Mathematics and Visualization

Series Editors

Gerald Farin
Hans-Christian Hege
David Hoffman
Christopher R. Johnson
Konrad Polthier

Springer
Berlin
Heidelberg
New York
Hong Kong
London
Milan
Paris
Tokyo

Gerald Farin
Bernd Hamann
Hans Hagen
Editors

Hierarchical and Geometrical Methods in Scientific Visualization

With 187 Figures, 30 in Color

 Springer

Editors

Gerald Farin
Arizona State University
Department of Computer Science
and Engineering
Tempe, AZ 85287-540, USA
e-mail: farin@asu.edu

Hans Hagen
Technical University Kaiserslautern
Department of Computer Science
67653 Kaiserslautern, Germany
e-mail: hagen@informatik.uni-kl.de

Bernd Hamann
University of California, Davis
Department of Computer Science
Davis, CA 95616-8562, USA
e-mail: hamann@cs.ucdavis.edu

Cataloging-in-Publication Data applied for
Bibliographic information published by Die Deutsche Bibliothek
Die Deutsche Bibliothek lists this publication in the Deutsche Nationalbibliografie;
detailed bibliographic data is available in the Internet at <http://dnb.ddb.de>.

Mathematics Subject Classification (2000): 76M27, 68U05, 68U07

ISBN 3-540-43313-9 Springer-Verlag Berlin Heidelberg New York

This work is subject to copyright. All rights are reserved, whether the whole or part of the material is concerned, specifically the rights of translation, reprinting, reuse of illustrations, recitation, broadcasting, reproduction on microfilm or in any other way, and storage in data banks. Duplication of this publication or parts thereof is permitted only under the provisions of the German Copyright Law of September 9, 1965, in its current version, and permission for use must always be obtained from Springer-Verlag. Violations are liable for prosecution under the German Copyright Law.

Springer-Verlag Berlin Heidelberg New York
a member of BertelsmannSpringer Science+Business Media GmbH

http://www.springer.de

© Springer-Verlag Berlin Heidelberg 2003
Printed in Germany

The use of general descriptive names, registered names, trademarks, etc. in this publication does not imply, even in the absence of a specific statement, that such names are exempt from the relevant protective laws and regulations and therefore free for general use.

Typeset by the authors. Edited by Kurt Mattes, Heidelberg, using a Springer TeX macro package
Cover design: *design & production* GmbH, Heidelberg

SPIN 10868921   46/3142LK - 5 4 3 2 1 0 - Printed on acid-free paper

Table of Contents

Dataflow and Remapping for Wavelet Compression
and View-dependent Optimization of Billion-triangle Isosurfaces . . 1
    Mark A. Duchaineau, Serban D. Porumbescu, Martin Bertram,
    Bernd Hamann, and Kenneth I. Joy

Extraction of Crack-free Isosurfaces
from Adaptive Mesh Refinement Data 19
    Gunther H. Weber, Oliver Kreylos, Terry J. Ligocki,
    John M. Shalf, Hans Hagen, Bernd Hamann, and Kenneth I. Joy

Edgebreaker on a Corner Table:
A Simple Technique for Representing
and Compressing Triangulated Surfaces 41
    Jarek Rossignac, Alla Safonova, and Andrzej Szymczak

Efficient Error Calculation for Multiresolution Texture-based
Volume Visualization 51
    Eric LaMar, Bernd Hamann, and Kenneth I. Joy

Hierarchical Spline Approximations 63
    David F. Wiley, Martin Bertram, Benjamin W. Jordan,
    Bernd Hamann, Kenneth I. Joy, Nelson L. Max,
    and Gerik Scheuermann

Terrain Modeling Using Voronoi Hierarchies 89
    Martin Bertram, Shirley E. Konkle, Hans Hagen, Bernd Hamann,
    and Kenneth I. Joy

Multiresolution Representation of Datasets with Material Interfaces   99
    Benjamin F. Gregorski, David E. Sigeti, John Ambrosiano,
    Gerald Graham, Murray Wolinsky, Mark A. Duchaineau,
    Bernd Hamann, and Kenneth I. Joy

Approaches to Interactive Visualization of Large-scale
Dynamic Astrophysical Environments 119
    Andrew J. Hanson and Philip Chi-Wing Fu

Data Structures for Multiresolution Representation
of Unstructured Meshes 143
    Kenneth I. Joy, Justin Legakis, and Ron MacCracken

Scaling the Topology of Symmetric, Second-Order Planar
Tensor Fields 171
    Xavier Tricoche, Gerik Scheuermann, and Hans Hagen

Simplification of Nonconvex Tetrahedral Meshes 185
    Martin Kraus and Thomas Ertl

A Framework for Visualizing Hierarchical Computations 197
    Terry J. Ligocki, Brian Van Straalen, John M. Shalf,
    Gunther H. Weber, and Bernd Hamann

Virtual-Reality Based Interactive Exploration
of Multiresolution Data 205
    Oliver Kreylos, E. Wes Bethel, Terry J. Ligocki,
    and Bernd Hamann

Hierarchical Indexing for Out-of-Core Access
to Multi-Resolution Data 225
    Valerio Pascucci and Randall J. Frank

Mesh Fairing Based on Harmonic Mean Curvature Surfaces . . . 243
    Robert Schneider, Leif Kobbelt, and Hans-Peter Seidel

Shape Feature Extraction 269
    Georgios Stylianou and Gerald Farin

Network-based Rendering Techniques for Large-scale
Volume Data Sets 283
    Joerg Meyer, Ragnar Borg, Bernd Hamann, Kenneth I. Joy,
    and Arthur J. Olson

A Data Model for Distributed Multiresolution Multisource
Scientific Data 297
    Philip J. Rhodes, R. Daniel Bergeron, and Ted M. Sparr

Adaptive Subdivision Schemes for Triangular Meshes 319
    Ashish Amresh, Gerald Farin, and Anshuman Razdan

Hierarchical Image-based and Polygon-based Rendering
for Large-Scale Visualizations 329
    Chu-Fei Chang, Zhiyun Li, Amitabh Varshney,
    and Qiaode Jeffrey Ge

Appendix: Color Plates 347

Dataflow and Remapping for Wavelet Compression and View-dependent Optimization of Billion-triangle Isosurfaces

Mark A. Duchaineau[1], Serban D. Porumbescu[1,3], Martin Bertram[1,2], Bernd Hamann[3], and Kenneth I. Joy[3]

[1] Center for Applied Scientific Computing (CASC), Lawrence Livermore National Laboratory, P.O. Box 808, L-561, Livermore, CA 94551, USA,
duchaineau1@llnl.gov
[2] AG Computergraphik und Computergeometrie, Universität Kaiserslautern, Postfach 3049, D-67653 Kaiserslautern, Germany,
bertram@informatik.uni-kl.de
[3] Center for Image Processing and Integrated Computing (CIPIC), Department of Computer Science, University of California at Davis, Davis, CA 95616-8562, USA, {hamann,joy,porumbes}@cs.ucdavis.edu

Abstract. Currently, large physics simulations produce 3D discretized field data whose individual isosurfaces, after conventional extraction processes, contain upwards of hundreds of millions of triangles. Detailed interactive viewing of these surfaces requires (a) powerful compression to minimize storage, and (b) fast view-dependent optimization of display triangulations to most effectively utilize high-performance graphics hardware. In this work, we introduce the first end-to-end multiresolution dataflow strategy that can effectively combine the top performing subdivision-surface wavelet compression and view-dependent optimization methods, thus increasing efficiency by several orders of magnitude over conventional processing pipelines. In addition to the general development and analysis of the dataflow, we present new algorithms at two steps in the pipeline that provide the "glue" that makes an integrated large-scale data visualization approach possible. A shrink-wrapping step converts highly detailed unstructured surfaces of arbitrary topology to the semi-structured meshes needed for wavelet compression. Remapping to triangle bintrees minimizes disturbing "pops" during realtime display-triangulation optimization and provides effective selective-transmission compression for out-of-core and remote access to extremely large surfaces. Overall, this is the first effort to exploit semi-structured surface representations for a complete large-data visualization pipeline.

1 Background

Two formerly distinct multi-resolution methods—subdivision-surface wavelet compression, and realtime display through view-dependent optimization— must be integrated to achieve a highly scalable and storage-efficient system

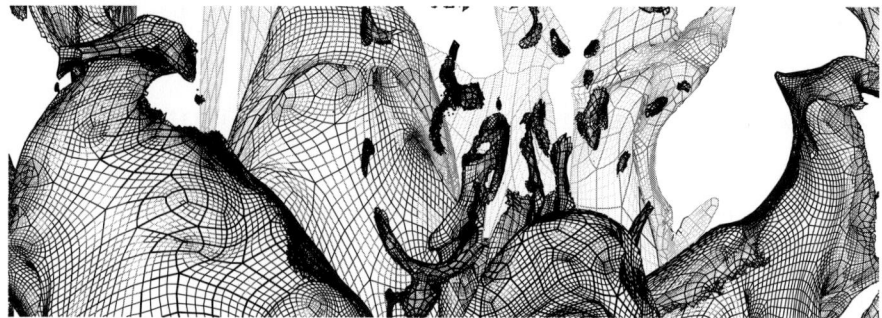

Fig. 1. Shrink-wrap result for .3% of a terascale-simulation isosurface. This conversion from irregular to a semi-regular mesh enables us to apply wavelet compression and view-dependent optimization.

for visualizing the results of recent large-scale 3D simulations. These semi-structured methods have been shown individually to have the highest performance of other respective approaches, but have required a number of missing re-mapping and optimization steps to be effectively scaled to huge data sets and to be linked together. This paper fills in these gaps and demonstrates the potential of the semi-structured approach.

Terascale physics simulations are now producing tens of terabytes of output for a several-day run on the largest computer systems. For example, the Gordon Bell Prize-winning simulation of a Richtmyer-Meshkov instability in a shock-tube experiment [12] produced isosurfaces of the mixing interface consisting of 460 million unstructured triangles using conventional isosurface extraction methods. New parallel systems are three times as powerful as the one used for this run, so billion-triangle surfaces are to be expected. Since we are interested in user interaction, especially sliding through time, surfaces are pre-computed; if they were not, the 100 kilotriangle-per-processor rates for the fastest isosurface extractors would result in several minutes per surface on 25 processors. Using 32-bit values for coordinates, normals and indices requires 16 gigabytes to store a single surface, and several terabytes for a single surface tracking through all 274 time steps of the simulation. This already exceeds the compressed storage of the 3D fields from which the surfaces are derived, and adding additional isolevels or fields per time step would make this approach infeasible. With the gigabyte-per-second read rates of current RAID storage, it would require 16 seconds to read a single surface. A compression factor of 100 with no loss of visual quality would cleanly solve both the storage and load-rate issues. This may be possible with new finite bicubic subdivision-surface wavelets that we introduced in [1]. Methods based on non-finite wavelet decomposition filters [10] are not likely to scale well to the massively parallel co-processing setting that we find critical for the future.

Another bottleneck occurs with high-performance graphics hardware. The fastest commercial systems today can effectively draw around 20 million triangles per second, i.e., around 1 million triangles per frame at 20 frames per second. Thus around a thousand-fold reduction in triangle count is needed. This reduction is too aggressive to be done without taking the interactively-changing viewpoint into account. As the scientist moves close to an area of interest, that area should immediately and seamlessly be fully resolved while staying within the interactive triangle budget. This can be formulated as the optimization of an adaptive triangulation of the surface to minimize a view-dependent error measure, such as the projected geometric distortion on the screen. Since the viewpoint changes 20 times per second, and the error measure changes with the viewpoint, the million-element adaptation must be re-optimized continuously and quickly. The fastest time in which any algorithm can perform this optimization is $O(\Delta output)$, an amount of time proportional to the number of element changes in the adaptation per frame. The Realtime Optimally Adapting Meshes (ROAM) algorithm achieves this optimal time using adaptations built from triangle bintrees [6], implemented for height-map surfaces. Using this approach in a flight-simulation example, it was found that around 3% of the elements change per frame, which resulted in a 30-fold speedup in optimization rates. This is critically important since the bottleneck in fine-grained view-dependent optimizers is processing time rather than graphics hardware rates. In this work we extend the ROAM optmizer to work on general surface shapes formulated as a kind of subdivision-surface.

Wavelet compression and view-dependent optimization are two powerful tools that are part of the larger dataflow from 3D simulation to interactive rendering. These are by no means the only challenging components of a terascale visualization system. For example, conversion is required to turn the irregular extracted surfaces into a semi-structured form appropriate for further processing, see Figure 1. Before turning to our focus on the two remapping steps that tie together the overall dataflow, we provide an overview of the complete data pipeline for surface interaction.

2 End-to-end Multiresolution Dataflow

The processing required to go from massive 3D data to the interactive, optimal display triangulations may be decomposed into six steps:

Extract: Obtain unstructured triangles through accelerated isosurface extraction methods or through material-boundary extraction from volume-fraction data [2]. The extraction for our purposes actually produces two intermediate results, 1) a signed-distance field near the desired surface, along with 2) a polygonal mesh of the surface ready for simplification.

Shrink-wrap: Convert the high-resolution triangulation to a similarly detailed surface that has subdivision-surface connectivity (i.e., is *semi-regular*), has high parametric quality, and minimizes the number of structured blocks. This step has three phases: (1) compute the complete signed-distance transform of the surface from the near-distance field obtained in the Extract step; (2) simplify the surface to a base mesh; and (3) iteratively *subdivide*, *smooth* and *snap* the new mesh to the fine-resolution surface until a specified tolerance is met (see Section 3).

Wavelet compress: For texture storage, geometry archiving, and initial transmission from the massively-parallel runs, use bicubic wavelets based on subdivision surfaces for nearly lossless compression [1]. Good shrink-wrap optimization can make a significant improvement in the wavelet compression performance of this step.

Triangle bintree remap: Remap the shrink-wrap parameterization to be optimal for subsequent view-dependent optimization (this is different than being optimal for high-quality wavelet compression). The bintree hierarchies use piecewise-linear "wavelets" without any vanishing moments, but where most wavelet coefficients can be a single scalar value in a derived normal direction, and where highly localized changes are supported during selective refinement (see Section 4).

Selective decompress: Asynchronously feed a stream of compressed detail where the view-dependent adaptation is most likely to find missing values during selective refinement in the near future. This stream can be efficiently stored in chunks for efficient I/O or network transmission (further details are provided later in this Section).

Display-list optimization: Perform the realtime optimization of the adaptive display-list triangulation each frame, taking maximal advantage of frame-to-frame coherence to accelerate frustum culling, element priority computation, and local refinement and coarsening to achieve the optimum per-view geometry [6].

These six processing steps occur in the order listed for terascale surface visualization. The first three steps—extraction, shrink-wrapping and wavelet compression—typically occur in batch mode either as a co-process of the massively parallel physics simulation, or in a subsequent parallel post-processing phase. Given the selective access during interaction, and given the already existing wavelet hierarchy, the remapping for ROAM interaction can be supported on demand by modest parallel resources. Because the ROAM remapping works in time O(output) in a coarse-to-fine progression, the amount of computation per minute of interaction is independent of the size of the physics grid or the fine-resolution surfaces. The ROAM remapper is envisioned as a runtime data service residing on a small farm of processors and disks, that reads, remaps and feeds a highly compact data stream on demand to the client ROAM display-list optimizer. This server-to-client link could be across a LAN or over high-speed WAN to provide remote access. The ROAM algorithm works at high speed per frame to optimize the display triangulation, and thus will reside proximate to the graphics hardware.

The decompression and remapping services can be abstracted into an asynchronous client-server interface between the ROAM optimizer client running at high frame rates, and a *loader* server performing the reads, remote transmission, decompression, remapping and local transmission. In the ROAM algorithm, refinement and coarsening of the displayed geometry occurs on triangle bintrees as depicted in Figure 2. The unit operation is the *simple split*, wherein two triangles T and its base neighbor T_B are replaced with four children T_0, T_1, T_{B0} and T_{B1}. The key change to the ROAM algorithm is that the simple split operation is allowed to fail in the case that the decompressed/remapped information for the four kids, such as geometry and textures, is not available yet. In that case, the priority of the split becomes temporarily set to zero until an event handler notes the arrival of the required information. Similarly, the chains of simple split operations resulting from a *force split* (depicted in Figure 3), require that the force split fail if any simple split in the chain fails.

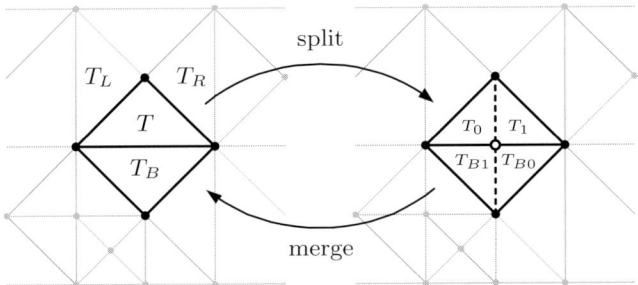

Fig. 2. Simple split and merge operations on a bintree triangulation. A typical neighborhood is shown for triangle T on the left. Splits can fail if the required details have not arrived from the loader.

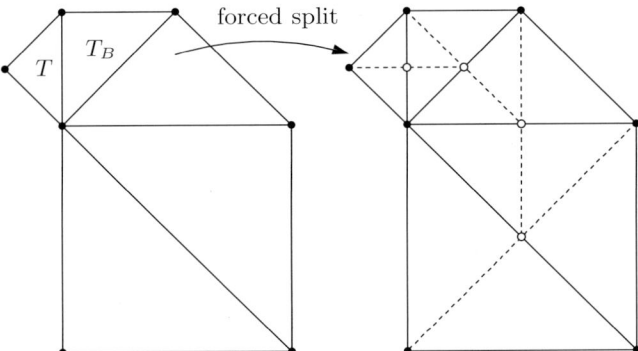

Fig. 3. A chain of simple splits, known as a *force split*, is required during mesh optimization to ensure continuity. Any simple split failure results in failure for the remainder of the chain.

A cache is kept of the remapped/decompressed information (the geometry and textures). A hash scheme is used to access the remapped information, keyed on the unique split-center vertex index. If the required information for the four kids exists, the split operation will succeed in finding them in the cache. Else the split-center key is put on the loader's decompress/remap request queue and the split fails. After a simple split succeeds, the information record for the four kids is taken off the re-use queue. On merge, the four-kid record is placed on the "most recent" end of the re-use queue. To conserve limited memory or graphics hardware resources, the least-recently-used records may be re-used.

The loader performs remapping on any awaiting remap requests in the order received. Upon completion, it generates an event to the ROAM optimizer that the four-kid record should be stored in the cache and made available for future simple splits. The priority of the associated splits are reset to their normal values at this time.

In order for the loading scheme to avoid causing delays to the per-frame processing, the display optimizer does not stop and wait for loads to occur, but continues to perform splits and merges in response to viewpoint changes. The delayed (failed) splits will eventually succeed after some frames, resulting in brief sub-optimal display accuracy but at full frame rates. However, due to frame-to-frame coherence these inaccuracies are expected to be minor and quickly corrected. To avoid the inaccuracies altogether, a form of eager evaluation is included that anticipates future split requirements. It works by putting the loader requests associated with the grandkids on the request queue each time a split occurs. These requests should be given lower priority than the ones immediately needed by currently-attempted split operations. Finally, to avoid stale entries in the loader request queue from accumulating, queue entries may be removed if the request was not repeated on the current frame.

The remainder of this paper focuses on the two remapping steps in this dataflow. The first step prepares a surface for wavelet compression, the second converts the parameterization for ROAM triangle bintree optimization.

3 Shrink-wrapping Large Isosurfaces

Subdivision surfaces originated with the methods of Catmull-Clark [3] and Doo-Sabin [4], with greatly renewed research activity in the last few years. Wavelets for these types of surfaces were first devised by Lounsbery and collaborators [11,7,14], with additional work on wavelets for Loop and Butterfly subdivision surfaces and compression by Khodakovsky *et al.* [10]. Work on finite filters with Catmull-Clark subdivision-surface wavelets was performed by us in Bertram *et al.* [1], which also introduced an early variant on the shrink-wrap procedure described in this section, which is applicable to isosurfaces. The shrink-wrapping method we introduced in [1] was used to demonstrate wavelet transforms of complex isosurfaces in a non-parallel, topology-preserving setting. Here we elaborate on extensions to allow parallel and topology-simplifying shrink-wrapping. An alternate approach to remapping for isosurfaces was presented by Wood *et al.* [15], but appears to be difficult to parallelize or to extend for topology simplification.

Before subdivision-surface wavelet compression can be applied to an isosurface, it must be remapped to a mesh with subdivision-surface connectivity. Minimizing the number of base-mesh elements increases the number of levels in a wavelet transform, thus increasing the potential for compression. Because of the extreme size of the surfaces encountered, and the large number of them, the remapper must be fast, work in parallel, and be entirely automated. The compression using wavelets is improved by generating high-quality meshes during the remapping step. For this, highly smooth and nearly square parameterizations result in the smallest magnitude and count of wavelet coefficient magnitudes, respectively, and yields small outputs after entropy coding. In addition, we would like to allow the new mesh to optionally have simplified topology, akin to actual physical shrink-wrapping of complex 3D objects with a few connected sheets of plastic. The strategy pursued here is to first produce a highly simplified version of the mesh, followed by the re-introduction of detail in a semi-regular refinement through a coarse-to-fine iterative optimization process.

The algorithm uses as input a scalar field on a 3D mesh and an isolevel, and provides as output a surface mesh with subdivision-surface connectivity, i.e., a collection of logically-square patches of size $(2^n+1) \times (2^n+1)$ connected on mutual edges. The algorithm, on a high level, is organized in three steps:

Signed distance transform: For each grid point in the 3D mesh, compute the signed-distance field, i.e., the distance to the closest surface point, negated if in the region of scalar field less than the isolevel. Starting with the vertices of the 3D mesh elements containing isosurface components, the transform is computed using a kind of breadth-first propagation. Data parallelism is readily achieved by queueing up the propagating values on the boundary of a block subdomain, and communicating to the block-face neighbor when no further local propagation is possible.

Determine base mesh: To preserve topology, edge-collapse simplification is used on the full-resolution isosurface extracted from the distance field using conventional techniques. This phase is followed by an edge-removal phase (edges but not vertices are deleted) that improves the vertex and face degrees to be as close to four as possible. Parallelism for this mode of simplification is problematic, since edge-collapse methods are inherently serial using a single priority queue to order collapses.

To allow topology reduction, the 3D distance field is simplified before the isosurface is extracted. Simplification can be performed by using wavelet low-pass filtering on regular grids, or after resampling to a regular grid for curvilinear or unstructured 3D meshes. The use of the signed-distance field improves simplified-surface quality compared to working directly from the original scalar field. To achieve the analog of physical shrink-wrapping, a min operation can be used in place of the wavelet filtering. For each cell in an octree, the minimum of all the child cells is recursively computed. Assuming the solid of interest has values greater than an isolevel, the isosurfaces of the coarse octree levels will form a topologically-simplified surface that is guaranteed to completely contain the fully-resolved solid. This form of simplification is easily parallelized in a distributed setting. It is not clear at this time how to achieve topology preservation and significant distributed-data parallelism.

Subdivide and fit: The base mesh is iteratively fit and the parameterization optimized by repeating three phases: (1) subdivide using Catmull-Clark rules; (2) perform edge-length-weighted Laplacian smoothing; and (3) snap the mesh vertices onto the original full-resolution surface with the help of the signed-distance field. Snapping involves a hunt for the nearest fine-resolution surface position that lies on a line passing through the mesh point in an estimated normal direction of the shrink-wrap mesh. The estimated normal is used, instead of (for example) the distance-field gradient, to help spread the shrink-wrap vertices evenly over high-curvature regions. The signed-distance field is used to provide Newton-Raphson-iteration convergence when the snap hunt is close to the original surface, and to eliminate nearest-position candidates whose gradients are not facing in the directional hemisphere centered on the estimated normal. Steps 2 and 3 may be repeated several times after each subdivision step to improve the quality of the parameterization and fit. In the case of topology simplification, portions of the shrink-wrap surface with no appropriate snap target are left at their minimal-energy position determined by the smoothing and the boundary conditions of those points that do snap. Distributed computation is straightforward since all operations are local and independent for a given resolution.

The distance-field computation uses a regular cubic grid (for unstructured finite element input, a fast 3D variant of triangle scan conversion is used to resample to a regular cubic grid). Each vertex in the 3D grid contains the field value and the coordinates and distance to the closest point found so far, where the distances are initialized to a very large number. There is also a circular array of queued vertex indices (a single integer per vertex), of size no larger than the number of vertices in the 3D grid. This circular array is used as a simple first-in, first-out queue used to order the propagation of distance values through the domain. It is initialized to be empty.

The propagation is started at the isosurface of interest. For the purposes of this paper, material-fraction boundaries are treated one material at a time, using the same formulation as for isosurfaces by considering that material's volume fraction as a field and thresholding it at 50%. All the vertices of all the cells containing the surface are put onto the queue, and their closest points and distances are determined directly from the field values of their cell. These nearest points are computed by approximating the field as a linear function in the neighborhood of the cubic-grid vertex v:

$$\hat{f}(v) = f(v_0) + \nabla f(v_0) \cdot (v - v_0)$$

where we define the gradient at a vertex using central differences:

$$\nabla f(v_0) = \begin{bmatrix} (f(v_0 + d_x) - f(v_0 - d_x))/2 \\ (f(v_0 + d_y) - f(v_0 - d_y))/2 \\ (f(v_0 + d_z) - f(v_0 - d_z))/2 \end{bmatrix}$$

for $d_x = [1,0,0]^T$, $d_y = [0,1,0]^T$ and $d_z = [0,0,1]^T$. The closest point in the local linear approximation will be

$$\hat{v}_c = v_0 + (f_0 - f(v_0)) \frac{\nabla f(v_0)}{\|\nabla f(v_0)\|_2^2}$$

where $f(v)$ if the trilinearly-interpolated field for the cubic grid, and f_0 is the desired isolevel. Optionally, the precise distance to the trilinear field isosurface is obtained by performing Newton-Raphson iterations starting at $\hat{v}_c$, using the gradient and field values from the trilinear interpolant.

The distance-field propagation now proceeds by repeatedly processing the output end of the queue until no queue entries are left. The vertex index i is removed from the queue, and its 26 vertex neighbors are examined. If the current closest point $v_c(i)$ to vertex i is closer to neighbor $v(j)$ than j's current closest point $v_c(j)$, then we set $v_c(j) \leftarrow v_c(i)$, update the current distance $d(j)$, and put j on the input end of the queue if it is not already on the queue. Overall our tests show that a vertex will be updated between about 1.2 and 1.8 times during the propagation, meaning that the algorithm is observed empirically to run in time O(N) for N vertices. The distance transform on the 256 × 256 × 384 volume used for the large example in Figure 1 took 63 seconds to compute on a 1.8GHz Pentium 4 processor with 1GB of 800MHz RDRAM memory.

Note that closest points by definition travel perpendicular to the isosurface in straight lines. This, combined with the fact that the propagation updates are highly localized, leads to a simple and effective means of parallelization. Each processor is assigned a random subset of small blocks (e.g. 64^3 cells) as a passive load-balancing mechanism. Each processor initializes the queue for the vertices near the isosurface on all its blocks. The processing then proceeds as follows. The processor chooses a block with a nonempty queue. It processes all entries of the queue until the queue is empty, and accumulates a list of updated vertices on each of the six block faces. For each face that has updates, a message containing the list of updates is sent to the processor that owns the neighboring block. The processor receives any available update messages, puts the updated vertices on the indicated block queue, and sends an acknowledgement reply message. Also the processor receives any available acknowledgements to its previous update transmissions. If at any time the processor transitions to having all empty block queues and no pending acknowledgements to receive, then it increments a global completion counter. If it transitions out of the completion state (by receipt of further updates), then it decrements the global counter before sending acknowledgements. The entire distance-transform computation is complete when the global counter equals the number of processors. The processor continues with another block if available, otherwise it waits idle for acknowledgements, update messages or global completion.

Once the signed distance field has been computed, a coarse base mesh can be constructed using the octree minima values obtained through a simple, parallelizable, coarse-to-fine recursion on the octree cells. The octree adaptation is determined by the average deviation from the trilinear approximation that the full-resolution data takes on over an octree cell. For example, in a top-down recursive splitting, any octree cell where the average deviation is more than 0.2 times the cell width is split further. Octree cell neighbors are recursively split if they are more than one level coarser than the current cell. After the recursion completes, the faces from the leaf octree cell boundaries form the base mesh. In the case of edges that are on the boundary of octree leaves of different resolutions, the mutual refinement of the edge is used, leading faces to have between four and eight edges (each edge may be split if any edge-neighbor is at finer resolution; neighbors are, by construction, at most one level finer). This base mesh is now ready for the iterative application of smoothing, snapping and refinement to create an optimized fit of the detailed surface.

Smoothing is defined by a simple edge-weighted local averaging procedure, with an optional correction to improve processing in high-curvature regions. The simple weighting formula to update a vertex p is

$$p' = (1 - \alpha)p + \alpha \bar{p}$$

where p is the old vertex position, $\alpha = 0.2$ (approximately), and

$$\bar{p} = \frac{\sum_i r_i p_i}{\sum_i r_i}$$

where p_i are the edge-neighbor vertices of p, and $r_i = \|p_i - p\|_2$ are the edge lengths to them. The smoothing neighborhood is depicted in Figure 4.

This smoothing tends to shrink spherical shapes progressively to a point, which can lead to stability and quality problems in high-curvature regions. An option to correct for this is to project p' in the normal direction n onto a local least-squares-optimal quadratic height map. This map is computed with respect to the estimated normal at p and its tangent plane, passing through p, and otherwise minimizing the least-squares error with respect to the p_i's.

Snapping involves tracing a ray from the current vertex position p, in the normal direction n estimated from the current shrink-wrap mesh neighborhood, and updating the vertex position if a suitable intersection is found on

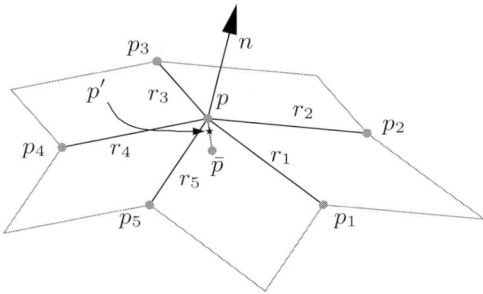

Fig. 4. A neighborhood of vertex p undergoing smoothing. The simple version of smoothing involves edge-weighted averaging with neighbors. An optional correction projects to a local quadratic height map where the height direction is given by the local normal n.

the high-resolution isosurface. As with the smoothing procedure, the normal estimate is computed by averaging the set of cross products of successive edges neighboring p, and normalizing to unit length. Intersections are found in two stages, first by using binary search for zeros of the signed distance field, and later by using Newton-Raphson iterations that again use the distance field. The transition to the Newton-Raphson procedure is attempted at each step of the binary search, and if it fails to converge in five steps then the binary search is continued.

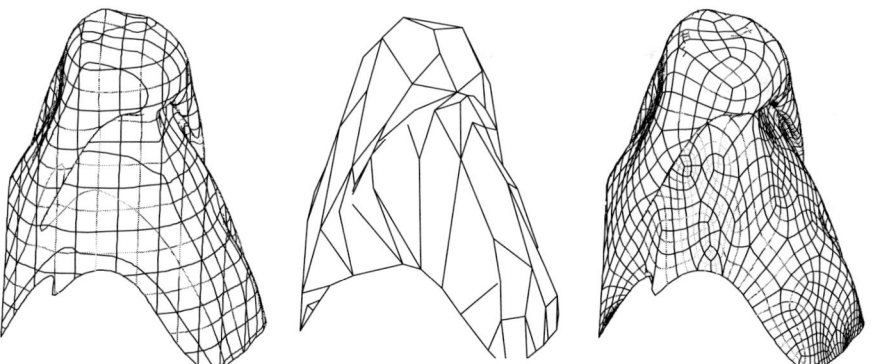

Fig. 5. Shrink wrapping steps (left to right): full-resolution isosurface, base mesh constructed from edge collapses, and final shrink-wrap with subdivision-surface connectivity.

The shrink-wrap process is depicted in Figure 5 for approximately .016% of the 460 million-triangle Richtmyer-Meshkov mixing interface in topology-preserving mode. The original isosurface fragment contains 37,335 vertices, the base mesh 93 vertices, and the shrink-wrap result 75,777 vertices. A larger portion of the same isosurface is shown in Figure 1. The extracted, full resolution unstructured mesh contains 976,321 vertices, the base mesh 19,527 vertices, and the fine-resolution shrink-wrap mesh 1,187,277 vertices. The shrink-wrap refinement process in both cases was continued until the fit was accurate to a tiny fraction of a cell width. Indeed, the fit surface takes on the artifacts of the isosurface extraction process, slight "staircase" patterns at the grid resolution.

The shrink-wrap processing succeeded in our goal of creating a parameterization that results in high compression rates. With only 1.6% of the wavelet coefficients, the root mean squared error is a fraction of the width of a finite element in the grid, implying effective compression even before considering improvements through quantization and entropy coding.

4 Remapping for ROAM

The ROAM algorithm typically exploits a piecewise block-structured surface grid to provide efficient selective refinement for view-dependent optimization. A triangle bintree structure is used, with split and merge operations for selective refinement and coarsening respectively, as described in Section 2 (see Figures 2-3). In this section we introduce a second re-mapping procedure that solves the problem of *tangential motion* that arises when performing selective refinement on the "curvy" parameterizations that result from the shrink-wrapping algorithm.

A simple demonstration of the problem is shown in Figure 6. Note that even for a perfectly planar region, significant "pops" will occur during selective refinement due to the nonlinear parameterization. A natural choice for remapping is to project the original surface geometry in an estimated normal direction onto the coarse bintree triangles acting as the new domain. Occasionally the projection causes excessive distortions or folds, but typically it can be applied without problems. When projection can be applied, it reduces the new split-point updates to all be in the local "height" or normal direction, and can thus completely avoid tangential motion. Indeed, this can be seen as a form of compression, since only a single scalar Δ-height need be stored instead of a 3-component vector.

Although it would be simpler to have a single type of mapping, the local-heightmap scheme for triangle bintrees is not optimal for bicubic subdivision-surface wavelet compression, and so we choose to keep two distinct remappings. The curved parameterizations are ideal for the higher-order wavelet compression. The ROAM-optimized mappings often appear jagged and less pleasing to a human observer *who is looking at the mesh lines*, but nevertheless are much better performing for actual use during view-dependent display, where the mesh lines are *not* shown.

Several lines of previous research serve as a backdrop for the methods we have devised. The earliest known local hierarchical displacement frames were introduced by Forsey and Bartels in the context of hierarchical B-spline

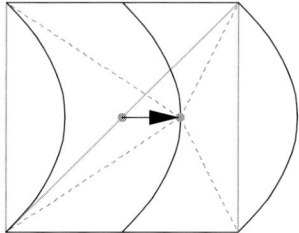

Fig. 6. The highly curved parameterization from shrink-wrapping is shown in black, whereas the natural parameterization for triangle bintrees is shown in gray. Without remapping, the large tangential motion (arrow) will occur when splitting.

editing [8]. A more general method, including integration with wavelets, was introduced in [5]. The recent work of Guskov *et al.* [9] has many similarities to our approach, but in the context of geometric modeling and compression rather than view-dependent display optimization.

The ROAM remapping works from coarse to fine resolutions, following the simple split operations in whatever order they appear during selective refinement. The vertices of the base mesh are left fixed at their original positions, and are stored as full three-component coordinates. For every edge-bisection vertex formed in one simple split step, estimated normals are computed by averaging the normals of the two triangles whose common edge is being bisected, and making the averaged normal unit length. All vertices carry with them a reference to the $2^n + 1$-squared patch that they reside in, along with the two parameter coordinates $[u, v]$ within that patch. During edge bisection, the *parametric midpoint* is computed by "topologically gluing" at most two patches together from the original mesh, computing the mid-parameter values in this glued space, then converting those parameters back to unglued parameters. The gluing and remapping is depicted in Figure 7.

Given the constraints on our procedure, it is not possible for bisecting-edge endpoints to cross more than one patch boundary. A ray-trace intersection is performed from the midpoint of the line segment being bisected, in the estimated normal direction. Since we expect the intersection to be near the parametric midpoint in most cases, it is efficient to begin ray intersection tests there for early candidates.

Since the surface being ray-traced stays fixed throughout the remapping, the construction of typical ray-surface intersection-acceleration structures can be amortized and overall offer time savings. However, we have found that, overall, the hunt using the parametric midpoint as a starting point is very fast in practice and further acceleration methods are not needed. For the final precise intersection evaluation, we use interval-Newton iterations [13]. Intersections are rejected if they are not within a parametric window defined by the four remapped vertices of the two triangles being split, shrunk by some factor (e.g., reduced by 50%) around the parametric midpoint. The

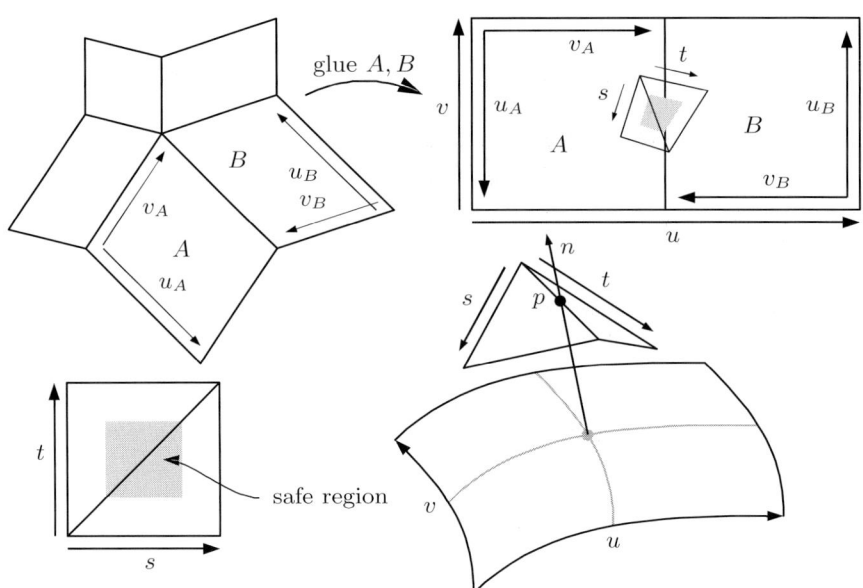

Fig. 7. Two patches parameterized by $[u_A, v_A]$ and $[u_B, v_B]$ are glued to form a single patch parameterized by $[u, v]$. A ray-tracing procedure finds the intersection of the ray through midpoint p in the estimated normal direction n. The $[u, v]$ of the intersection is acceptable if it appears in the image of the safe region of $[s, t]$-space in the glued space.

closest acceptable intersection is chosen (closest with respect to the distance from p in world space). If an acceptable intersection exists, then the single scalar value is given specifying the distance from p along the normal ray n. If no intersections exist or are acceptable, then parametric midpoint is chosen and is specified with a three-component vector of the displacement from edge midpoint p to the parametric midpoint.

The result of remapping is shown in Figure 8 for a test object produced by Catmull-Clark subdivision with semi-sharp features. The original parameterization on the left is optimal for compression by bicubic subdivision-surface wavelets, but produces extreme and unnecessary tangential motions during triangle-bintree refinement. The remapped surface, shown on the right, has bisection-midpoint displacements ("poor man's wavelets") of length zero in the flat regions of the disk, and displacements of minimized length elsewhere. We note that while the main motivation for this procedure is to increase accuracy and reduce the "pops" during realtime display-mesh optimization, the typical reduction to a single scalar value of the displacement vectors (wavelet coefficients) gives a fair amount of compression. This is desirable when the remap from high-quality wavelet parameterization and compression is too time-consuming to allow some clients to perform it themselves, so that a

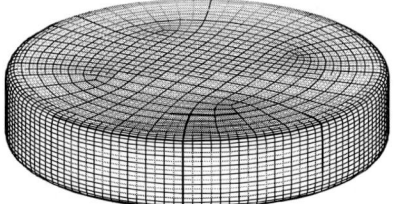 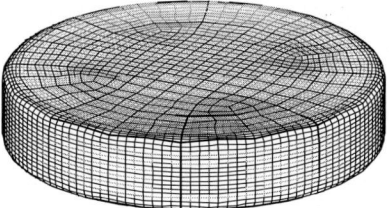

Fig. 8. Before (left) and after remapping for triangle bintree hierarchies. Tangential motion during subdivision is eliminated.

more capable server performs the remap yet there is still a fair amount of compression over the server-to-client communications link.

Generally the ROAM remapping decreases the error for a given level of approximation. However, tiny discrepancies exist in the finest resolution mesh due to remapping. It is certainly possible to continue remapping beyond the resolution of the input mesh, in effect fully resolving the exact artifacts and other details of the original mesh. In most applications this is not necessary.

The ROAM algorithm naturally requires only a tiny fraction of the current optimal display mesh to be updated each frame. Combined with caching and the compression potential of remapping, this promises to provide an effective mechanism for out-of-core and remote access to the surfaces on demand during interaction.

5 Future Work

We have demonstrated, for the first time, the end-to-end use of semi-structured representations for large-data compression and view-dependent display. The primary lesson learned is that two remapping steps are needed for best performance, rather than one, and that a runtime on-demand caching scheme is needed to best perform the second remapping. Overall the software is still very much a research prototype with many challenges remaining to create a production-ready system:

1. For topology-preserving simplification, the inherently serial nature of the queue-based schemes must be overcome to harness parallelism.
2. Means such as transparent textures must be devised to handle the non-snapping shink-wrap regions that result from topology simplification.
3. The shrink-wrapping procedure can fail to produce one-to-one, onto mappings in some cases even when such mappings exist. Perhaps it is possible to revert to expensive simplification schemes that carry one-to-one, onto mappings only in problematic neighborhoods.
4. Shrink-wrapping needs to be extended to produce time-coherent mappings for time-dependent surfaces. This is a great challenge due to the complex evolution that surfaces undergo in time-dependent physics simulations.

Acknowledgments

This work was performed under the auspices of the U.S. Department of Energy by University of California Lawrence Livermore National Laboratory under contract No. W-7405-Eng-48. We thank LLNL for support through the Science and Technology Education Program, the Student-Employee Graduate Research Fellowship Program, the Laboratory-Directed Research and Development Program, the Institute for Scientific Computing Research, and the Accelerated Strategic Computing Initiative.

In addition, this work was supported by the National Science Foundation under contract ACI 9624034 (CAREER Award), through the Large Scientific and Software Data Set Visualization (LSSDSV) program under contract ACI 9982251, and through the National Partnership for Advanced Computational Infrastructure (NPACI); the Office of Naval Research under contract N00014-97-1-0222; the Army Research Office under contract ARO 36598-MA-RIP; the NASA Ames Research Center through an NRA award under contract NAG2-1216; the Lawrence Livermore National Laboratory under ASCI ASAP Level-2 Memorandum Agreement B347878 and under Memorandum Agreement B503159; the Lawrence Berkeley National Laboratory; the Los Alamos National Laboratory; and the North Atlantic Treaty Organization (NATO) under contract CRG.971628. We also acknowledge the support of ALSTOM Schilling Robotics and SGI. We thank the members of the Visualization and Graphics Research Group at the Center for Image Processing and Integrated Computing (CIPIC) at the University of California, Davis.

References

1. Martin Bertram, Mark A. Duchaineau, Bernd Hamann, and Kenneth I. Joy. Bicubic subdivision-surface wavelets for large-scale isosurface representation and visualization. *Proceedings of IEEE Vis00*, October 2000.
2. Kathleen S. Bonnell, Daniel R. Schikore, Kenneth I. Joy, Mark Duchaineau, and Bernd Hamann. Constructing material interfaces from data sets with volume-fraction information. *Proceedings of IEEE Vis00*, October 2000.
3. E. Catmull and J. Clark. Recursively generated b-spline surfaces on arbitrary topological meshes. *Computer-Aided Design*, 10:350—355, September 1978.
4. D. Doo and M. Sabin. Behaviour of recursive division surfaces near extraordinary points. *Computer-Aided Design*, 10:356—360, September 1978.
5. Mark A. Duchaineau. *Dyadic Splines*. PhD thesis, University of California, Davis, 1996.
6. Mark A. Duchaineau, Murray Wolinsky, David E. Sigeti, Mark C. Miller, Charles Aldrich, and Mark B. Mineev-Weinstein. ROAMing terrain: Real-time optimally adapting meshes. *IEEE Visualization '97*, pages 81–88, November 1997. ISBN 0-58113-011-2.
7. Matthias Eck, Tony DeRose, Tom Duchamp, Hugues Hoppe, Michael Lounsbery, and Werner Stuetzle. Multiresolution analysis of arbitrary meshes. In

Robert Cook, editor, *SIGGRAPH 95 Conference Proceedings*, pages 173–182. ACM SIGGRAPH, August 1995.
8. David R. Forsey and Richard H. Bartels. Hierarchical b-spline refinement. In *Computer Graphics (Proceedings of SIGGRAPH 88)*, volume 22, pages 205–212, Atlanta, Georgia, August 1988.
9. Igor Guskov, Kiril Vidimce, Wim Sweldens, and Peter Schröder. Normal meshes. *Proceedings of SIGGRAPH 2000*, pages 95–102, July 2000.
10. Andrei Khodakovsky, Peter Schröder, and Wim Sweldens. Progressive geometry compression. In Kurt Akeley, editor, *Siggraph 2000, Computer Graphics Proceedings*, Annual Conference Series, pages 271–278. ACM Press / ACM SIGGRAPH / Addison Wesley Longman, 2000.
11. Michael Lounsbery. *Multiresolution Analysis for Surfaces of Arbitrary Topological Type*. PhD thesis, Dept. of Computer Science and Engineering, U. of Washington, 1994.
12. Arthur A. Mirin, Ron H. Cohen, Bruce C. Curtis, William P. Dannevik, Andris M. Dimits, Mark A. Duchaineau, Don E. Eliason, Daniel R. Schikore, Sarah E. Anderson, David H. Porter, Paul R. Woodward, L. J. Shieh, and Steve W. White. Very high resolution simulation of compressible turbulence on the IBM-SP system. *Supercomputing 99 Conference*, November 1999.
13. Tomoyuki Nishita, Thomas W. Sederberg, and Masanori Kakimoto. Ray tracing trimmed rational surface patches. *Computer Graphics (SIGGRAPH '90 Proc.)*, 24(4):337–345, August 1990.
14. Eric J. Stollnitz, Tony D. DeRose, and David H. Salesin. *Wavelets for Computer Graphics: Theory and Applications*. Morgann Kaufmann, San Francisco, CA, 1996.
15. Zoë J. Wood, Mathieu Desbrun, Peter Schröder, and David Breen. Semi-regular mesh extraction from volumes. In T. Ertl, B. Hamann, and A. Varshney, editors, *Proceedings Visualization 2000*, pages 275–282. IEEE Computer Society Technical Committee on Computer Graphics, 2000.

Extraction of Crack-free Isosurfaces from Adaptive Mesh Refinement Data

Gunther H. Weber[1,2,3], Oliver Kreylos[1,3], Terry J. Ligocki[3], John M. Shalf[3,4], Hans Hagen[2], Bernd Hamann[1,3], and Kenneth I. Joy[1]

[1] Center for Image Processing and Integrated Computing (CIPIC), Department of Computer Science, One Shields Avenue, University of California, Davis, CA 95616-8562, U.S.A.
[2] AG Computergraphik, FB Informatik, University of Kaiserslautern, Erwin-Schrödinger Straße, D-67653 Kaiserslautern, Germany
[3] National Energy Research Scientific Computing Center (NERSC), Lawrence Berkeley National Laboratory, 1 Cyclotron Road, Berkeley, CA 94720, U.S.A.
[4] National Center for Supercomputing Applications (NCSA), University of Illinois, Urbana-Champaign, IL 61801, U.S.A.

Abstract. Adaptive mesh refinement (AMR) is a numerical simulation technique used in computational fluid dynamics (CFD). This technique permits efficient simulation of phenomena characterized by substantially varying scales in complexity. By using a set of nested grids of different resolutions, AMR combines the simplicity of structured rectilinear grids with the possibility to adapt to local changes in complexity within the domain of a numerical simulation that otherwise requires the use of unstructured grids. Without proper interpolation at the boundaries of the nested grids of different levels of a hierarchy, discontinuities can arise. These discontinuities can lead, for example, to cracks in an extracted isosurface. Treating locations of data values given at the cell centers of AMR grids as vertices of a dual grid allows us to use the original data values of the cell-centered AMR data in a marching-cubes (MC) isosurface extraction scheme that expects vertex-centered data. The use of dual grids also induces gaps between grids of different hierarchy levels. We use an index-based tessellation approach to fill these gaps with "stitch cells." By extending the standard MC approach to a finite set of stitch cells, we can define an isosurface extraction scheme that avoids cracks at level boundaries.

1 Introduction

AMR was introduced to computational physics by Berger and Oliger [3] in 1984. A modified version of their algorithm was published by Berger and Colella [2]. AMR has become increasingly popular in the computational physics community, and it is used in a variety of applications. For example, Bryan [4] uses a hybrid approach of AMR and particle simulations to simulate astrophysical phenomena.

Fig. 1 shows a simple two-dimensional (2D) AMR hierarchy produced by the Berger–Colella method. The basic building block of a d–dimensional Berger-Colella AMR hierarchy is an axis-aligned, structured rectilinear grid.

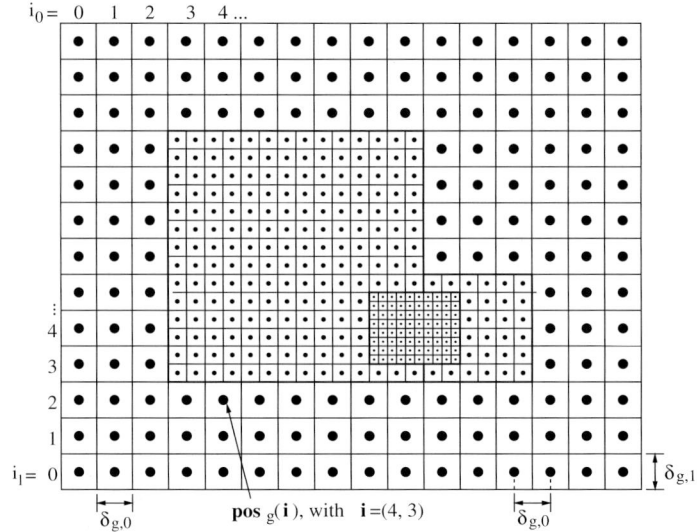

Fig. 1. AMR hierarchy consisting of four grids in three levels. The root level consists of one grid. This grid is refined by a second level consisting of two grids. A fourth grid refines the second level. It overlaps both grids of the second level. Boundaries of the grids are drawn as bold lines. Locations at which dependent variables are given are indicated by solid discs

Each grid g consists of n_j hexahedral cells in each axial direction. We treat this number as an integer resolution vector $\boldsymbol{n}$[1]. The grid spacing, i.e., the widths of grid cell in each dimension, is constant in a specific direction and given as a vector $\boldsymbol{\delta}_g$, see Fig. 1. Fig. 1 illustrates that the distance between two samples is equal to the grid spacing. Each grid can be positioned by specifying its origin $\boldsymbol{o}_g$. The simulation method typically applied to AMR grids is a finite-difference method. Typically, a *cell-centered* data format is used, i.e., dependent function values are associated with the centers of cells. Thus, the dependent function value associated with a cell $\boldsymbol{i}_g$, with $0 \leq i_{g,j} < n_j$, is located at

$$pos_{g,j}(\boldsymbol{i_g}) = o_{g,j} + \left(i_j + \frac{1}{2}\right)\delta_j \quad, \tag{1}$$

see Fig. 1. Since sample locations are implicitly given by the regular grid structure, it suffices to store dependent data values in a simple array using a fixed ordering scheme, e.g., row-major order. We denote the region covered by a grid g by Γ_g.

An AMR hierarchy consists of several levels Λ_l comprising one or multiple grids. All grids in the same level have the same resolution, i.e., all grids in

---

[1] For convenience, we denote the j–th component of a vector $\boldsymbol{x}$ as x_j.

a level share the same cell width vector $\boldsymbol{\delta}_g = \boldsymbol{\delta}_{\Gamma_l}$. The region covered by a level Γ_{Λ_l} is the union of regions covered by the grids of that level. In most AMR data sets, only the root level covers a contiguous region in space, while all other levels typically consist of several disjoint regions.

The hierarchy starts with the *root level* Λ_0, the coarsest level. Each level Λ_l may be refined by a finer level Λ_{l+1}. A grid of the refined level is commonly referred to as a *coarse grid* and a grid of the refining level as a *fine grid*. The *refinement ratio* r_l specifies how many cells of a fine grid contained in level l fit into a coarse-grid cell along each axial direction. The refinement ratio is specified as a positive integer rather than a vector, as it is usually the same for all axial directions. A refining grid can only refine complete grid cells of the parent level, i.e., it must start and end at the boundaries of grid cells of the parent level. A refining grid refines an entire level Λ_l, i.e., it is completely contained in Γ_{Λ_l} but not necessarily in the region covered by a single grid of that level. (This is illustrated in Fig. 1, where the grid comprising the second level overlaps both grids of the first level.) Thus, in many cases it is convenient to access grid cells on a per-level basis instead of a per-grid basis.

To obtain the index of a cell within a level, we assume that a *level grid* with a cell width equal to $\boldsymbol{\delta}_{\Lambda_l}$ covers the entire domain Γ_{Λ_0}, see Fig. 2. This level grid starts at the minimum extent of a bounding box surrounding the

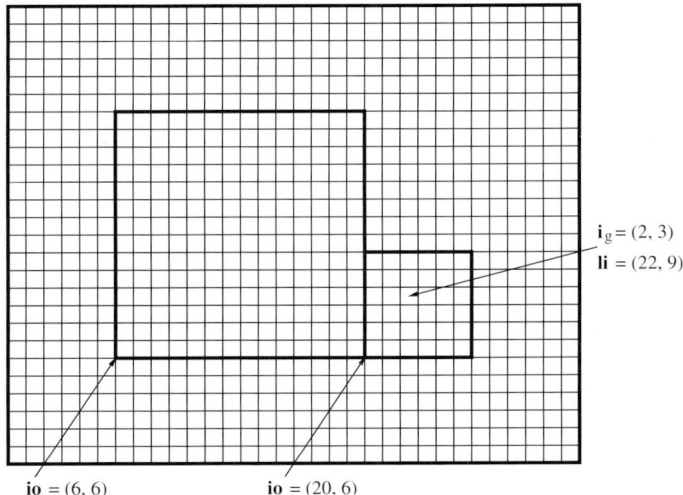

Fig. 2. To specify the index of a cell with respect to a level rather than a single grid, we assume that an entire level is covered by a level grid with a cell width equal to $\boldsymbol{\delta}_{\Gamma_l}$. Individual grids within a level cover rectangular sub-regions of that level grid. An index of a grid cell can either be specified with respect to the level grid ($\boldsymbol{li}$) or with respect to a grid containing that cell ($\boldsymbol{i}_g$). The integer origin of a grid is the level index of the grid cell with index **0** within the grid

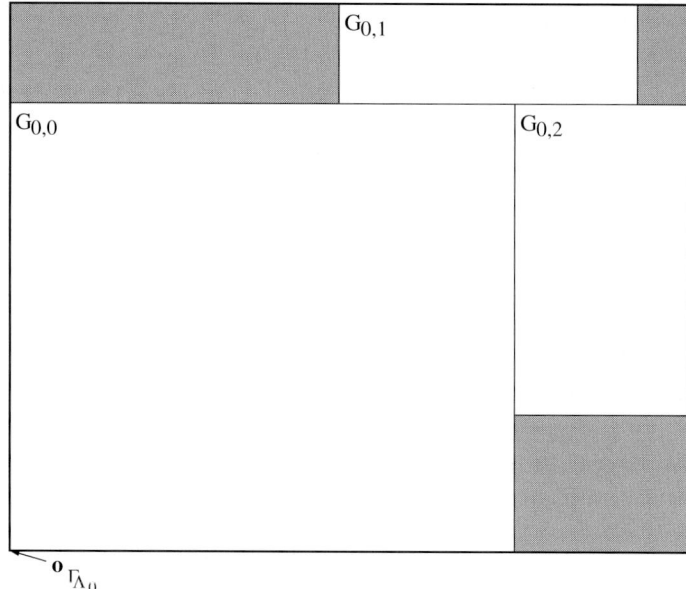

Fig. 3. The AMR hierarchy shown in this figure contains three grids in the root level. Regions that are not covered by any grid are shaded. The origin of a level grid is the minimum extent of the bounding box enclosing all grids in the root level

root level o_{Λ_0}, where

$$o_{\Lambda_0, j} = \min \{o_{g,j} | o_g \in \Lambda_0\} \quad , \tag{2}$$

as shown in Fig. 3. Since all grids in a level are placed with respect to boundaries of grid cells in a parent level, grid cells in the level grid coincide with grid cells in a grid of that level or are outside the region covered by the level Γ_{Λ_l}. The *level index* li of a grid cell is its index in the level grid. Using the level index, the origin of a grid in a level can be defined as an *integer origin*. The integer origin io_g of a grid g is the level index of the cell with index $\mathbf{0}$ within the grid, see Fig. 2. Its components are defined by

$$io_{g,j} = \frac{(o_{g,j} - o_{\Lambda_0, j})}{\delta_{\Gamma_l}} \quad , \text{ where } g \in \Lambda_l \quad . \tag{3}$$

Since all grid cells of a fine grid must start and end at boundaries of cells of a coarse level, the components of the integer origin $io_{g,j}$ and the number of cells $n_{g,j}$ along an arbitrary axial direction of a grid g belonging to level l are always integer multiples of r_l. Individual grids in a level correspond to rectangular sub-regions of the level grid. To access a cell with a given level index li, first the grid g that contains that level index has to be found, provided that such a grid exists. Second, the index i_g of the cell within that grid is obtained by subtracting this grid's integer origin io_g from li.

Due to the hierarchical nature of AMR simulations, AMR data lend themselves to hierarchical visualization. One of the problems encountered when isosurfaces are extracted using an MC method is the cell-centered AMR format. MC methods expect dependent data values at a cell's vertices instead of data associated with a cell's center. If a re-sampling step is used to replace the values at a cell's center with values at its vertices, "dangling" nodes arise at level boundaries. These dangling nodes can cause cracks in the isosurface when using an MC method, even if a consistent interpolation scheme is used, i.e., one that assigns the same value to a dangling node as is assigned to its location in the coarse level. To avoid re-sampling, we interpret the locations of the samples, see Eq. 1, as vertices of a new grid that is "dual" to the original one. This dual grid is then used in a standard MC approach.

The use of dual grids creates gaps between grids of different hierarchy levels. We fill those gaps in an index-based stitching step. Vertices, edges and faces of a fine grid are connected to vertices in the coarse level using a look-up table (LUT) for the possible refinement configurations. The resulting stitch mesh consists of tetrahedra, pyramids, triangular prisms and hexahedral cells. By extending an MC method to these additional cell types, we define an isosurface extraction scheme that avoids cracks in an isosurface at level boundaries.

The original Berger–Colella scheme [2] requires a layer with a width of at least one grid cell between a refining grid and the boundary of the refined level. Even though Bryan [4] eliminates this requirement, we still require it. This is necessary, as this requirement ensures that only transitions between a coarse level and the next finer level occur in an AMR hierarchy. Allowing transitions between arbitrary levels would force us to consider a large number of cases during the stitching process. (The number of cases would be limited only by the number of levels in an AMR hierarchy, since transitions between arbitrary levels are possible.) These requirements are equivalent to requirements described by Gross et al. [7] who also do not permit transitions between arbitrary levels.

2 Related Work

Relatively little research has been published regarding the visualization of AMR data. Norman et al. [14] convert an AMR hierarchy into finite-element hexahedral cells with cell-centered data that can be handled by standard visualization tools (like AVS [1], IDL [8], or VTK [15]), while preserving the hierarchical nature of the data. Ma [10] describes a parallel rendering approach for AMR data. Even though he re-samples the data to vertex-centered data, he still uses the hierarchical nature of AMR data and contrasts it to re-sampling it to the highest level of resolution available. Max [11] describes a sorting scheme for cells for volume rendering, and uses AMR data as one application of his method. Weber et al. [17] present two volume rendering schemes for AMR data. One scheme is a hardware-accelerated renderer for data previewing. This renderer partitions an AMR hierarchy into blocks of equal resolution and renders the complete data set by rendering blocks in back-to-front order. Their other scheme is based on cell projection and allows progressive rendering of AMR hierarchies. It is possible to render an AMR hierarchy starting with a coarse representation and refining it by subsequently integrating the results from rendering finer grids.

Isosurface extraction is a commonly used technique for visual exploration of scalar fields. Our work is based on the MC method, introduced by Lorensen and Cline [9], where a volume is traversed cell-by-cell, and the part of the isosurface within each cell is constructed using an LUT. The LUT in the original paper by Lorensen and Cline contained a minor error that can produce cracks in the extracted isosurface. This is due to ambiguous cases where different isosurface triangulations in a cell are possible. Nielson and Hamann [12], among others, addressed this problem and proposed a solution to it. Van Gelder and Wilhelms [6] provided a survey of various solutions that were proposed for this problem. They showed that, in order to extract a topologically correct isosurface, more than one cell must be considered at a time. If topological correctness of an isosurface is not required, it is possible to avoid cracks without looking at surrounding cells. In our implementation, we use the LUT from VTK [15] that avoids cracks by taking special care during LUT generation.

Octree-based methods can be used to speed up the extraction of isosurfaces. Shekhar et al. [16] use an octree for hierarchical data representation. By adaptively traversing the octree and merging cells that satisfy certain criteria, they reduce the amount of triangles generated for an isosurface. Their scheme removes cracks in a resulting isosurface by moving fine-level vertices at boundaries to a coarser level to match up contours with the coarse level. Westermann et al. [19] modified this approach by adjusting the traversal criteria and improving the crack-removal strategy. Nielson et al. [13] use coons patches to avoid cracks between adjacent regions of different resolution. Instead of avoiding T-intersections or matching up contours they develop a Coons-patch based interpolation scheme for coarse-level cells that avoids cracks. Gross et al. [7] use a combination of wavelets and quadtrees to approximate surfaces, e.g., from terrain data. Considering an estimate based on their wavelet transform, their approach determines a level in the quadtree structure to represent a given region with a specific precision. Handling transitions between quadtree levels is similar to handling those between levels in an AMR hierarchy. Weber et al. [18] have introduced grid stitching to extract isosurfaces from AMR data sets. In this paper, we extent and elaborate on this work.

3 Dual Grids

The MC method assumes that data values are associated with cell vertices, but the prototypical AMR method produces values at cell centers. One possibility to deal with this incompatibility problem is to re-sample a given data set to a vertex-centered format. However, re-sampling causes "dangling nodes" in the fine level. Even if the same values are assigned to the dangling nodes as the interpolation scheme assigns to their location in the coarse level, dangling nodes can cause cracks when the MC method is applied, see [19]. We solve these problems by using a *dual grid* for isosurface extraction. This dual grid is defined by the function values at the cell centers, see Eq. 1. The implied connectivity information between these centers is given by the neighborhood configuration of the original cells. Cell centers become the vertices of the vertex-centered dual grid. The indices of a cell with respect to a grid or a level (level index), defined in Section 1, become indices of vertices in the dual grid.

The dual grids for the first two levels of the AMR hierarchy shown in Fig. 1 are shown in Fig. 4. We note that the dual grids have "shrunk" by one cell in each axial direction with respect to the original grid. The result is a gap between the coarse grid and the embedded fine grids. Due to the existence of this gap, there are no dangling nodes causing discontinuities in an isosurface. However, to avoid cracks in extracted isosurfaces as a result of gaps between grids, a tessellation scheme is needed that "stitches" grids of two different hierarchy levels.

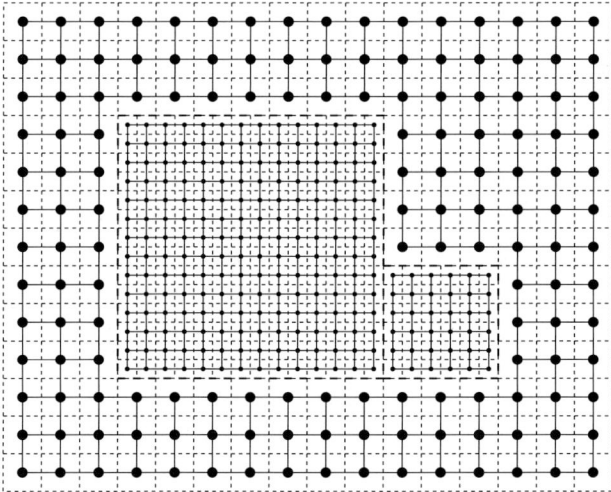

Fig. 4. Dual grids for the three AMR grids comprising the first two hierarchy levels shown in Fig. 1. The original AMR grids are drawn in dashed lines and the dual grids in solid lines

4 Stitching 2D Grids

A *stitch mesh* used to fill gaps between different levels in the hierarchy is constrained by the boundaries of the coarse and the fine grids. In order to merge levels seamlessly, the stitch mesh must not subdivide any boundary elements of the existing grids. In the 2D case, this is achieved by requiring that only existing vertices are used and no new vertices generated. Since one of the reasons for using the dual grids is to avoid the insertion of new vertices, whenever possible, this causes no problems.

In the 2D case, a constrained Delaunay triangulation, see, for example, Chew [5], can be used to fill the gaps between grids. For two reasons, we do not do this. While in the 2D case only edges must be shared between the stitching grid and the dual grids, entire faces must be shared in the 3D case. The boundary faces of rectilinear grids are rectangles that cannot be shared by tetrahedra without being subdivided, thus causing cracks when used in an MC-based isosurface extraction scheme. Furthermore, an index-based approach is more efficient, since it takes advantage of the regular structure of the boundaries while avoiding problems that might be caused by this regular structure when using a Delaunay-based approach.

The stitching process for a refinement ratio of two is shown in Fig. 5. Stitch cells are generated for edges along the boundary and for the vertices of the fine grid. The stitch cells generated for the edges are shown in dark grey, while the stitch cells generated for the vertices are drawn in light grey. For the transition between one fine and one coarse grid, each edge of the fine grid is connected alternatingly to either a vertex or an edge of the coarse grid. This yields triangles and quadrilaterals as additional cells. The quadrilaterals

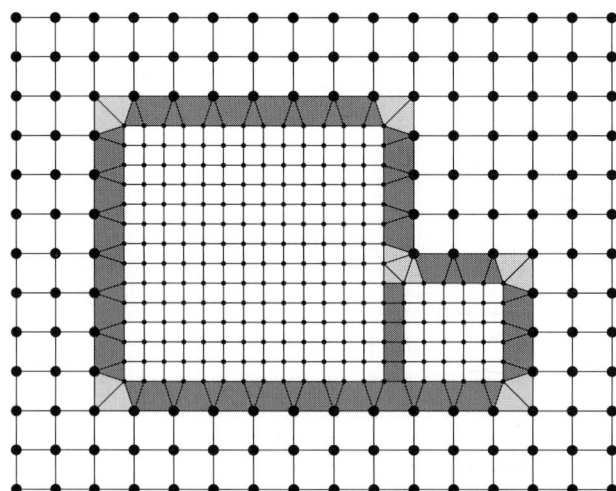

Fig. 5. Stitch cells for first two levels of AMR data set shown in Fig. 4

are not subdivided, since subdivision is not unique. (This in turn would cause problems in the 3D case when quadrilaterals become boundary faces shared between cells.) The vertices are connected to the coarse grid via two triangles. A consistent partition of the deformed quadrilateral is possible. The obvious choice is to connect each edge to the two coarse edges that are "visible" from it.

In the case of multiple grids, a check must be performed: Are the grid points in the coarse grid refined or not? If a fine edge is connected to a coarse point, this check is simple. If the coarse point is refined, the fine edge must be connected to another fine edge; this yields a rectilinear instead of a triangular cell. The case of connecting to a coarse edge is more complicated and is illustrated in Figs. 6 (i)–(iv). If both points are refined, see Fig. 6(iv), the fine edge is connected to another fine edge. As a result, adjacent fine grids yield the same cells as a "continuous" fine grid. Problem cases occur when only one of the points is refined, see Figs. 6(ii) and 6(iii). Even though it is possible to skip these cases and handle them as vertex cases of the other grid, a more consistent approach is to include them in the possible edge cases. However, the same tessellations should be generated for both cases, as shown in Fig. 6.

The cases arising from connecting a vertex are illustrated in Figs. 6 (v)–(vii). In addition to replacing refined coarse grid points by the nearest fine-grid point, adjoining grids must be merged. If either of the coarse grid points

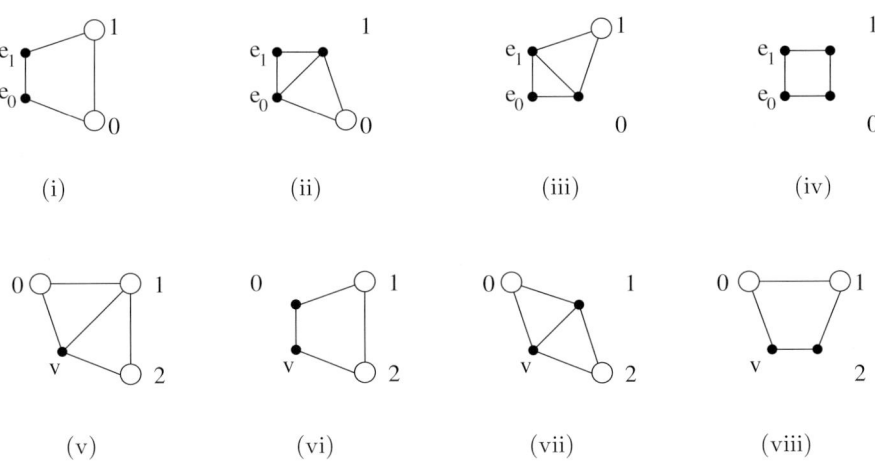

Fig. 6. Possible cases for connecting a boundary edge $\overline{e_0 e_1}$, cases (i)–(iv), or a boundary vertex v, cases (v)–(viii), to a coarse grid. If cells of the coarse grid are refined, the coarse grid points (circles) are replaced by the corresponding refining point (solid black discs)

0, see Fig. 6(v), or 2, see Fig. 6(v), is refined, it is possible to change the border vertex to a border-edge segment by connecting it to the other refined grid point and treating it as an edge, and using the connection configurations from the previous paragraph, i.e., those shown in Figs. 6 (i)–(iv). (This case occurs along the bottom edge of the fine grids shown in Fig. 5.)

Even though arbitrary integer-refinement ratios r_l are possible for AMR grids, refinement ratios of two and four are the most common ones used. The stitching process can be generalized to more general refinement ratios. Instead of connecting edge segments of the refining grid alternatingly to a coarse-grid edge segment and point, $(r_l - 1)$ consecutive edge segments must be connected to one common coarse-grid point. Every r_l-th fine edge must be connected to a coarse edge. The same connection strategy results from connecting each fine grid edge to a parallel "phantom edge" that would exist if the grid continued in that direction. If both end points of the phantom edge are within the same grid cell in the parent level, the edge is connected to the coarse grid point within that cell. If the phantom edge crosses a boundary between two coarse grid cells, the fine edge is connected to the edge formed by the two grid points in those cells.

Even though the valence of the grid points of the coarse grid is increased, this is not a problem with the commonly used refinement ratios. Furthermore, it is important to note that arbitrary refinement ratios would not add more refinement configurations. The fundamental connection strategies remain the same. A fine-grid edge is connected to a coarse-grid vertex or a coarse-grid edge. A fine-grid vertex is connected to two coarse-grid edges. The cell subdivisions shown in Fig. 6 can be used for arbitrary refinement ratios.

5 Stitching 3D Grids

Our index-based approach can be generalized to 3D AMR grids. In the simple case of one fine grid embedded in a coarse grid, boundary faces, edges and vertices of the fine grid must be connected to the coarse grid.

Each of the six boundary faces of a grid consists of a number of rectangles defined by four adjacent grid points on the face. A boundary face is connected to the coarse mesh by connecting each of its comprising rectangles to the coarse grid. For each quadrilateral, the level indices of the four grid points that would extend the grid in normal direction are computed. These are transformed into level indices of the parent level by dividing them by the refinement ratio r_l of the fine level. In each of the two directions implied by a rectangle, these transformed points may have the same level index component. If they have the same index for a direction, the fine rectangle must be connected to one vertex in this direction; otherwise, it must be connected to an edge in that direction. The result is the same as the combination of two 2D edge cases. The various combinations result in rectangles being connected to either a vertex, a line segment (in the two possible directions) or another rectangle. The cell types resulting from these connections are pyramids, see Fig. 7(i), deformed triangle prisms, see Fig. 7(ii), and deformed hexahedral cells, see Fig. 7(iii).

An edge is connected to the coarse grid by connecting its comprising edge segments to the grids in the parent level. For each segment, the level indices of the six grid points that would extend the grid beyond the edge are computed. These indices are also transformed into level indices of the parent level. Depending on whether the edge segment crosses a boundary face of the original AMR grid or not, the edge must either be connected to three perpendicular edges or two rectangles of the coarse grid. This is equivalent to a combination of the vertex and edge connection types of the 2D case. If the viewing direction is parallel to the edge segment (such that it appears to the viewer as a point), it must always be connected to two perpendicular edges of the coarse grid. In the direction along the edge, one connects it to a point or a parallel edge. Connecting an edge segment to the coarse grid results in two tetrahedra, shown in Fig. 7(iv), or two deformed triangle prisms, shown in Fig. 7(v), as connecting cells.

A vertex is connected by calculating the level indices of the seven points that would extend the grid. These are transformed into level indices of the parent level. The result is the same as the combination of two 2D vertex cases. The vertex is connected to three rectangles of the coarse grid via pyramid cells, as shown in Fig. 7(vi).

When the coarse level is refined by more than one fine grid, one must check each coarse-grid point for refinement and adapt the generated tessellation accordingly. The simplest case is given when a fine grid boundary rectangle is connected to a coarse point, see Fig. 7(i). If this coarse grid point is refined an adjacent fine grid exists and the fine grid boundary rectangle must be connected to the other fine grid's boundary rectangle. This case illustrates that it is helpful to retain the indices within the fine level in addition to converting them to the coarse level. In the unrefined case, the rectangle was connected to a coarse point, because transforming the four level indices from the fine level to the coarse level yielded the same coarse level index. If the grid point corresponding to that coarse level index is refined, the "correct" fine boundary rectangle can be determined using the original fine level indices. (For a refinement ration of $r_l = 2$ the correct choice of the fine-level rectangle is also implied by the connection type, but for general refinement ratios the fine-level index must be retained.)

Connecting a fine rectangle to a coarse edge, see Fig. 7(ii), is slightly more difficult. Each of the endpoints of a coarse edge can either be refined or unrefined. The resulting refinement configurations and their tessellations are shown in Fig. 8. Refinement configurations, the cases in Fig. 8 and subsequent figures are numbered as follows: For each connection type shown in Fig. 7, the coarse-grid grid points that are connected to a fine grid element are numbered

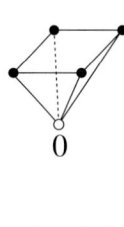

(i) Rectangle-to-vertex

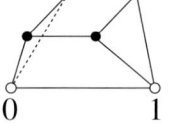

(ii) Rectangle-to-edge

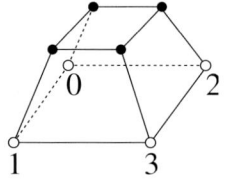

(iii) Rectangle-to-rectangle

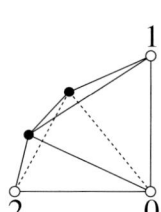

(iv) Edge-to-edges

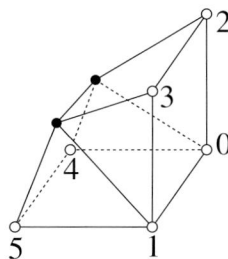

(v) Edge-to-rectangles

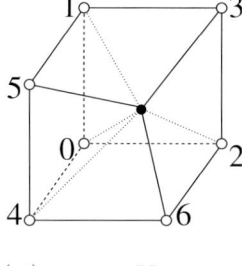

(vi) Vertex-to-rectangles

Fig. 7. Possible connection types for quadrilateral, edge and vertex in 3D case

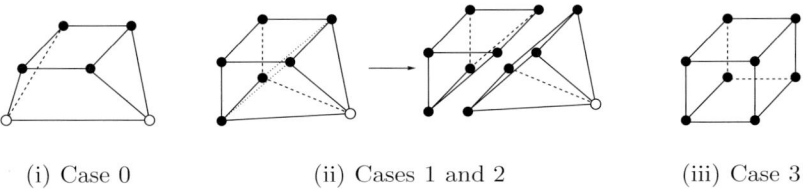

(i) Case 0 (ii) Cases 1 and 2 (iii) Case 3

Fig. 8. Refinement configurations for connecting a fine-grid rectangle to a coarse-grid edge

according to the corresponding sub-figure. A case number is obtained by starting with case 0. For each refined coarse-grid vertex k, 2^k is added to the number associated with this case. To determine to which refining grid points the fine rectangle should be connected we retain the fine-level indices in addition to converting them to coarse level indices.

When connecting a fine-grid edge to two perpendicular coarse-grid edges, eight refinement configurations arise. If all coarse-grid vertices are unrefined, two tetrahedra are generated, as illustrated in Fig. 7(iv). If either coarse-grid vertex 1 or 2 is refined, the fine-grid edge must be upgraded to a coarse-grid rectangle and the corresponding tessellation function called with appropriate vertex ordering. (This procedure ensures that adjacent fine grids produce the same stitch tessellation as a continuous fine grid.) In the remaining case, when only coarse-grid vertex 0 is refined, two pyramids are generated instead of tetrahedra, see Fig. 9.

For connection types where a fine-grid quadrilateral, see Fig. 7(iii), edge, see Fig. 7(v), or vertex, see Fig. 7(v), is connected to coarse-grid rectangles, eight points are considered. These points form a deformed hexahedral cell. One must consider 16 possible refinement configurations when a fine-grid rectangle is connected, since the four vertices belonging to the fine-grid rectangle are always refined, see Fig. 10. More cases arise when a fine-grid edge or a vertex is connected to the coarse level. It is important to devise an

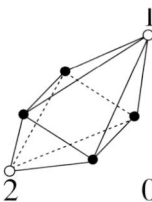

Fig. 9. If coarse-grid vertex 0 is refined, when a fine-grid edge is connected to two coarse-grid edges, see Fig. 7(iv), two pyramids are generated instead of two tetrahedra

efficient scheme to determine the tessellation for a given refinement configuration. Each cell face corresponds to a possible 2D refinement configuration as shown in Fig. 6. It is important to note that the 2D refinement configurations that produce subdivided quadrilaterals are the same configurations that yield non-planar cell boundaries, i.e., boundaries that we must subdivide. The subdivision information alone is sufficient to determine a tessellation for any given refinement configuration. It is not necessary to consider the positions of the points. Each cell face Fig. 10 and subsequent figures is subdivided using the canonical tessellations depicted in Fig. 6, illustrated by the dotted lines in a figure illustrating tessellations of a connection type, e.g., Fig. 10.

The subdivision of cell faces implies subdivisions of hexahedral cells into pyramids, triangular prisms and tetrahedra. In certain cases, see, for example, Fig. 10(iv), a cell type arises that does not correspond to the standard cells (hexahedra, pyramids, triangle prisms and tetrahedra), and that cannot be subdivided further without introducing additional vertices. Even though it is possible to generate a case table to extend MC to this cell type, the asymmetric form of this cell makes this extension difficult. Symmetry considerations that are used to reduce the number of cases that need to be considered cannot be applied. Therefore, we handle cells of this type by generating an additional vertex at the centroid of the cell. By connecting the

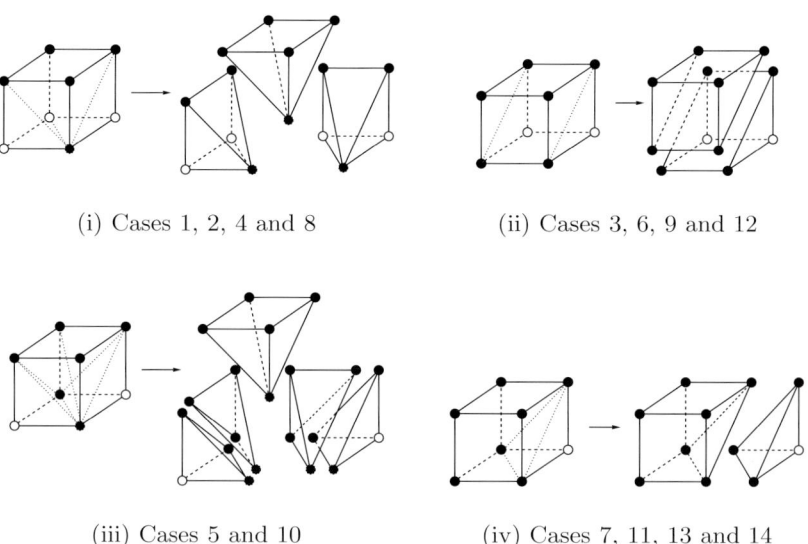

(i) Cases 1, 2, 4 and 8   (ii) Cases 3, 6, 9 and 12

(iii) Cases 5 and 10   (iv) Cases 7, 11, 13 and 14

Fig. 10. Refinement configurations for connecting a fine-grid rectangle to a coarse-grid rectangle

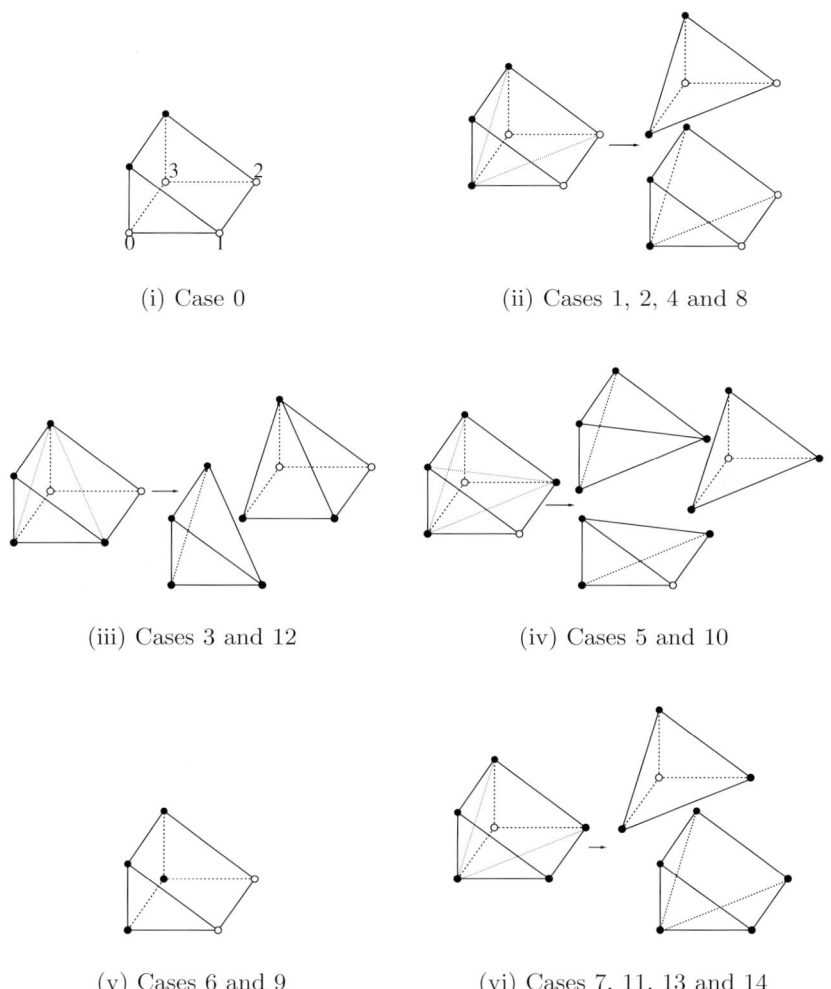

Fig. 11. Possible refinement configurations and corresponding tessellations for triangle prism cell

centroid to all cell vertices we obtain a tessellation consisting of pyramids and tetrahedra.

Edges are connected to the coarse grid by considering eight vertices forming a deformed hexahedral cell. When an edge is connected, two of these eight points belong to the edge. The other six vertices are coarse-grid vertices and can be either refined or unrefined. Thus, it is necessary to consider 64 possible refinement configurations. It is necessary to consider all six coarse grid points at once, since the boundary faces need not always be subdivided faces, as Fig. 7(v) implies. If, for example, all coarse grid points are refined, a single, not-tessellated cell must be produced. However, in certain refinement situations (Cases 0–3, 5, 7, 10, 11, 15, 17, 19, 27, 34, 35, 39, 51), the two cell faces perpendicular to the fine-grid edge segment must be divided as shown in Fig. 7(v). In these cases it is possible to connect the fine-grid edge segment to the coarse grid by handling each of the triangle prisms separately. Each triangle prism is tessellated according to the refinement configurations shown in Fig. 11. When connecting a fine-grid edge to two coarse-grid rectangles, see Fig. 7(iii), it must be upgraded to to the rectangle case if either grid points 2 and 3 or grid points 4 and 5 are refined. In these cases (Cases 12–15, 28–31 and 44–63) the refinement configurations for connecting a fine-grid rectangle are used according to Fig. 10. The tessellations for the remaining cases are shown in Fig. 12. In Fig. 12, we only considered symmetry with respect to a plane perpendicular to the edge to reduce the number of cases. The cases in Figs. 12(ix) and 12(x) differ only by rotation but yield the same tessellation. We note that only the partition of the boundary faces matters in determining a valid tessellation. Even though the refinement configurations shown Figs. 12(iv) and 12(vi) differ, they yield the same partition of a cell's boundary faces and thus the same tessellation.

Vertices can be upgraded to edges, or even quadrilaterals, when more than two grids meet at a given location. The fine vertex shown in Fig. 7(vi) can be changed to an edge, if any of the coarse grid points 3, 5 or 6 is refined. In these cases, the procedure used to connect an edge segment is called with appropriate vertex ordering. As a result, the same tessellations are used as in the case of a continuing edge. Furthermore, this procedure ensures that an additional upgrade to the rectangle case is handled automatically when needed. In the remaining cases, each of the pyramids of the unrefined case can be handled independently. If the base face of a pyramid is not planer, it is subdivided using the corresponding configuration from Fig. 6, and the pyramid is split into two tetrahedra.

6 Isosurface Extraction

Within individual grids, we apply a slightly modified MC approach for isosurface extraction. Instead of considering all cells of a grid for isosurface generation, we consider only those cells that are not refined by a finer grid. We do this by pre-computing a map with refinement information for each grid. For each grid cell, this map contains an index of a refining grid or an entry indicating that the cell is unrefined. This enables us to quickly skip refined portions of the grid. For the generation of an isosurface within stitch cells, the MC method must be extended to handle the cell types generated during the stitching process. This extension is achieved by generating case tables for each of the additional cell types. These new case tables must be

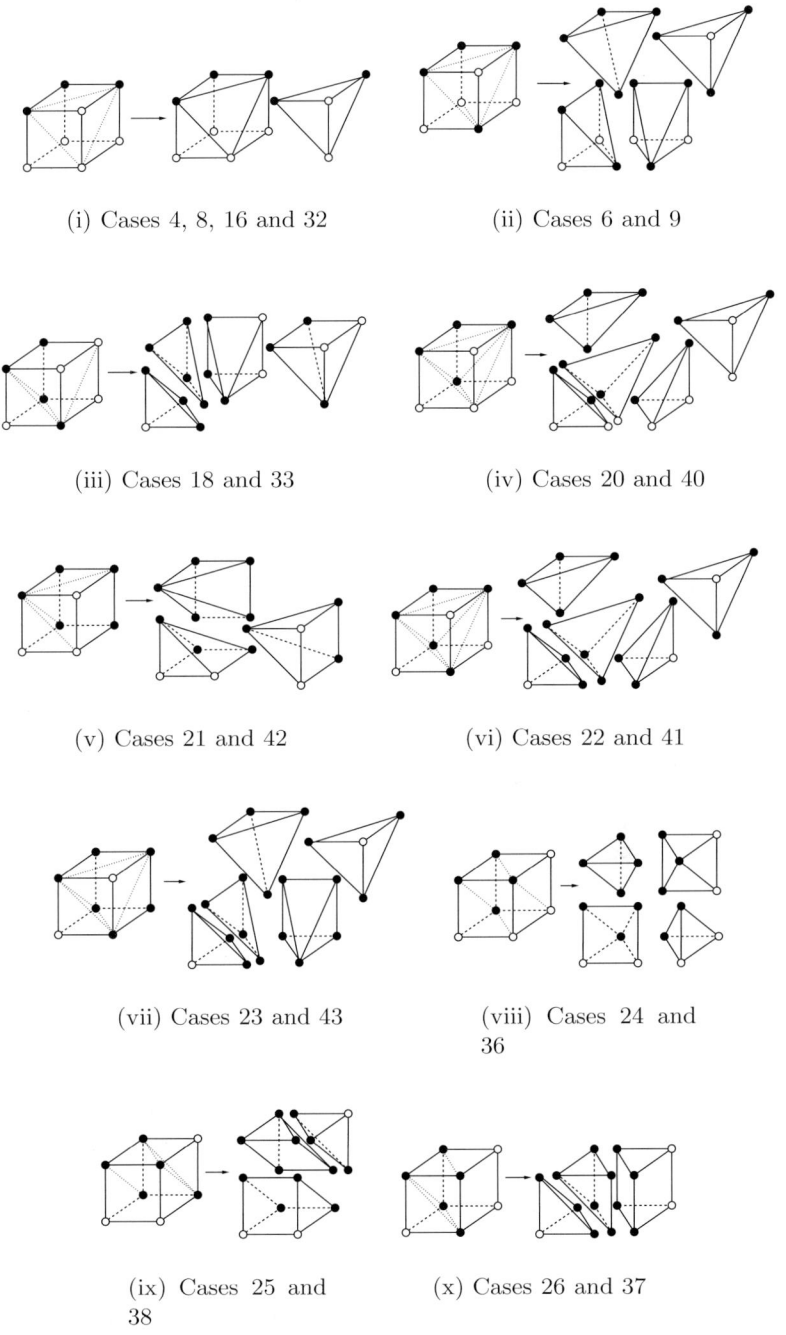

(i) Cases 4, 8, 16 and 32

(ii) Cases 6 and 9

(iii) Cases 18 and 33

(iv) Cases 20 and 40

(v) Cases 21 and 42

(vi) Cases 22 and 41

(vii) Cases 23 and 43

(viii) Cases 24 and 36

(ix) Cases 25 and 38

(x) Cases 26 and 37

Fig. 12. Remaining refinement configurations for connecting fine-grid edge to coarse-grid rectangles

compatible with the one used in the standard MC approach, i.e., ambiguous cases, see Section 2 must be handled in exactly the same way as handled for typical hexahedral cells.

7 Results

Fig. 13 shows isosurfaces extracted from an AMR data set. This data set is the result from an astrophyiscal simulation of star clusters performed by Bryan [4]. The isosurface in Fig. 13(i) shows an isosurface extracted from two levels of the hierarchy, and Fig. 13(ii) one extracted from three levels. To highlight the transitions between levels, the parts of the isosurface extracted from different levels of the hierarchy are colored differently. Isosurface parts extracted from the root, the first and the second levels are colored red, orange and light blue, respectively. Portions extracted from the stitch meshes between the root and the first level are colored in green, and portions extracted from the stitch meshes between the first and second levels are colored in yellow. The root level and the first level of the AMR hierarchy each consist of one $32 \times 32 \times 32$ grid. The second level consists of 12 grids of resolutions $6 \times 12 \times 6$, $6 \times 4 \times 2$, $8 \times 12 \times 10$, $6 \times 4 \times 4$, $14 \times 4 \times 10$, $6 \times 6 \times 12$, $12 \times 10 \times 12$, $10 \times 4 \times 8$, $6 \times 6 \times 2$, $16 \times 26 \times 52$, $14 \times 16 \times 12$ and $36 \times 52 \times 36$. All measurements were performed on a standard PC with a 700MHz Pentium III processor.

8 Conclusions and Future Work

We have presented a method for the extraction of crack-free isosurfaces from AMR data. By using a dual-grid approach and filling gaps with stitch cells we avoid re-sampling of data and dangling nodes. By extending the standard MC method to the cell types resulting from grid stitching, we have developed an isosurfacing scheme that produces consistent and seamless isosurfaces. Theoretically any continuous (scattered data) interpolation scheme could also lead to a crack free isosurface. However, the use of dual grids and stitch cells ensures that the resolution of extracted isosurfaces automatically varies with the resolution of the data given in a region.

Several extensions to our method are possible. The use of a generic triangulation scheme would allow the use of our method for other, more general AMR data, where grids might not necessarily be axis-aligned, e.g., data sets produced by the AMR method of Berger and Oliger [3]. There are also possible improvements to our index-based scheme. The original AMR scheme by Berger and Colella [2] requires a layer of width of at least one grid cell between a refining grid and the boundary of a refined level. Even though Bryan [4] eliminates this requirement, we still require it. This is necessary, because this requirement ensures that only transitions between a coarse level and the next finer level occur in an AMR hierarchy. Allowing transitions between arbitrary levels would require us to consider too many cases during

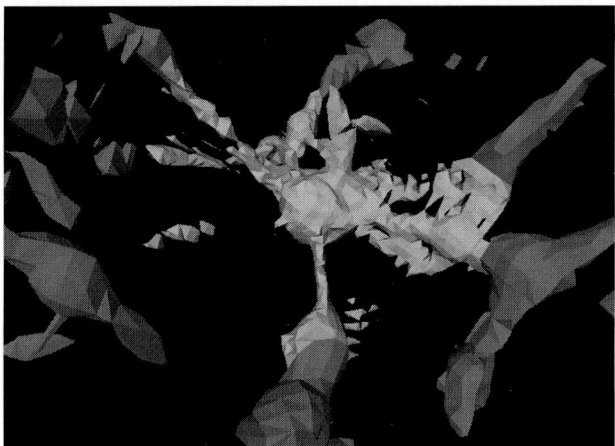

(i) Isosurface obtained when using two of seven levels of AMR hierarchy. Stitch cell generation required approximately 55ms, and isosurface generation required approximately 250ms

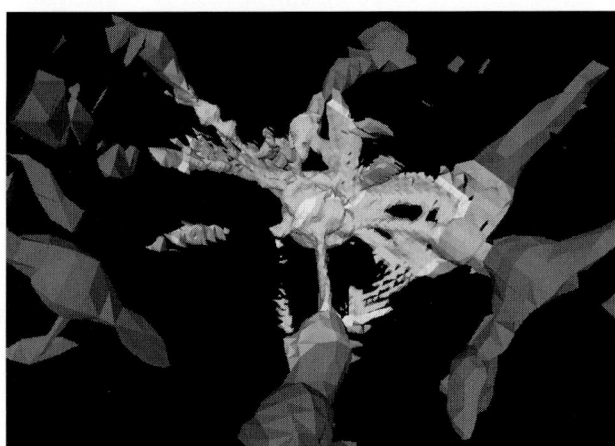

(ii) Isosurface obtained when using three of seven levels of AMR hierarchy. Stitch cell generation required approximately 340ms, and isosurface generation required approximately 600ms

Fig. 13. Isosurface extracted from AMR hierarchy simulating star clusters (data set courtesy of Greg Bryan, Massachusetts Institute of Technology, Theoretical Cosmology Group, Cambridge, Massachusetts). (See Color Plate 1 on page 349.)

the stitching process. (The number of cases would be limited only by the number of levels in an AMR hierarchy, since within this hierarchy transitions between arbitrary levels are possible.) These requirements are equivalent to the requirements described by Gross et al. [7], where transitions between arbitrary levels are also prevented. Unfortunately, this requirement does not allow us to handle the full range of AMR data sets in use today, e.g., those produced by the methods of Bryan [4].

To handle the full range of AMR grid structures, our grid-stitching approach must be extended. Any LUT-based approach has inherent problems, since the number of possible level transitions is bounded only by the number of levels in a hierarchy. For the transformation of level coordinates to grid coordinates, we currently examine each grid in a level whether it contains a given grid point. This is efficient enough for moderately sized data sets; but for larger data sets, a space subdivision-based search scheme should be used. It should be possible to use a modification of the generalized k–D trees from Weber et al. [17] for this purpose.

9 Acknowledgments

This work was supported by the Directory, Office of Science, Office of Basic Energy Sciences, of the U.S. Department of Energy under contract no. DE-AC03-76SF00098, awarded to the Lawrence Berkeley National Laboratory; the National Science Foundation under contract ACI 9624034 (CAREER Award), through the Large Scientific and Software Data Set Visualization (LSSDSV) program under contract ACI 9982251, and through the National Partnership for Advanced Computational Infrastructure (NPACI); the Office of Naval Research under contract N00014-97-1-0222; the Army Research Office under contract ARO 36598-MA-RIP; the NASA Ames Research Center through an NRA award under contract NAG2-1216; the Lawrence Livermore National Laboratory under ASCI ASAP Level-2 Memorandum Agreement B347878 and under Memorandum Agreement B503159; the Los Alamos National Laboratory; and the North Atlantic Treaty Organization (NATO) under contract CRG.971628.

We also acknowledge the support of ALSTOM Schilling Robotics and SGI. We thank the members of the NERSC/LBNL Visualization Group; the LBNL Applied Numerical Algorithms Group; the Visualization and Graphics Research Group at the Center for Image Processing and Integrated Computing (CIPIC) at the University of California, Davis, and the AG Graphische Datenverarbeitung und Computergeometrie at the University of Kaiserslautern, Germany.

References

1. AVS5. Product of Advanced Visual Systems, see http://www.avs.com/products/AVS5/avs5.htm.

2. Marsha Berger and Phillip Colella. Local adaptive mesh refinement for shock hydrodynamics. *Journal of Computational Physics*, 82:64–84, May 1989. Lawrence Livermore National Laboratory, Technical Report No. UCRL-97196.
3. Marsha Berger and Joseph Oliger. Adaptive mesh refinement for hyperbolic partial differential equations. *Journal of Computational Physics*, 53:484–512, March 1984.
4. Greg L. Bryan. Fluids in the universe: Adaptive mesh refinement in cosmology. *Computing in Science and Engineering*, 1(2):46–53, March/April 1999.
5. L. Paul Chew. Constrained delaunay triangulations. *Algorithmica*, 4(1):97–108, 1989.
6. Allen Van Gelder and Jane Wilhelms. Topological considerations in isosurface generation. *ACM Transactions on Graphics*, 13(4):337–375, October 1994.
7. Markus H. Gross, Oliver G. Staadt, and Roger Gatti. Efficient triangular surface approximations using wavelets and quadtree data structures. *IEEE Transactions on Visualization and Computer Graphics*, 2(2):130–143, June 1996.
8. Interactive Data Language (IDL). Product of Research Systems, Inc., see `http://www.rsinc.com/idl/index.cfm`.
9. William E. Lorensen and Harvey E. Cline. Marching cubes: A high resolution 3D surface construction algorithm. *Computer Graphics (SIGGRAPH '87 Proceedings)*, 21(4):163–169, July 1987.
10. Kwan-Liu Ma. Parallel rendering of 3D AMR data on the SGI/Cray T3E. In: *Proceedings of Frontiers '99 the Seventh Symposium on the Frontiers of Massively Parallel Computation*, pages 138–145, IEEE Computer Society Press, Los Alamitos, California, February 1999.
11. Nelson L. Max. Sorting for polyhedron compositing. In: Hans Hagen, Heinrich Müller, and Gregory M. Nielson, editors, *Focus on Scientific Visualization*, pages 259–268. Springer-Verlag, New York, New York, 1993.
12. Gregory M. Nielson and Bernd Hamann. The asymptotic decider: Removing the ambiguity in marching cubes. In: Gregory M. Nielson and Larry J. Rosenblum, editors, *IEEE Visualization '91*, pages 83–91, IEEE Computer Society Press, Los Alamitos, California, 1991.
13. Gregory M. Nielson, Dave Holiday, and Tom Roxborough. Cracking the cracking problem with coons patches. In: David Ebert, Markus Gross, and Bernd Hamann, editors, *IEEE Visualization '99*, pages 285–290,535, IEEE Computer Society Press, Los Alamitos, California, 1999.
14. Michael L. Norman, John M. Shalf, Stuart Levy, and Greg Daues. Diving deep: Data management and visualization strategies for adaptive mesh refinement simulations. *Computing in Science and Engineering*, 1(4):36–47, July/August 1999.
15. William J. Schroeder, Kenneth M. Martin, and William E. Lorensen. *The Visualization Toolkit*, second edition, 1998. Prentice-Hall, Upper Saddle River, New Jersey.
16. Raj Shekhar, Elias Fayyad, Roni Yagel, and J. Fredrick Cornhill. Octree-based decimation of marching cubes surface. In: Roni Yagel and Gregory M. Nielson, editors, *IEEE Visualization '96*, pages 335–342, 499, IEEE Computer Society Press, Los Alamitos, California, October 1998.
17. Gunther H. Weber, Hans Hagen, Bernd Hamann, Kenneth I. Joy, Terry J. Ligocki, Kwan-Liu Ma, and John M. Shalf. Visualization of adaptive mesh refinement data. In: Robert F. Erbacher, Philip C. Chen, Jonathan C. Roberts,

Craig M. Wittenbrink, and Matti Groehn, editors, *Proceedings of the SPIE (Visual Data Exploration and Analysis VIII, San Jose, CA, USA, Jan 22–23)*, volume 4302, pages 121–132, SPIE – The International Society for Optical Engineering, Bellingham, WA, January 2001.
18. Gunther H. Weber, Oliver Kreylos, Terry J. Ligocki, John M. Shalf, Hans Hagen, Bernd Hamann, and Kenneth I. Joy. Extraction of crack-free isosurfaces from adaptive mesh refinement data. In: David Ebert, Jean M. Favre, and Ronny Peikert, editors, *Proceedings of the Joint EUROGRAPHICS and IEEE TCVG Symposium on Visualization, Ascona, Switzerland, May 28–31, 2001*, pages 25–34, 335, Springer Verlag, Wien, Austria, May 2001.
19. Rüdiger Westermann, Leif Kobbelt, and Thomas Ertl. Real-time exploration of regular volume data by adaptive reconstruction of isosurfaces. *The Visual Computer*, 15(2):100–111, 1999.

Edgebreaker on a Corner Table: A Simple Technique for Representing and Compressing Triangulated Surfaces

Jarek Rossignac, Alla Safonova, and Andrzej Szymczak

College of Computing and GVU Center, Georgia Institute of Technology

Abstract. A triangulated surface S with V vertices is sometimes stored as a list of T independent triangles, each described by the 3 floating-point coordinates of its vertices. This representation requires about 576V bits and provides no explicit information regarding the adjacency between neighboring triangles or vertices. A variety of boundary-graph data structures may be derived from such a representation in order to make explicit the various adjacency and incidence relations between triangles, edges, and vertices. These relations are stored to accelerate algorithms that visit the surface in a systematic manner and access the neighbors of each vertex or triangle. Instead of these complex data structures, we advocate a simple Corner Table, which explicitly represents the triangle/vertex incidence and the triangle/triangle adjacency of any manifold or pseudo-manifold triangle mesh, as two tables of integers. The Corner Table requires about $12V\log_2 V$ bits and must be accompanied by a vertex table, which requires 96V bits, if Floats are used. The Corner Table may be derived from the list of independent triangles. For meshes homeomorphic to a sphere, it may be compressed to less that 4V bits by storing the "clers" sequence of triangle-labels from the set {C,L,E,R,S}. Further compression to 3.6V bits may be guaranteed by using context-based codes for the clers symbols. Entropy codes reduce the storage for large meshes to less than 2V bits. Meshes with more complex topologies may require $O(\log_2 V)$ additional bits per handle or hole. We present here a publicly available, simple, state-machine implementation of the Edgebreaker compression, which traverses the corner table, computes the CLERS symbols, and constructs an ordered list of vertex references. Vertices are encoded, in the order in which they appear on the list, as corrective displacements between their predicted and actual locations. Quantizing vertex coordinates to 12 bits and predicting each vertex as a linear combinations of its previously encoded neighbors leads to short displacements, for which entropy codes drop the total vertex location storage for heavily sampled typical meshes below 16V bits.

1 Introduction

3D graphics plays an increasingly important role in applications where 3D models are accessed through the Internet. Due to improved design and model acquisition tools, to the wider acceptance of this technology, and to the need for higher accuracy, the number and complexity of these models are growing more rapidly than phone and network bandwidth. Consequently, it is imperative to continue increasing the terseness of 3D data transmission formats and the performance and reliability of the associated compression and decompression algorithms.

Although many representations have been proposed for 3D models, polygon and triangle meshes are the de facto standard for exchanging and viewing 3D models. A triangle mesh may be represented by its vertex data and by its connectivity. *Vertex data* comprises coordinates of all the vertices and optionally the vertex colors and the coordinates of the associated normal vectors and textures. In its simplest form, *connectivity* captures the incidence relation between the triangles of the mesh and their bounding vertices. It may be represented by a triangle-vertex incidence table, which associates with each triangle the references to its three bounding vertices.

In practice, the number of triangles is roughly twice the number of vertices. Consequently, when pointers or integer indices are used as vertex-references and when floating point coordinates are used to encode vertex locations, uncompressed connectivity data consumes twice more storage than vertex coordinates.

Vertex coordinates may be compressed through various forms of vector quantization. Most vertex compression approaches exploit the coherence in vertex locations by using local or global predictors to encode corrections instead of absolute vertex data. Both the encoder and the decoder use the same prediction formula. The encoder transmits the difference between the predicted and the correct vertex data. It uses variable length codes for the corrections. The better the prediction – the shorter the codes. The decoder receives the correction, decodes it and adds it to the predicted to obtain the correct information for the next vertex. Thus the prediction can only exploit data that has been previously received and decoded. Most predictive schemes require only local connectivity between the next vertex and its previously decoded neighbors. Some global predictors require having the connectivity of the entire mesh. Thus it is imperative to optimize connectivity compression techniques that are independent of vertex data.

The Edgebreaker compression scheme, discussed here, has been extended to manifold meshes with handles and holes [Ross99], to triangulated boundaries of non-manifold solids [RoCa99], and to meshes that contain only quadrilaterals or a combination of simply-connected polygonal faces with an arbitrary number of sides [King99b]. It was also optimized for meshes with nearly regular connectivity [SKR00, SKR00b]. Nevertheless, for sake of simplicity, in this chapter, we restrict our focus to meshes that are each homeomorphic to a sphere.

As several other compression schemes [TaRo98, ToGo98, Gust98], Edgebreaker visits the triangles in a spiraling (depth-first) triangle-spanning-tree order and generates a string of descriptors, one per triangle, which indicate how the mesh can be rebuilt by attaching new triangles to previously reconstructed ones. The popularity of Edgebreaker lies in the fact that all descriptors are symbols from the set {C,L,E,R,S}. No other parameter is needed. Because half of the descriptors are Cs, a trivial code (C = 0, L = 110, E = 111, R = 101, S = 100) guarantees that storage will not exceed 2 bits per triangle. A slightly more complex code guarantees 1.73 bits per triangle [King99]. This upper bound on storage does not rely on statistic-based entropy or arithmetic coding schemes, which in general perform poorly on small or irregular meshes. Consequently, the simple encoding of Edgebreaker is particularly attractive for large catalogs of small models and for crude simplifications of complex meshes. For large meshes, entropy codes may be used to further reduce the storage to less than a bit per triangle [RoSz99]. The string of descriptors produced by Edgebreaker is called the *clers* string. An efficient decompression algorithm for the *clers* sequence [RoSz99] interprets the symbols to build a simply connected triangulated polygon, which represents the triangle-spanning tree. Then, it zips up the borders of that polygon by matching pairs of its bounding edges in a bottom-up order with respect to the vertex-spanning-tree that is the dual of the triangle-spanning-tree. We describe here a compact implementation of this decompression. A previously proposed alternative, called Spirale Reversi [IsSo99], interprets the reversed *clers* string and builds the triangle tree from the leaves.

The contributions of this chapter are a simple data structure, called the Corner-Table, for representing the connectivity of triangle meshes and a very compact (single page) description of the complete Edgebreaker compression and decompression algorithms, which trivializes their implementation. The data structure, examples, and source codes for this implementation are publicly available [EB01]. Because the corner table is nothing more than two arrays of integers and because the decompression is simple and fast, the scheme may be suitable for hardware implementation. We first define our notation and introduce the Corner-Table, then we present the simplified Compression and Decompression algorithms.

2 Notation and Corner-Table

Vertices are identified using positive integers. Their locations are stored in an array called G for "geometry". Each entry of G is a 3D point that encodes the location of a vertex. (Other vertex attributes are ignored for simplicity in this chapter.) We have overloaded the "+" and "–" operators to perform additions and subtraction of points and vectors. Thus G[1]–G[0] returns the vector from the first vertex to the second. Edgebreaker compression stores a point and a sequence of corrective vectors in the string called *delta*, using WRITE(*delta*, D) statements, where D is a point or vector. The corrective vectors will be encoded using a variable length binary format in a separate post-processing entropy-compression step. During decompression, the first call READ(*delta*) returns a decoded version of the first vertex. Subsequent calls to READ(*delta*) return corrective vectors, which are added to the vertex estimates to produce correct vertices.

Compression stores, in a string called *clers*, a sequence of symbols from the set {C,L,E,R,S}, encoded using a simple binary format: {0, 110, 111, 101, 100}. Better codes may be substituted easily, if desired.

During decompression, the symbols (i.e., their binary format) are read and used to switch to the correct operation. We assume that the READ instruction knows to read two more bits when the first one is a 1.

The Corner Table data structure used by Edgebreaker is composed of two global arrays (the V and O tables) and of two temporary tables (M, U), which are only used during compression. V, O, and U have 3 times as many entries as there are triangles. M has as many entries as vertices. V and O hold the integer references to vertices and to opposite corners. M and U hold binary flags indicating whether the corresponding vertex or triangle has already been visited.

Although Edgebreaker manipulates integer indices, we use (*our own*) object-oriented notation to increase the readability of the algorithms that follow. We use lower-case letters that follow a period to refer to table entries or functions with the corresponding uppercase name. For example, if c is an integer, c.v stands for V[c] and c.o stands for O[c]. However, when we assign values to specific entries in these tables, we still write V[c] = b, rather than c.v = b, to remind the reader that we are updating an entry in the V table. We use left-to-right expansion of this "object-oriented" notation, thus c.o.v stands for V[O[c]].

We also introduce the "next corner around triangle" functions: N(c), which will be written c.n and which returns c–2, if c MOD 3 is 2, and c+1 otherwise. This functions permits to move from one corner of a triangle to the next according to the agreed-upon orientation of the triangle, which we assume to be consistent throughout the mesh. The "previous corner around triangle" function, written as c.p stands for N(N(c)). For example, the statement V[a.p] = b.n.v translates to V[N(N(a))] = V[N(b)].

A **corner** c is the association of a triangle c.t with one of its bounding vertices c.v (see Fig. 1). The entries in V and O are consecutive for the 3 corners (c.p, c, c.n) of each triangle. Thus, c.t returns the integer division of c by 3 and the corner-triangle relation needs not be stored explicitly. For example, when c is 4, c.t is 1 and thus c is a corner of the second triangle. We use c.t only to mark previously visited triangles in the U table.

The notation c.v returns the id of the vertex associated with corner c. We use this id to mark previously visited vertices in the M table or to access the geometry of the vertex (c.v.g). The notation c.o returns the id of the corner opposite to c. To be precise, c.o is the only integer b for which: c.n.v == b.p.v and c.p.v == b.n.v. For convenience, we also define c.l as c.p.o and c.r as c.n.o. These relations are illustrated in the figure below. We assume a counter-clockwise orientation.

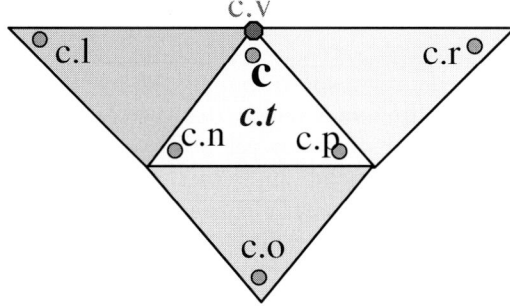

Fig. 1. Using the V and O tables, given a corner, c, we can access: its triangle, c.t; its vertex, c.v; the previous and next corners, c.p and c.n, in c.t; the opposite corner, c.o; and the corners of the left and right neighbors, c.l and c.r.

3 Compression

Edgebreaker is a state machine. At each state it moves from a triangle Y to an adjacent triangle X. It marks all visited triangles and their bounding vertices. Let Left and Right denote the other two triangles that are incident upon X. Let **v** be the vertex common to X, Left, and Right. If **v** has not yet been visited, then neither have Left and Right. This is case C. If **v** has been visited, we distinguish four other cases, which corresponds to four situations where one, both, or neither of the Left and Right triangles have been visited. These situations and the associated *clers* symbols are shown in Fig. 2. The arrow indicates the direction to the next triangle. Previously visited triangles are not shown. Note that in the S case, Edgebreaker moves to the right, using a recursive call, and then to the left.

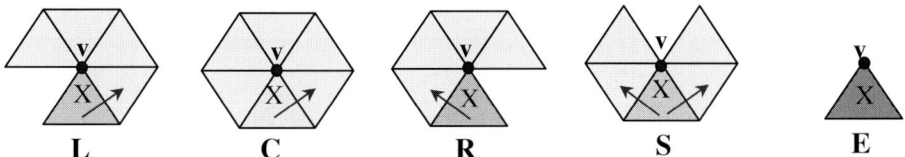

Fig. 2. Using the V and O tables, given a corner, c, we can access: its triangle, c.t; its vertex, c.v; the previous and next corners, c.p and c.n, in c.t; the opposite corner, c.o; and the corners of the left and right neighbors, c.l and c.r.

The compression algorithm is composed of an initialization followed by a call to *Compress*. The initial corner c may be chosen randomly. The initialization encodes and marks the three vertices of the first triangle, marks the triangle as visited, and calls compress.

Compress is a recursive procedure that traverses the mesh along a spiraling triangle-spanning tree. A recursion starts at triangles that are of type S.

It compresses the branch adjacent to the right edge of such a triangle. When the corresponding E triangle is reached, the branch traversal is complete and we "RETURN" from the recursion to pursue the left branch. The encounter of an E that does not match an S terminates the compression process. If the tip vertex of a new triangles has not yet been visited ("IF $c.v.m! = 1$"), we are on a C triangle and we encode in *delta* the corrective vector for the tip of the current triangle using a parallelogram rule [ToGo98]. We also encode a C symbol (for example code(C) may return a 0) in the *clers* string. When the tip of the new triangle has been visited, we distinguish amongst the four other cases, based on the status of the neighboring (left and right) triangles. The compression pseudo-code is provided in the frame below.

```
PROCEDURE initCompression (c){
    GLOBAL M[]={0...}, U[]={0...};        # init tables for marking visited vertices and triangles
    WRITE(delta, c.p.v.g);                # store first vertex as a point
    WRITE(delta, c.v.g – c.p.v.g);        # store second vertex as a difference vector with first
    WRITE(delta, c.n.v.g – c.v.g);        # store third vertex as a difference vector with second
    M[c.v] = 1; M[c.n.v] = 1; M[c.p.v] = 1;  # mark these 3 vertices
    U[c.t] = 1;                           # paint the triangle and go to opposite corner
    Compress (c.o); }                     # start the compression process

RECURSIVE PROCEDURE Compress (c) {        # compressed simple t-meshes
    REPEAT {                              # start traversal for triangle tree
        U[c.t] = 1;                       # mark the triangle as visited
        IF c.v.m != 1                     # test whether tip vertex was visited
        THEN {WRITE(delta, c.v.g – c.p.v.g – c.n.v.g + c.o.v.g);  # append correction for c.v
            WRITE(clers, code(C));        # append encoding of C to clers
            M[c.v] = 1;                   # mark tip vertex as visited
            c = c.r}                      # continue with the right neighbor
        ELSE IF c.r.t.u == 1              # test whether right triangle was visited
            THEN IF c.l.t.u == 1          # test whether left triangle was visited
                THEN {WRITE(clers, code(E)); RETURN}# append code for E and pop
                ELSE {WRITE(clers, code(R)); c = c.l }  # append code for R, move to left triangle
            ELSE IF c.l.t.u == 1          # test whether left triangle was visited
                THEN {WRITE(clers, code(L)); c = c.r }  # append code for L, move to right triangle
                ELSE {WRITE(clers, code(S));           # append code for S
                    Compress(c.r);        # recursive call to visit right branch first
                    c = c.l } } }         # move to left triangle
```

The Fig. 3, below, shows the labels for triangles that have been visited during a typical initial phase of compression.

Figure 4 shows the final steps of compression for a branch or for the whole mesh. It appends the symbols CRSRLECRRRLE to *clers*. The first triangle is marked by an arrow.

4 Decompression

The decompression algorithm builds the two arrays, V and O, of the corner Corner-Table and the G table of vertex locations. After initializing the first triangle in *initDecompression*, the recursive procedure *Decompress* is called for corner 1. In each iteration, Edgebreaker appends a new triangle to a

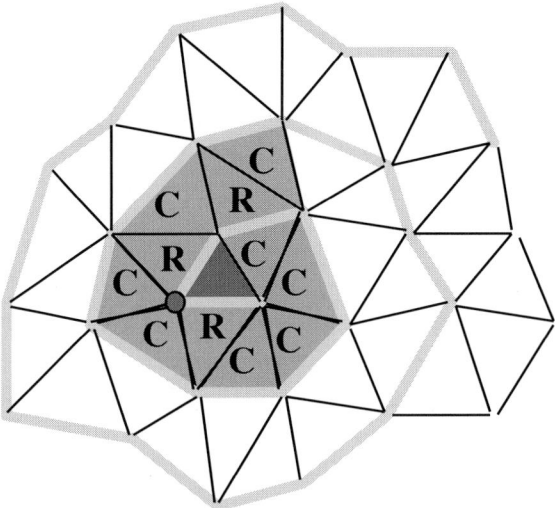

Fig. 3. The clers sequence CCCCRCCRCRC is produced as Edgebreaker starts compressing this mesh.

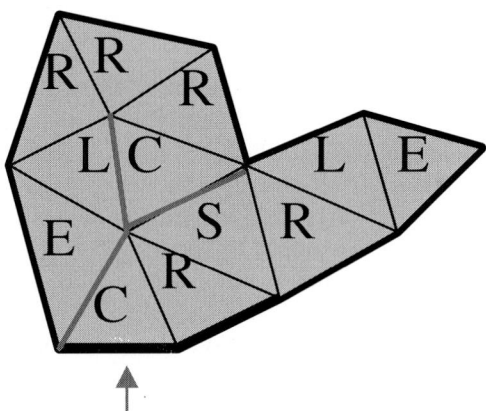

Fig. 4. The clers sequence CRSRLECRRRLE is produced as Edgebreaker finishes compressing this mesh.

previously visited one. It interprets the binary encoding of the next symbol from the *clers* string. If it is a C, Edgebreaker associates the label −1 with the corner opposite the left edge and stores if in O. This temporary marking will be replaced with the correct reference to the opposite corner by a zip. If the symbol is an L, Edgebreaker associates a different label (−2) with the opposite edge and tries to zip, by identifying it with the adjacent edge on the left. When an R symbol is encountered, the opposite edge is labeled −2. No zipping takes place. When an E symbol is encountered, both edges are labeled

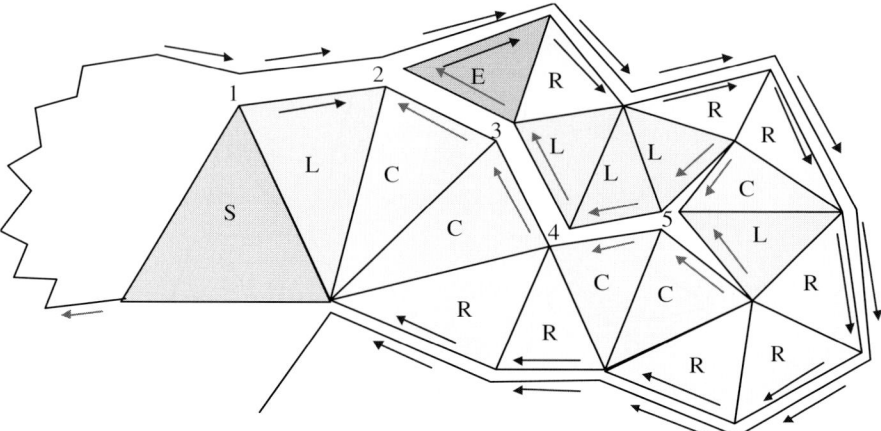

Fig. 5. The clers string SLCCRRCCRRRLCRRLLLRE will generate the mesh shown here. The left edge of the first L triangle is not zipped immediately. The left edge of the second L triangle is zipped reaching vertex 5. Then, as we encounter the subsequent three L triangles, their left edges are zipped right away. The first left edge of the E triangle is also zipped. The rest will be zipped later, when the left branch of the S triangle is decoded.

–2, and an iterative zipping is attempted. This zipping will continue as long as the free edge on the right of the last zipped vertex is marked with –2 and the free edge on the left is marked –1. An S symbols forks a recursive call to *Decompress*, which will construct and zip a subset of the mesh that is incident to the right edge of the current triangle. Then the reconstruction proceeds to decode and build the branch attached to the left edge of the current triangle. Typically about 2% of the triangles are of type S. The process is illustrated in Fig. 5.

The pseudocode for the decompression algorithm is shown in the frame below.

5 Conclusion

3D mesh compression and planar graph encoding techniques have been the subject of numerous publications (see [Ross99] for a review of prior art). All these approaches have been presented at a high level. Many are complex and difficult to implement. In comparison, the proposed compression and decompression algorithms are trivial to implement. More importantly, the source code is extremely small and uses simple arrays of integers as a data structure. This simplicity makes them suitable for many Internet and possibly even hardware applications.

```
PROCEDURE initDecompression {
    GLOBAL V[] = { 0,1,2,0,0,0,0,0,...};         # table of vertex Ids for each corner
    GLOBAL O[] = {-1,-3,-1,-3,-3,-3...};          # table of opposite corner Ids for each corner
    GLOBAL T = 0;                                 # id of the last triangle decompressed so far
    GLOBAL N = 2;                                 # id of the last vertex encountered
    DecompressConnectivity(1);                    # starts connectivity decompression

    GLOBAL M[]={0...}, U[]={0...};                # init tables for marking visited vertices and triangles
    G[0] = READ(delta);                           # read first vertex
    G[1] = G[0]+ READ(delta);                     # set second vertex using first plus delta
    G[2] = G[1]+ READ(delta);                     # set third vertex using second plus new delta
    GLOBAL N = 2;                                 # id of the last vertex encountered
    M[0] = 1;  M[1] = 1; M[2] = 1;                # mark these 3 vertices
    U[0] = 1;                                     # paint the triangle and go to opposite corner
    DecompressVertices(O[1]); }                   # starts vertices decompression

RECURSIVE PROCEDURE DecompressConnectivity(c) {
    REPEAT {                                      # Loop builds triangle tree and zips it up
        T++;                                      # new triangle
        O[c] = 3T; O[3T] = c;                     # attach new triangle, link opposite corners
        V[3T+1] = c.p.v; V[3T+2] = c.n.v;         # enter vertex Ids for shared vertices
        c = c.o.n;                                # move corner to new triangle
        Switch decode(READ(clers)) {              # select operation based on next symbol
            Case C: { O[c.n] = -1; V[3T] = ++N;}  # C: left edge is free, store ref to new vertex
            Case L: { O[c.n] = -2; zip(c.n); }    # L: orient free edge, try to zip once
            Case R: { O[c] = -2; c = c.n }        # R: orient free edge, go left
            Case S: { DecompressConnectivity(c); c = c.n }   # S: recursion going right, then go left
            Case E: { O[c] = -2; O[c.n] = -2; zip(c.n); RETURN }}}  # E: zip, try more, pop

RECURSIVE PROCEDURE Zip(c) {
    b = c.n; WHILE b.o>=0 DO b=b.o.n;             # tries to zip free edges opposite c
                                                  # search clockwise for free edge
    IF b.o != -1 THEN RETURN;                     # pop if no zip possible
    O[c]=b; O[b]=c;                               # link opposite corners
    a = c.p; V[a.p] = b.p.v;                      # assign co-incident corners
    WHILE a.o>=0 && b!=a DO {a=a.o.p; V[a.p]=b.p.v};
    c = c.p; WHILE c.o >= 0 && c!= b DO c = c.o.p;   # find corner of next free edge on right
    IF c.o == -2 THEN Zip(c) }                    # try to zip again

RECURSIVE PROCEDURE DecompressVertices(c) {
    REPEAT {                                      # start traversal for triangle tree
        U[c.t] = 1;                               # mark the triangle as visited
        IF c.v.m != 1                             # test whether tip vertex was visited
            THEN {G[++N] = c.p.v.g+c.n.v.g-c.o.v.g+READ(delta);   # update new vertex
                M[c.v] = 1;                       # mark tip vertex as visited
                c = c.r;}                         # continue with the right neighbor
        ELSE IF c.r.t.u == 1                      # test whether right triangle was visited
            THEN IF c.l.t.u == 1                  # test whether left triangle was visited
                THEN RETURN                       # pop
                ELSE { c = c.l }                  # move to left triangle
            ELSE IF c.l.t.u == 1                  # test whether left triangle was visited
                THEN { c = c.r }                  # move to right triangle
                ELSE { DecompressVertices (c.r);  # recursive call to visit right branch first
                    c = c.l } } }                 # move to left triangle
```

Acknowledgements

This material is based upon work supported by the National Science Foundation under Grant 9721358.

References

[GuSt98] S. Gumhold and W. Strasser, "Real Time Compression of Triangle Mesh Connectivity", Proc. ACM Siggraph, pp. 133–140, July 1998.

[IsSo99]   M. Isenburg and J. Snoeyink, "Spirale Reversi: Reverse decoding of the Edgebreaker encoding", Tech. Report TR-99-08, Computer Science, UBC, 1999.
[King99]   D. King and J. Rossignac, "Guaranteed 3.67V bit encoding of planar triangle graphs", 11th Canadian Conference on Computational Geometry (CCCG'99), pp. 146–149, Vancouver, CA, August 15–18, 1999.
[King99b] D. King and J. Rossignac, "Connectivity Compression for Irregular Quadrilateral Meshes" Research Report GIT-GVU-99-29, Dec 1999.
[RoCa99]   J. Rossignac and D. Cardoze, "Matchmaker: Manifold Breps for non-manifold r-sets", Proceedings of the ACM Symposium on Solid Modeling, pp. 31–41, June 1999.
[Ross99]   J. Rossignac, "Edgebreaker: Connectivity compression for triangle meshes", IEEE Transactions on Visualization and Computer Graphics, 5(1), 47–61, Jan–Mar 1999. *(Sigma Xi award: Best Paper from Georgia Tech.)*
[RoSz99]   J. Rossignac and A. Szymczak, "Wrap&Zip decompression of the connectivity of triangle meshes compressed with Edgebreaker", Computational Geometry, Theory and Applications, 14(1/3), 119–135, November 1999.
[SKR00]   A. Szymczak,. D. King, J. Rossignac, "An Edgebreaker-based efficient compression scheme for regular meshes", Proc of the 12th Canadian Conference on Computational Geometry, Fredericton, New Brunswick, August 16–19, 2000.
[SKR00b]   A. Szymczak, D. King, J. Rossignac, "An Edgebreaker-based Efficient Compression Scheme for Connectivity of Regular Meshes", Journal of Computational Geometry: Theory and Applications, 2000.
[TaRo98]   G. Taubin and J. Rossignac, "Geometric Compression through Topological Surgery", ACM Transactions on Graphics, 17(2), 84–115, April 1998. *(IBM award: Best Computer Science Paper from IBM.)*
[ToGo98]   C. Touma and C. Gotsman, "Triangle Mesh Compression", Proceedings Graphics Interface 98, pp. 26–34, 1998.

Efficient Error Calculation for Multiresolution Texture-based Volume Visualization

Eric LaMar*, Bernd Hamann, and Kenneth I. Joy**

Center for Applied Scientific Computing (CASC)
Lawrence Livermore National Laboratory
Livermore, CA 94550, USA

Abstract. Multiresolution texture-based volume visualization is an excellent technique to enable interactive rendering of massive data sets. Interactive manipulation of a transfer function is necessary for proper exploration of a data set. However, multiresolution techniques require assessing the accuracy of the resulting images, and re-computing the error after each change in a transfer function is very expensive. We extend our existing multiresolution volume visualization method by introducing a method for accelerating error calculations for multiresolution volume approximations. Computing the error for an approximation requires adding individual error terms. One error value must be computed once for each original voxel and its corresponding approximating voxel. For byte data, i.e., data sets where integer function values between 0 and 255 are given, we observe that the set of "error pairs" can be quite large, yet the set of *unique* error pairs is small. Instead of evaluating the error function for each original voxel, we construct a table of the unique combinations and the number of their occurrences. To evaluate the error, we add the products of the error function for each unique error pair and the frequency of each error pair. This approach dramatically reduces the amount of computation time involved and allows us to re-compute the error associated with a new transfer function quickly.

1 Introduction

When rendering images from approximations, it is necessary to know how close a generated image is to the original data. For multiresolution volume visualization, it is not possible to compare the images generated from original data to all possible images generated from approximations. The reason for using the approximations is to substantially reduce the amount of time required to render the data. If we assume that there is a reasonable amount of correlation between the data and the resulting imagery, we can compute an

---

* lamar1@llnl.gov, Center for Applied Scientific Computing (CASC), Lawrence Livermore National Laboratory, Box 808, L-661, Livermore, CA 94550 U.S.A.
** {hamann,joy}@cs.ucdavis.edu, Center for Image Processing and Integrated Computing (CIPIC), Dept. of Computer Science, University of California, Davis, CA 95616-8562

error value between the approximations and the original data (in 3D object space). We can then use that value to estimate the error in the 2D imagery.

Even so, the amount of time required to evaluate the error over an entire data hierarchy can be significant. This consideration is especially important when it is necessary for a user to interactively modify a transfer function: each change in the transfer function requires us to re-compute the error for the entire hierarchy.

We introduce a solution based on the observation that, for many data sets, the range size of scalar data sets is often many orders smaller that the domain size (physical extension) – and that, instead of evaluating an error function for each original voxel, it suffices to count the frequencies of unique pairs of error terms. In the case of 8-bit (byte) integer data, there are only 256^2 combinations (each term is a single byte, or 256 possible values). To compute the error, instead of adding individual error terms, we add the products of the error (for a unique pair of error terms) and the frequency of that unique pair.

For example, a typical 512^3 voxel data set, with one byte per voxel, contains 2^{27} bytes. To compute the error, a naive method would evaluate an error function for each of the original 2^{27} voxels. However, there are only 256^2 *unique* pairs of error terms. Thus, to compute the error, we evaluate the error function for each unique pair of error terms, or 2^{16} times – which is 2^{11} times faster than the naive method. This algorithm requires us to examine the entire data set for unique pairs of error terms – this is a preprocessing cost that can be performed off-line.

This work is a direct extension of earlier work reported in [LJH99], [LHJ00], and [LDHJ00]. We first review the generation and rendering of a volume hierarchy in Section 2, discuss the error criterion we use in Section 3, cover some basic optimizations of the general approach in Section 4, provide performance statistics in Section 5, and discuss directions for future work in Section 6.

2 Multiresolution Volume Visualization

Our multiresolution volume visualization system uses hardware-accelerated 3D texturing for rendering a volume; it represents a volume with index textures. A transfer function is applied by first building a look-up table, transferring it into the graphics system, and then transferring the texture tile. The translation of index values to display values (luminance or color) is performed in hardware. In the following discussion, we use the term *tile* to refer to the data and *node* to refer to the element of a binary tree or octree. In some passages, we will use *tile* to refer to both. *Voxels* are attributes of *tiles* and spatial location and extent are attributes of *nodes*.

2.1 Generating the Hierarchy

First, we review certain aspects of our multiresolution data representation and how it influences the decisions on how to evaluate and store error. The underlying assumption is that data sets are too large to fit into texture memory. Data sets are often too large even to fit in main memory. Given a volumetric data set, we produce a hierarchy of approximations. Each level in the approximation hierarchy is half the size of the next level. Each level is broken into constant-sized tiles – tiles that are small enough to fit in their entirety in texture memory, see [LJH99]. This is also called "bricking" in [GHY98]. Figure 1 shows the decomposition of a block consisting of 29 pix-

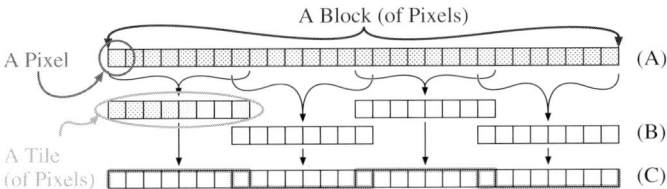

Fig. 1. Line (A) shows a block of pixels, is broken into four tiles of eight pixels on line (B). Line (C) is our shorthand notation for line (B).

els (line A) into four tiles of eight pixels (line B), with one shared pixel in the regions where the tiles overlap. The shared pixel is necessary because we linearly interpolate pixel centers. The alternating thicker borders in line (C) show the position of the individual tiles of a block – this is our "shorthand notation" for line (B).

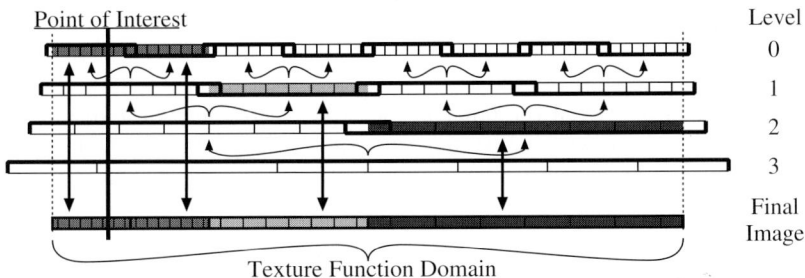

Fig. 2. Selecting from a texture hierarchy of four levels. Level 0 is the original texture, broken into eight tiles. The dashed lines show the domain of the texture function over the hierarchy. The bold vertical line represents a point p of interest. Tile selection depends on the width of the tile and the distance from the point. The red, green, and blue shaded regions in levels 0, 1, and 2, respectively, and then in the Final Image, show from which levels of detail the data is used to create the final image. (See Color Plate 2 on page 350.)

Figure 2 shows a one-dimensional texture hierarchy of four levels. The top level, level 0, is the original texture (of 57 pixels), broken into eight tiles (of eight pixels). Level 1 contains four tiles at half of the original resolution (29 pixels), and so on. The dashed vertical lines on either side show the domain of the texture function over the hierarchy. Arrows indicate the parent-child relationship of the hierarchy, defining a binary tree, rooted at the coarsest tile, level 3. The bold vertical line denotes a point of interest, p, and tiles are selected when the distance from p to the center of the tile is greater than the width of the tile. One starts with the root tile and performs selection until all tiles meet this distance-to-width criterion, or no smaller tiles exist[1]. The double-headed vertical arrows show selected tiles and their correspondence in the final image.

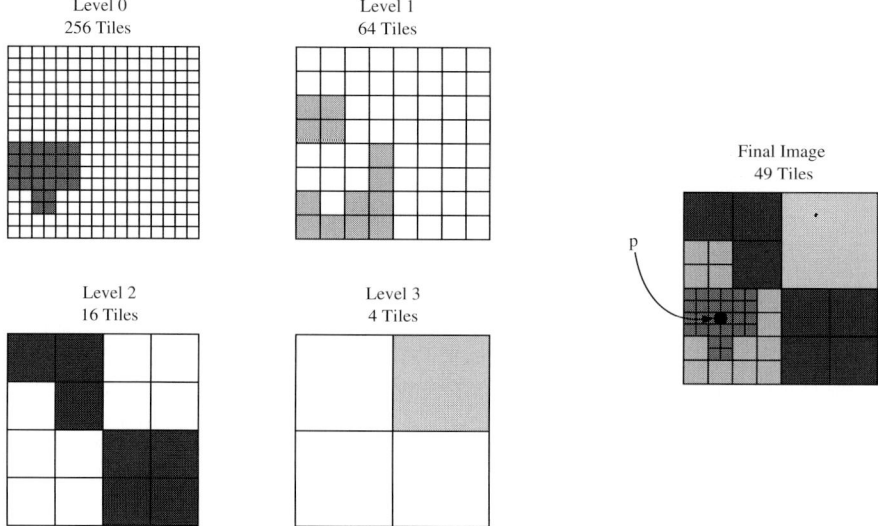

Fig. 3. Selecting tiles in two dimensions from a texture hierarchy of five levels (Level 4 not shown). Given the point p, tiles are selected when the distance from the center of the tile to p is greater than the length of the diagonal of the tile. Tiles selected from each level are shaded.

Figure 3 shows a two-dimensional quadtree example. The original texture, level 0, has 256 tiles. The shaded regions in each level show the portion of that level used to approximate the final image. The selection method in the two-dimensional case is similar to the one for the one-dimensional case: A node is selected when the distance from the center of the node to the point p is greater than the length of the diagonal of the node. The original texture is divided into 256 tiles. The adaptive rendering scheme uses 49 tiles or, roughly,

---

[1] This is the case on the left side of Figure 2.

one-fifth of the data. This scheme can easily be used for three-dimensional textures using an octree.

2.2 Node Selection and Rendering

The first rendering step determines which nodes will be rendered. This means finding a set of non-overlaying nodes that meet some error criterion. The general logic is to subdivide nodes, starting at the root node, until the error criterion is met, always subdividing the node with the greatest error.

We use a priority queue of nodes that is sorted by descending error, *i.e.*, the first node in the queue is the one with the highest error. We first push the root node onto the queue. We then iterate the following: we examine to top node (the node with the highest error). If the error criterion is not met (which we define below), that node is subdivided, or removed from the queue and all of its children are added to the queue. This continues until the error criterion is met. This is guaranteed to terminate as all leaf nodes have zero error. All nodes still in the queue meet the error criterion and are rendered.

Our primary selection filter is based on one of these two error criterion:

- L-infinity (l_∞): Subdivide the node if the node's associated l_∞ error is greater than some maximum value, *i.e.*, all rendered nodes must have an error less than this maximum value.
- Root-mean-square (RMS): Subdivide the node if the root-mean-square error over all selected nodes is greater than some maximum value, *i.e.*, the root of the sum of the squared differences considering all rendered nodes must be less than this maximum value. Rather than re-calculate the error each time a node is subdivided, we keep a running error total, subtracting a node's error when it is removed from the queue and adding a node's error when it is added to the queue.

We note that many other error criterion can be used and that our algorithm is not limited to the two selected here. Nodes are sorted and composited in back-to-front order. We order nodes with respect to the view direction such that, when drawn in this order, no node is drawn behind a rendered node. The order is fixed for the entire tree for orthogonal projections and has to be computed just once for each new rendering [GHY98]. For perspective projections, the order must be computed at each node.

3 Error Calculation

We calculate error on a per-node basis: When a node meets the error criterion, it is rendered; otherwise, its child (higher-resolution) nodes are considered for rendering. We currently assume a piece-wise constant function implied by the set of given voxels, but use trilinear interpolation for the texture. This approach simplifies the error calculation.

We use two error norms: the L-infinity and root-mean-square (RMS) norms. Again, we note that other error norms may be better, but these are sufficiently simple as to not complicate the following discussion. Given two sets of function values, $\{f_i\}$ and $\{g_i\}$, $i = 0,\ldots,n-1$, the L-infinity error norm is defined as $l_\infty = max_i\{|x_i|\}$, and the RMS is defined as $E_{rms} = \sqrt{\frac{1}{n}\sum_i(x_i^2)}$, where $x_i = T[f_i] - T[g_i]$ and $T[x]$ is a transfer function. For the purposes of this discussion, we assume that $T[x]$ is a simple scalar function, mapping density to gray-scale luminance; the issue of error in color space is beyond the scope of this discussion. We evaluate the error function once for each Level-0 voxel.

A data set of size 512^3 contains 2^{27} voxels, with the same number of pairs of error terms for each level of the hierarchy. However, when using byte data, we observe that, though there are 2^{27} pairs of values, there are only 2^{16} (256^2) unique pairs of error values. This means, on average, that each unique pair is evaluated 2^{11} times.

Our solution is to add a two-dimensional table Q to each internal (e.g., approximating) node of the octree, see Figure 5. The elements $Q_{a,b}$ store the numbers of occurrences for each (f_i, g_i) pair, where a and b are the table indices corresponding to f_i and g_i, respectively. Thus

$$Q_{a,b} = \sum_i \begin{pmatrix} 1 \text{ if } f_i = a \text{ and } g_i = b \\ 0 \text{ otherwise} \end{pmatrix}.$$

The tables are created only once, when the data is loaded, and count the number of occurrences of unique error terms that the node "covers" of the original data.

To calculate the L-infinity error for a node, we search for the largest value, with the requirement that this value corresponds to a real pair of values:

$$l_\infty = max_{a,b | Q_{a,b} \neq 0}\{|T[a] - T[b]|\}.$$

We compute the RMS error for a node as

$$E_{rms} = \sqrt{\frac{1}{n}\sum_{a,b}(T[a] - T[b])^2 \times Q_{a,b}},$$

where $n = \sum_{a,b}(Q_{a,b})$. For a set of t nodes, we compute the RMS value

$$E_{rms} = \sqrt{\frac{1}{N}\sum_t\left(\sum_{a,b}(T[a] - T[b])^2 \times Q_{a,b}\right)},$$

where $N = \sum_t\left(\sum_{a,b} Q_{a,b}\right)$.

Image 4(B), shown in the upper right-hand corner of Figure 4, shows a graphical representation of the frequency relationship of the original and

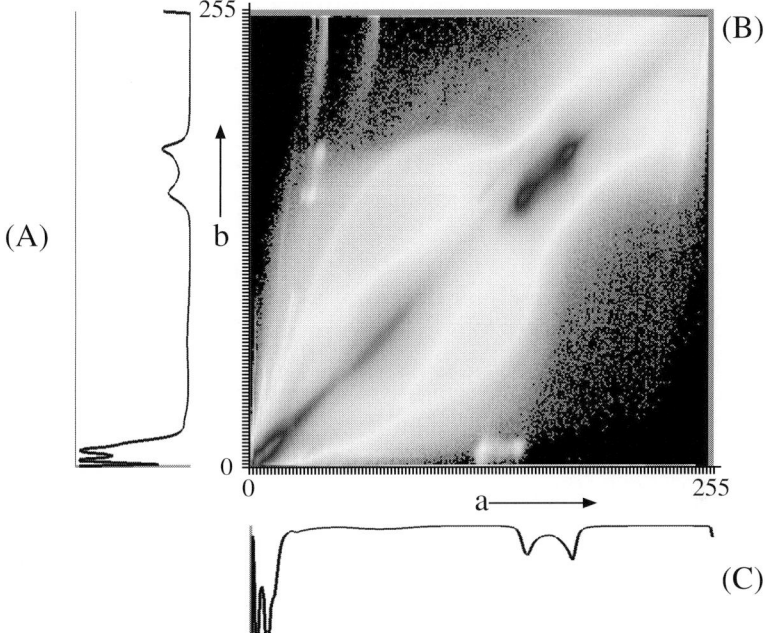

Fig. 4. The Visible Female CT data set. Image (B) in the upper-right corner shows the frequency relationship of the original and first approximation of the Visible Female CT data set. The image consists of 256 × 256 pixels, with a on the horizontal axis and b on the vertical axis; each pixel corresponds to a $Q_{a,b}$ element. This particular table, covering all of the original domain, would not be produced in practice; normally, Q tables associated with the first-level approximation cover fairly small regions of the original domain. Image (B) is shown here to provide the reader with insight into the nature of a typical Q table. The colors are assigned by normalizing the logarithm of the number of occurrences of a $Q_{a,b}$ element, linearly mapped to a rainbow color sequence, where zero maps to red and one maps to violet: $pixel_{a,b} = RainbowColorMap\left[\ln\left(Q_{a,b}\right)/\ln\left(Max_{a,b}\{Q_{a,b}\}\right)\right]$. Graph (A), on the left, shows the histogram of the original data (Level 0), with positive frequency pointing left. Graph (C), on the bottom, shows the histogram of the first approximation (Level 1), with positive frequency pointing down. (See Color Plate 3 on page 350.)

first approximation of the Visible Female CT dataset. Figure 4 indicates that the approximation is generally good: Most of the error terms are along the diagonal. The "lines" in the upper-left corner of image 4(B) may correspond to high gradients present in the data set. (The Visible Female CT data set was produced such that sections of the body fill a 512^2 image: these different sections are scanned with different spatial scales. We have not accounted for these different spatial scales in our version of the data set, thus there exist several significant discontinuities in data values in the data set.)

4 Optimizations

Evaluating a table is still fairly expensive when interactive performance is required. We currently use three methods to reduce the time needed to evaluate the error.

First, one observes that the order of the error terms for calculating L-infinity and root-mean-square errors does not matter: $x_i = |f_i - g_i| = |g_i - f_i|$ and $x_i = (f_i - g_i)^2 = (g_i - f_i)^2$. If we order the indices in $Q_{a,b}$ such that $a \leq b$, or $a = min(f_i, g_i)$ and $b = max(f_i, g_i)$, we obtain a triangular matrix that has slightly more than half the terms of a full matrix (32896 vs. 65536 entries). This consideration allows us to roughly halve the evaluation time and storage requirements.

Second, one observes that, typically, there is a strong degree of correlation between approximation and original data. This means that the values of $Q_{a,b}$ are large when a is close to b (i.e., near the diagonal); and small, often zero, when $a \ll b$, (i.e., far away from the diagonal). We have observed that the number of non-zero entries in a Q table decreases (i.e., the table is becoming more sparse) for nodes closer to the leaves of the octree. This happens since the nodes closer to the leaf nodes correspond to high-resolution approximations and thus a better approximation - the correlation is strong, and the non-zero values in the Q table cluster close to the diagonal. Also, the nodes closer to the leaves cover a progressively smaller section of the domain. We perform a column-major scan, i.e., traverse the table first by column, then by row. Thus, the Q tables become sparse, and we can terminate the checking of elements if we remember the last non-zero entry for a row. We maintain a table, L_{row}, that contains the index of the last non-zero value for a row. It may even be possible to perform run-length encoding for a row to skip over regions of zero entries. Figure 4 shows that there are large, interior regions with zero entries.

Third, one can use "lazy evaluation" of the error. When we re-calculate the error for all nodes in a hierarchy, and few of the nodes are rendered, much of the error evaluation is done but not used. Thus, for each new transfer function, we only re-calculate the error value of those nodes that are being considered for rendering, see [LHJ00]. In Figure 5, nodes with a red circle around the Q do not satisfy the error criterion, nodes with green squares around the Q meet the error criterion, and nodes with blue diamonds around the Q are never visited, and thus no error calculation is performed.

5 Results

Performance results hierarchy generation, error table generation, and error computation time were obtained on an SGI Origin2000 with 10GB of memory, using one (of 16) 195MHz R10K processors. However, images were produced on an SGI Onyx2 Infinite Reality with 512MB of memory, using one (of

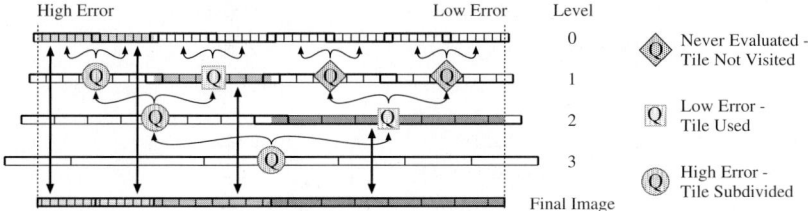

Fig. 5. Selecting from a texture hierarchy consisting of four levels with lazy evaluation of error. The error selection criterion selects high-resolution nodes in the high-error regions (left) and low resolution nodes in the low-error regions (right). The letter Q indicates nodes with a Q table, i.e., nodes that are all internal (approximating) nodes. Three classifications of tiles are shown. Blue-diamond tiles are never visited. Error is evaluated in red-round and green-square tiles; red-round tiles exceed the error requirement and are subdivided; green-square tiles meet the error requirement and are used.

four) 195MHz R10K processors. This was done for the following reasons: We are interested in the time required to process the hierarchy, free of memory limitations. The logical memory used during a visualization was 2.4GB, and thrashing completely dominated (by a factor of 10 or more) hierarchy/error-table generation and error calculation times when performed on the Onyx2. The SGI Origin2000 does not have a graphics subsystem, while the SGI Onyx2 does.

Table 1. Visible Female CT data set statistics.

Image	Level	Voxels	Nodes in Level	Nodes Rendered	Mem. (MB)	Time (sec.)	Error l_∞	E_{rms}
6a	0	$512^2 \times 1734$	2268	1560	390	83.4	0.000	0.00000
6b	1	$257^2 \times 868$	350	263	65	8.73	0.305	0.00678
6c	2	$129^2 \times 435$	63	49	13	2.38	0.305	0.00917
6d	3	$65^2 \times 218$	16	16	4	0.563	0.305	0.01059

We have used the Visible Female CT data set, consisting of $512^2 \times 1734$ voxels in our experiments. Figure 6 shows different-resolution images of this data set: Image 6(a) shows the original data, and images 6(b) to 6(d) are progressively 1/2 linear (1/8 total) size. Table 1 summarizes performance statistics for four (of six) levels of the hierarchy. The "Voxels" column shows the total size of the approximation in voxels, and the next column shows the total number of nodes associated with that level. The "Nodes Rendered" column shows how many nodes where used to render that level. The number of nodes rendered is actually less than one would expect: Many regions of the data set are constant, so there is no error when approximating these regions. Each tile contains 64^3 bytes = 256K. When transferring n tiles, the total

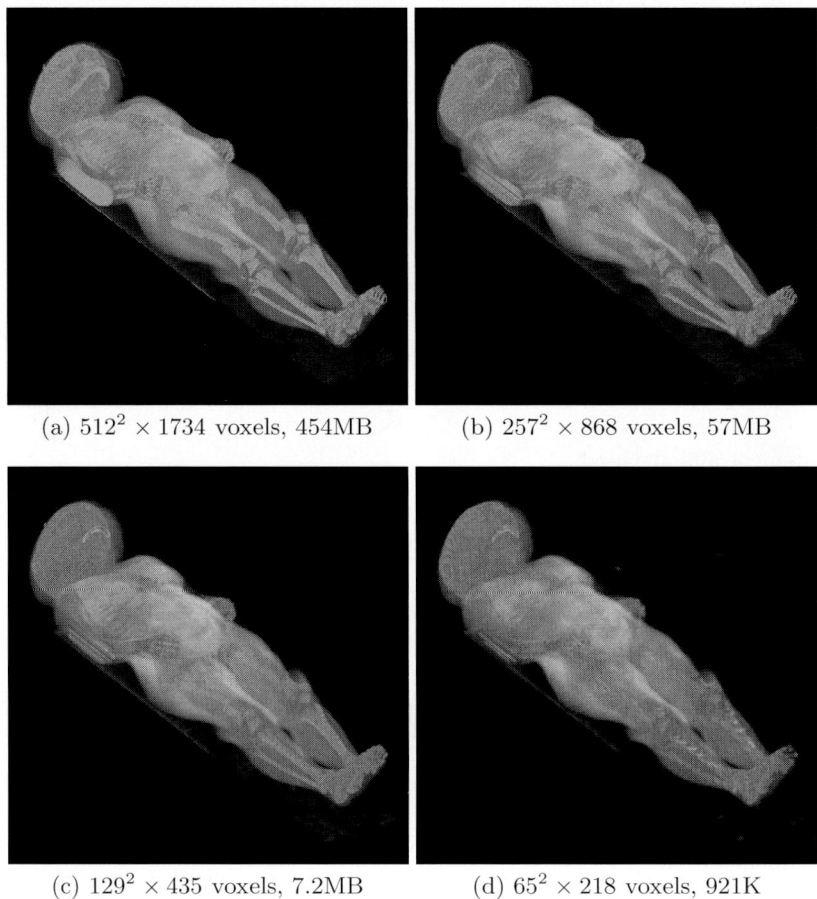

Fig. 6. Visible Female CT data set rendered at four resolutions (see Table 1). The transfer function shows bones in white, fat in yellow, muscle in red, and internal organs in green. Different spatial scales were used for different sections of the body during data acquisition. (See Color Plate 4 on page 351.)

memory transferred to the graphics subsystem for n tiles is 256K$\times n$, shown in the "Memory" column. The "Time" column lists the times, in seconds, required to produce the rendering for the various levels. The l_∞ and E_{rms} columns show the error values associated with those renderings.

Table 2 summarizes the times for various stages of our system. We note that the "Compute-Texture-Hierarchy" and "Compute-Error-Tables" stages are only performed once for a data set, and only the "Calculate-Error" step is performed for each new transfer function. The time for "Calculate-Error" is the times needed to re-compute the error for all 432 internal (i.e., approximating) nodes. The time per node is approximately 0.0028 seconds – and when coupled with a lazy evaluation scheme, "Calculate-Error" time is very

Table 2. Performance statistics for Visible Female CT data set.

Step	Index Texture
Compute Texture Hierarchy	1 min 46 sec
Compute Error Table	5 min 35 sec
Calculate Error	1.23 sec

small relative to the other parts of the rendering pipeline (see "Time" column in Table 1).

6 Conclusions and Future Work

We believe that there are several directions to continue this work. The first directions is to extend this technique to color images or more general vector-valued data hierarchies. The second direction is to consider data other than byte data: data sets with 12 and 16 bits per voxel are also common. However, a table would contain 4096^2 (2^{24}) or 65536^2 (2^{32}) entries. These tables would be larger than the actual volume data per node, and possibly require more time for evaluation. Could some quantizing approach work? Since these nodes should have a high degree of correlation, will these tables be sufficiently sparse to compress? A third direction is to apply our technique to time-varying data. The error between nodes in different time steps would be expressed in the same manner and could be encoded in a table.

Acknowledgments

This work was performed under the auspices of the U.S. Department of Energy by the University of California, Lawrence Livermore National Laboratory under contract No. W-7405-Eng-48. This work was supported by the National Science Foundation under contracts ACI 9624034 (CAREER Award), through the Large Scientific and Software Data Set Visualization (LSSDSV) program under contract ACI 9982251, and through the National Partnership for Advanced Computational Infrastructure (NPACI); the Office of Naval Research under contract N00014-97-1-0222; the Army Research Office under contract ARO 36598-MA-RIP; the NASA Ames Research Center through an NRA award under contract NAG2-1216; the Lawrence Livermore National Laboratory under ASCI ASAP Level-2 Memorandum Agreement B347878 and under Memorandum Agreement B503159; the Lawrence Berkeley National Laboratory; the Los Alamos National Laboratory; and the North Atlantic Treaty Organization (NATO) under contract CRG.971628. We also acknowledge the support of ALSTOM Schilling Robotics and SGI. We thank the members of the Visualization and Graphics Research Group at the Center for Image Processing and Integrated Computing (CIPIC) at the University of California, Davis, and the members of the Data Analysis and Exploration thrust of the Center for Applied Scientific Computing (CASC) at Lawrence Livermore National Laboratory.

References

[CCF94] Brian Cabral, Nancy Cam, and Jim Foran. Accelerated Volume Rendering and Tomographic Reconstruction Using Texture Mapping Hardware. In Arie Kaufman and Wolfgang Krueger, editors, *1994 Symposium on Volume Visualization*, pages 91–98. ACM SIGGRAPH, October 1994.

[GHY98] Robert Grzeszczuk, Chris Henn, and Roni Yagel. *SIGGRAPH '98 "Advanced Geometric Techniques for Ray Casting Volumes" course notes*. ACM, July 1998.

[LDHJ00] Eric C. LaMar, Mark A. Duchaineau, Bernd Hamann, and Kenneth I. Joy. Multiresolution Techniques for Interactive Texturing-based Rendering of Arbitrarily Oriented Cutting-Planes. In W. C. de Leeuw and R. van Liere, editors, *Data Visualization 2000*, pages 105–114, Vienna, Austria, May 29-30, 2000. Proceedings of VisSym '00 – The Joint Eurographics and IEEE TVCG Conference on Visualization, Springer-Verlag.

[Lev87] Marc Levoy. Display of Surfaces from Volume Data. *Computer Graphics and Applications*, 8(3):29–37, February 1987.

[LHJ00] Eric C. LaMar, Bernd Hamann, and Kenneth I. Joy. Multiresolution Techniques for Interactive Hardware Texturing-based Volume Visualization. In R. F. Erbacher, P. C. Chen, J. C. Roberts, and Craig M. Wittenbrink, editors, *Visual Data Exploration and Analysis*, pages 365–374, Bellingham, Washington, January 2000. SPIE – The International Society for Optical Engineering.

[LJH99] Eric C. LaMar, Kenneth I. Joy, and Bernd Hamann. Multi-Resolution techniques for Interactive Hardware Texturing-based Volume Visualization. In David Ebert, Markus Gross, and Bernd Hamann, editors, *IEEE Visualization 99*, pages 355–361. IEEE, ACM Press, October 25-29, 1999.

[WE98] Rüdiger Westermann and Thomas Ertl. Efficiently Using Graphics Hardware In Volume Rendering Applications. In Shiela Hoffmeyer, editor, *Proceedings of Siggraph 98*, pages 169–177. ACM, July 19-24, 1998.

Hierarchical Spline Approximations

David F. Wiley[1], Martin Bertram[2,1], Benjamin W. Jordan[3], Bernd Hamann[1], Kenneth I. Joy[1], Nelson L. Max[1,4], and Gerik Scheuermann[2]

[1] Center for Image Processing and Integrated Computing (CIPIC), Department of Computer Science, University of California, Davis, CA 95616-8562, U.S.A.; e-mail: {wiley, hamann, joy}@cs.ucdavis.edu
[2] University of Kaiserslautern, Fachbereich Informatik, P.O. Box 3049, D-67653 Kaiserslautern, Germany; e-mail: {bertram, scheuer}@informatik.uni-kl.de
[3] Pixar Feature Division, Pixar Animation Studios, 1001 West Cutting Blvd., Richmond, CA 94804, U.S.A.; e-mail: bjordan@pixar.com
[4] Center for Applied Scientific Computing (CASC), Lawrence Livermore National Laboratory, 7000 East Avenue, L-551, Livermore, CA 94550, U.S.A.; e-mail: max2@llnl.gov

Abstract. We discuss spline refinement methods that approximate multi-valued data defined over one, two, and three dimensions. The input to our method is a coarse decomposition of the compact domain of the function to be approximated consisting of intervals (univariate case), triangles (bivariate case), and tetrahedra (trivariate case). We first describe a best linear spline approximation scheme, understood in a least squares sense, and refine on initial mesh using repeated bisection of simplices (intervals, triangles, or tetrahedra) of maximal error. We discuss three enhancements that improve the performance and quality of our basic bisection approach. The enhancements we discuss are: (i) using a finite element approach that only considers original data sites during subdivision, (ii) including first-derivative information in the error functional and spline-coefficient computations, and (iii) using quadratic (deformed, "curved") rather than linear simplices to better approximate bivariate and trivariate data. We improve efficiency of our refinement algorithms by subdividing multiple simplices simultaneously and by using a sparse-matrix representation and system solver.

1 Introduction

Different methods are known and used for hierarchical representation of very large data sets. Unfortunately, only a small number of these methods are based on well developed mathematical theory. In the context of visualizing very large data sets in two and three dimensions, it is imperative to develop hierarchical data representations that allow us to visualize and analyze data at various levels of detail. General and efficient algorithms are needed to support the generation of hierarchical data representations and their applicability for visualization.

Our discussion deals with the construction of hierarchies of triangulations and spline approximations of functions. The main idea underlying the construction of our data hierarchy is repeated bisection of simplices (intervals, triangles, or tetrahedra). Bisection is chosen for its simplicity and the possibility to extend it to multiple dimensions easily. We construct an approximation hierarchy by repeatedly subdividing triangulations, "simplicial domain decompositions", and computing a best spline approximations. Our initial approximation is based on a coarse triangulation of the domain of interest, typically defined over the convex hull of all given data points/sites if the function to be approximated is known only at a finite number of locations. We identify regions of large error and subdivide simplices in these regions. These steps are required to perform our algorithm:

1. **Initial approximation.** Define a coarse initial triangulation of the domain of interest and compute the coefficients defining the best spline approximation for this mesh.
2. **Error estimation.** Analyze the error of this approximation by computing appropriate local and global error estimates relative to the input function.
3. **Refinement.** Subdivide the simplex (or set of simplices) with the largest local error estimate.
4. **Computation of best approximation.** Compute a new best linear spline approximation based on the new mesh.
5. **Iteration.** Repeat steps 2, 3, and 4 until a certain approximation error condition is met.

There are many hierarchical methods targeted at approximating large data sets. For example, wavelet methods are described in [2], [13], and [29]. The work described in [29] has the advantage of supporting both lossless and lossy compression. In general, wavelet methods work well for data lying on uniform, rectilinear, and power-of-two grids and provide fast and highly accurate compression.

Simplification methods using data elimination strategies are described in [3], [4], [11], [12], [16], [20], and [21]. These methods are more general than most wavelet methods since arbitrary input meshes can be converted to a form treatable by each method. Refinement methods similar to the ones we discuss here are described in [14], [17], and [27]. Most data-dependent refinement methods can also be adapted to arbitrary meshes.

The method described in [7] performs an iterative "thinning step" based on radial basis functions on scattered points while maintaining a Delaunay triangulation. The results of [7] support the notion that data-dependent triangulations better approximate functions in high-gradient regions.

Comparisons of wavelet, decimation, simplification, and data-dependent methods, including the meshes discussed in [13], [16], [20], and [27], are provided in [19]. This survey discusses the many approaches to surface simplification and also examines the complexity of some of the most commonly used methods.

The methods we discuss here apply to univariate, bivariate, and trivariate data. The underlying principles of our approach become evident in our discussion of the univariate case in Sect. 2. By discussing the univariate case in detail, generalizations to the bivariate and trivariate cases are more easily understood.

The hierarchies of approximations resulting from our methods can be used for visualization. Common visualization methods – including contouring, slicing, and volume visualization (i.e., ray-casting) – that can be applied to our hierarchical approximations are described in [15], [22], [24], [25], and [28].

2 Best Linear Spline Approximation and the Univariate Case

We begin the discussion with best linear spline approximation in the univariate case. Our method requires a few notations from linear algebra and approximation theory, and we discuss these briefly. We use the standard scalar product $\langle f, g \rangle$ of two functions $f(x)$ and $g(x)$, defined over the interval $[a, b]$,

$$\langle f, g \rangle = \int_a^b f(x)g(x)dx, \tag{1}$$

and the standard L^2 norm to measure a function $f(x)$,

$$\|f\| = \langle f, f \rangle^{1/2} = \left(\int_a^b \left(f(x) \right)^2 dx \right)^{1/2}. \tag{2}$$

It is well known from approximation theory, see, for example, [5], that the best approximation $f(x)$ of a given function $F(x)$, when approximating it by a linear combination

$$f(x) = \sum_{i=0}^{n-1} c_i f_i(x) \tag{3}$$

of independent functions $f_0(x), \ldots, f_{n-1}(x)$ (using f_i as abbreviation for $f_i(x)$), is defined by the normal equations

$$\begin{bmatrix} \langle f_0, f_0 \rangle & \cdots & \langle f_{n-1}, f_0 \rangle \\ \vdots & & \vdots \\ \langle f_0, f_{n-1} \rangle & \cdots & \langle f_{n-1}, f_{n-1} \rangle \end{bmatrix} \begin{bmatrix} c_0 \\ \vdots \\ c_{n-1} \end{bmatrix} = \begin{bmatrix} \langle F, f_0 \rangle \\ \vdots \\ \langle F, f_{n-1} \rangle \end{bmatrix}. \tag{4}$$

We can also write this linear system as $\boldsymbol{M}^{[n-1]} \boldsymbol{c}^{[n-1]} = \boldsymbol{F}^{[n-1]}$. This system is easily solved when dealing with a set of mutually orthogonal and normalized basis functions, i.e., in the case when $\langle f_i, f_j \rangle = \delta_{i,j}$ (*Kronecker delta*). In this case, only the diagonal entries of $\boldsymbol{M}^{[n-1]}$ are non-zero and the coefficients c_i are given by $c_i = \langle F, f_i \rangle$. For an arbitrary set of basis functions, one must investigate a means for an efficient solution of the linear system define by 4.

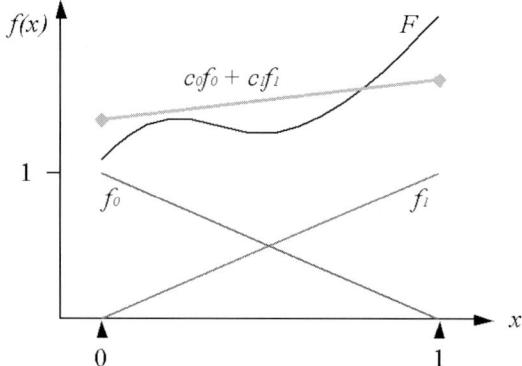

Fig. 1. Basis functions f_i, function F to be approximated, and approximation $f(x) = c_0 f_0 + c_1 f_1$. (See Color Plate 5 on page 352.)

Initially we approximate a given univariate function F by a single linear spline segment by computing $\boldsymbol{M}^{[1]} \boldsymbol{c}^{[1]} = \boldsymbol{F}^{[1]}$. We assume, without loss of generality, that F is defined over the interval $[0, 1]$ and that the basis functions are hat functions with the property $f_i(x_j) = \delta_{i,j}$, see Fig. 1.

When using hat functions as basis functions, the only non-zero elements of $\boldsymbol{M}^{[n-1]}$, for a particular row i, are the elements $\langle f_{i-1}, f_i \rangle$, $\langle f_i, f_i \rangle$, and $\langle f_{i+1}, f_i \rangle$. These scalar products are given by

$$\langle f_{i-1}, f_i \rangle = \frac{1}{6} \Delta_{i-1},$$

$$\langle f_i, f_i \rangle = \frac{1}{3} (\Delta_{i-1} + \Delta_i), \quad \text{and}$$

$$\langle f_{i+1}, f_i \rangle = \frac{1}{6} \Delta_i, \tag{5}$$

where $\Delta_i = x_{i+1} - x_i$. Thus, $\boldsymbol{M}^{[n-1]}$ is the tridiagonal matrix

$$\boldsymbol{M}^{[n-1]} = \begin{bmatrix} 2\Delta_0 & \Delta_0 & & & \\ \Delta_0 & 2(\Delta_0 + \Delta_1) & \Delta_1 & & \\ & \Delta_1 & 2(\Delta_1 + \Delta_2) & \Delta_2 & \\ & & \ddots & \ddots & \ddots \\ & & & \Delta_{n-2} & \Delta_{n-1} \end{bmatrix}. \tag{6}$$

Since it is necessary to re-compute the coefficients after each refinement due to the global nature of the best approximation problem, one can take advantage of efficient system solvers to solve this system in linear time, see, for example [6].

We want to compute a hierarchy of approximations of a function F by refining the initial approximation by adding more basis functions – or, in other words, by inserting more knots. We insert knots repeatedly until we have an approximation whose global error is smaller than some threshold. The error of the first approximation is defined as $E^{[1]} = \|F - (c_0 f_0 + c_1 f_1)\|$. If this value is larger than a specified tolerance, we refine by inserting a knot at $x = \frac{1}{2}$. The addition of this knot changes the basis function sequence and, therefore, we must compute a new best linear spline approximation for the new knot set by solving $\boldsymbol{M}^{[2]} \boldsymbol{c}^{[2]} = \boldsymbol{F}^{[2]}$; the error for this approximation is $E^{[2]} = \|F - \sum_{i=0}^{2} c_i f_i\|$. Should this new error still be too large, we refine the approximation further. We need to define a criterion that determines which segment to bisect. Local error estimates over each segment can be computed easily, so this information can be used to decide which interval to subdivide, in this case, either $[0, \frac{1}{2}]$ or $[\frac{1}{2}, 1]$.

We can define a global error estimate for an approximation based on the knot sequence $0 = x_0 < x_1 < x_2 < \ldots < x_{n-2} < x_{n-1} = 1$. We define the global error as

$$E^{[n-1]} = \|F - \sum_{i=0}^{n-1} c_i f_i\|. \tag{7}$$

In order to allow us to decide which segment to subdivide next, we define the local error for an interval $[x_i, x_{i+1}]$ as

$$e_i^{[n-1]} = \left(\int_{x_i}^{x_{i+1}} (F - (c_i f_i + c_{i+1} f_{i+1}))^2 \, dx \right)^{1/2}, \ i = 0, \ldots, n-2 . \tag{8}$$

We compute local errors for each of the segments and then bisect the segment with the maximum local error estimate at each iteration. If the maximum local error is not unique, it is sufficient to choose one segment randomly for subdivision. (Alternatively, one could subdivide all segments with the same maximum local error, thus leading to a unique solution). To improve efficiency, it is reasonable to select the m segments with the m largest local error estimates for subdivision. Such an approach seems to be more appropriate for very large data sets. An example of a hierarchical approximation of a univariate function is shown in Fig. 2.

The scalar products $\langle F, f_i \rangle$ and the error values $E^{[n-1]}$ and $e_i^{[n-1]}$ are computed by numerical integration. We use *Romberg integration* to perform these steps, see [1] and [17].

3 The Bivariate Case

Given an initial best linear spline approximation based on a small number of triangles, we compute the global error estimate for the approximation and, should this error be too large, bisect the longest edge of the triangle with maximal local error and insert a new knot at the midpoint of the longest edge of this triangle. If this edge is shared by another triangle, the neighboring triangle is also split.

Refinement leads to insertion of additional knots at each iteration. Again, we must re-compute spline coefficients after each knot insertion step due to the global nature of the best approximation problem.

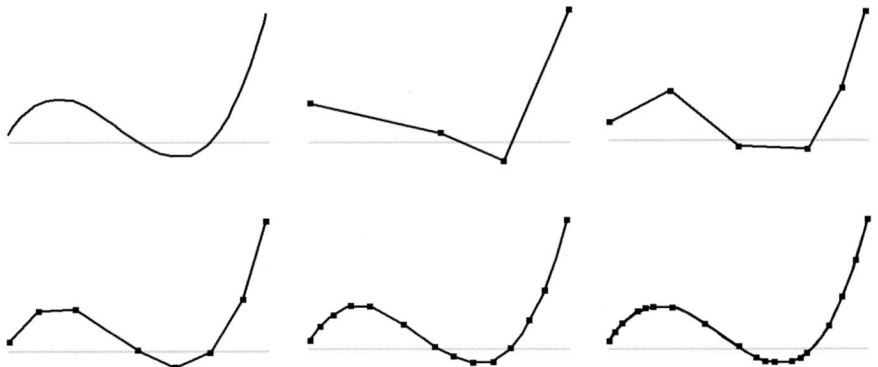

Fig. 2. Five approximations of $F(x) = 10x \left(x - \frac{1}{2}\right)\left(x - \frac{3}{4}\right)$, $x \in [0, 1]$. The original function is shown in the upper-left corner. Number of knots for approximations: 4, 6, 8, 14, and 19

Concerning the construction of an approximation hierarchy for function $F(x, y)$, we begin with a coarse initial triangulation of the domain. In the case of scattered data, a fitting step must be done to produce a smooth function for which the approximation hierarchy is constructed. Localized versions of *Hardy's multiquadric method*, see [18], or surface reconstruction techniques, similar to the one described in [9], can be used to yield smooth functions interpolating given discrete data. A minimal coarse triangulation of the convex hull of the scattered data sites suffices as an initial triangulation. Earlier work concerning scattered data interpolation and approximation is described in [10] and [23].

The approximation $f(x, y)$ is constructed from the given function $F(x, y)$, the underlying triangulation, which is a set of vertices $\boldsymbol{v}_i = [x_i, y_i]^T$, $i = 0, \ldots, n$, and hat basis functions $f_i = f_i(x, y)$. The basis function f_i – associated with vertex $\boldsymbol{v}_i$ – has a value of one at $\boldsymbol{v}_i$ and varies linearly to zero when going from $\boldsymbol{v}_i$ to all other vertices in the platelet of $\boldsymbol{v}_i$; f_i is zero outside the platelet of $\boldsymbol{v}_i$, see Fig. 3.

Univariate error estimates are easily extended to the bivariate case. Formally, the bivariate global error is the same as in the univariate case (7), and local errors for each triangle T_j are given by

$$e_j^{[n-1]} = \left(\int_{T_j} \left(F - \sum_{k=1}^{3} c_{j,k} f_{j,k}\right)^2 dxdy\right)^{\frac{1}{2}},$$
$$j = 0, \ldots, n_T - 1, \qquad (9)$$

where n_T is the number of triangles in a mesh and $f_{j,k}$ is the basis function associated with the kth vertex of the jth triangle.

The bivariate case requires integration over triangles. For this purpose, we make use of the *change-of-variables theorem*, which allows us to effectively

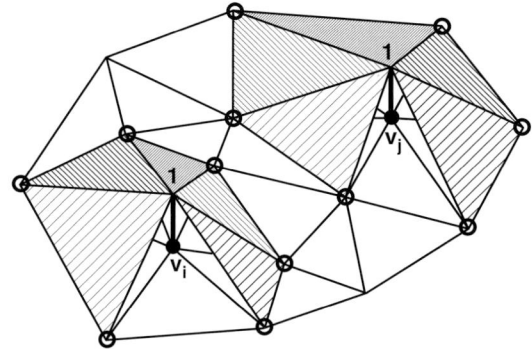

Fig. 3. Platelets of v_i and v_j and associated basis functions. (Triangles in front have been removed for clarity)

integrate functions over a triangle with vertices $v_0 = [x_0, y_0]^T$, $v_1 = [x_1, y_1]^T$, and $v_2 = [x_2, y_2]^T$:

Change-of-variables theorem

Let R and R^* be regions in the plane and let $M\colon R^* \to R$ be a C^1-continuous, one-to-one mapping such that $M(R^*) = R$. Then, for any bivariate integrable function f, the equation

$$\int_R f(x,y)\,dxdy = \int_{R^*} f(x(u,v), y(u,v))\, J\, dudv \qquad (10)$$

holds, where J is the *Jacobian* of M,

$$J = \det \begin{bmatrix} \frac{\partial}{\partial u} x(u,v) & \frac{\partial}{\partial v} x(u,v) \\ \frac{\partial}{\partial u} y(u,v) & \frac{\partial}{\partial v} y(u,v) \end{bmatrix}. \qquad (11)$$

Thus, to effectively compute integrals of functions over triangles we only need to consider the linear transformation

$$\begin{bmatrix} x(u,v) \\ y(u,v) \end{bmatrix} = \begin{bmatrix} x_1 - x_0 & x_2 - x_0 \\ y_1 - y_0 & y_2 - y_0 \end{bmatrix} \begin{bmatrix} u \\ v \end{bmatrix} + \begin{bmatrix} x_0 \\ y_0 \end{bmatrix}. \qquad (12)$$

This transformation maps the *standard triangle* T^* with vertices $u_0 = [0,0]^T$, $u_1 = [1,0]^T$, and $u_2 = [0,1]^T$ in the uv-plane to the arbitrary triangle T with vertices $v_0 = [x_0, y_0]^T$, $v_1 = [x_1, y_1]^T$, and $v_2 = [x_2, y_2]^T$ in the xy-plane. (Both triangles must be oriented counterclockwise.) For this linear mapping, the change-of-variables theorem yields

$$\int_T f(x,y)\,dxdy = J \int_{v=0}^{1} \int_{u=0}^{1-v} f(x(u,v), y(u,v))\, dudv, \qquad (13)$$

where the Jacobian is given by

$$J = \det \begin{bmatrix} x_1 - x_0 & x_2 - x_0 \\ y_1 - y_0 & y_2 - y_0 \end{bmatrix}. \qquad (14)$$

The only scalar products of basis function pairs one must consider in the bivariate case are $\langle N_0, N_0 \rangle$ and $\langle N_0, N_1 \rangle$, where $N_i(u_j, v_j) = \delta_{i,j}$ is a linear spline basis function defined over the standard triangle. (A scalar product $\langle f_i, f_j \rangle$ is only non-zero if the platelets of the vertices v_i and v_j intersect.) The values of these two scalar products are

$$\langle N_0, N_0 \rangle = \int_{v=0}^{1} \int_{u=0}^{1-v} (1-u-v)^2 du dv = \frac{1}{12} \quad \text{and}$$

$$\langle N_0, N_1 \rangle = \int_{v=0}^{1} \int_{u=0}^{1-v} (1-u-v)u \, du dv = \frac{1}{24}. \tag{15}$$

Thus, the scalar product $\langle f_i, f_i \rangle$ is given by

$$\langle f_i, f_i \rangle = \sum_{j=0}^{n_i-1} \int_{T_j} f_i f_i \, dx dy = \tfrac{1}{12} \sum_{j=0}^{n_i-1} J_j, \tag{16}$$

where n_i is the number of platelet triangles associated with vertex v_i and J_j is the Jacobian associated with the jth platelet triangle. The scalar product $\langle f_i, f_k \rangle$ of two basis functions, whose associated vertices v_i and v_k are connected by an edge, is given by

$$\langle f_i, f_k \rangle = \sum_{j=0}^{n_{i,k}-1} \int_{T_j} f_i f_k \, dx dy = \tfrac{1}{24} \sum_{j=0}^{n_{i,k}-1} J_j, \tag{17}$$

where $n_{i,k}$ is the number of platelet triangles shared by vertices v_i and v_k and J_j is the Jacobian associated with the jth platelet triangle. An example of a hierarchy of bivariate best linear spline approximations is shown in Fig. 4.

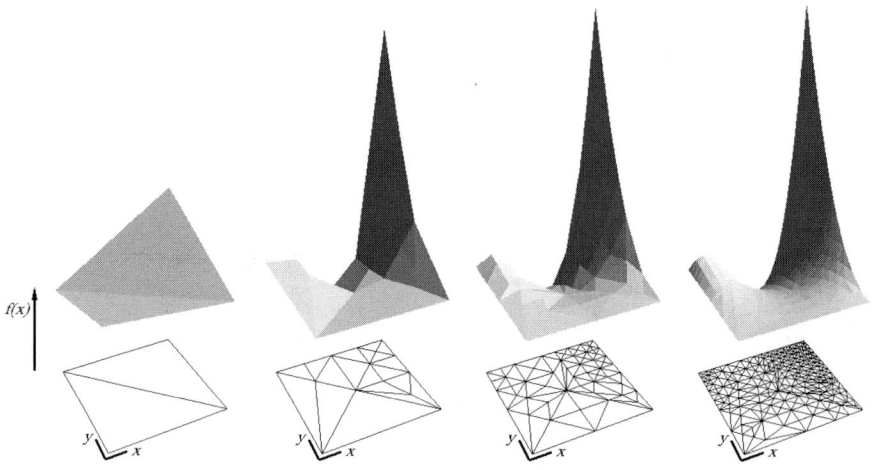

Fig. 4. Four approximations of $F(x,y) = 10x \left(x - \frac{1}{4}\right) \left(x - \frac{3}{4}\right) y^2$, $x, y \in [0,1]$; the origin is indicated by a pair of perpendicular line segments; number of knots for approximations: 4, 15, 58, and 267; corresponding global error estimates: 0.1789, 0.0250, 0.0057, and 0.0011

4 The Trivariate Case

The trivariate case is a straightforward generalization of the bivariate case. The only notable difference from the bivariate case is the use of the change-of-variables theorem. We consider the mapping of the *standard tetrahedron* with vertices $\boldsymbol{u}_0 = [0,0,0]^T$, $\boldsymbol{u}_1 = [1,0,0]^T$, $\boldsymbol{u}_2 = [0,1,0]^T$, and $\boldsymbol{u}_3 = [0,0,1]^T$ in uvw-space to the tetrahedron with vertices $\boldsymbol{v}_0 = [x_0, y_0, z_0]^T$, $\boldsymbol{v}_1 = [x_1, y_1, z_1]^T$, $\boldsymbol{v}_2 = [x_2, y_2, z_2]^T$, and $\boldsymbol{v}_3 = [x_3, y_3, z_3]^T$ in xyz-space. The resulting linear transformation is given by

$$\begin{bmatrix} x(u,v,w) \\ y(u,v,w) \\ z(u,v,w) \end{bmatrix} = \begin{bmatrix} x_1 - x_0 & x_2 - x_0 & x_3 - x_0 \\ y_1 - y_0 & y_2 - y_0 & y_3 - y_0 \\ z_1 - z_0 & z_2 - z_0 & z_3 - z_0 \end{bmatrix} \begin{bmatrix} u \\ v \\ w \end{bmatrix} + \begin{bmatrix} x_0 \\ y_0 \\ z_0 \end{bmatrix}. \quad (18)$$

In this case, the change-of-variables theorem implies that

$$\int_T f(x,y,z)\,dxdydz =$$
$$J \int_{w=0}^{1} \int_{v=0}^{1-w} \int_{u=0}^{1-v-w} f(x(u,v,w), y(u,v,w), z(u,v,w))\,dudvdw, \quad (19)$$

where the Jacobian is given by

$$J = \det \begin{bmatrix} x_1 - x_0 & x_2 - x_0 & x_3 - x_0 \\ y_1 - y_0 & y_2 - y_0 & y_3 - y_0 \\ z_1 - z_0 & z_2 - z_0 & z_3 - z_0 \end{bmatrix}. \quad (20)$$

As in the bivariate case, we need to consider only the scalar products of $\langle N_0, N_0 \rangle$ and $\langle N_0, N_1 \rangle$, where $N_i(u_j, v_j, w_j) = \delta_{i,j}$ is a linear spline basis function over the standard tetrahedron. The values of these two scalar products are

$$\langle N_0, N_0 \rangle = \int_{w=0}^{1} \int_{v=0}^{1-w} \int_{u=0}^{1-v-w} (1-u-v-w)^2\,dudvdw = \frac{1}{60}$$

and

$$\langle N_0, N_1 \rangle = \int_{w=0}^{1} \int_{v=0}^{1-w} \int_{u=0}^{1-v-w} (1-u-v-w)u\,dudvdw = \frac{1}{120}. \quad (21)$$

Thus, the scalar product $\langle f_i, f_i \rangle$ is given by

$$\langle f_i, f_i \rangle = \tfrac{1}{60} \sum_{j=0}^{n_i - 1} J_j, \quad (22)$$

where n_i is the number of platelet tetrahedra associated with vertex $\boldsymbol{v}_i$ and J_j is the Jacobian associated with the jth platelet tetrahedron. The scalar product $\langle f_i, f_k \rangle$ of two basis functions, whose associated vertices $\boldsymbol{v}_i$ and $\boldsymbol{v}_k$ are connected by an edge, is given by

$$\langle f_i, f_k \rangle = \tfrac{1}{120} \sum_{j=0}^{n_{i,k}-1} J_j, \quad (23)$$

where $n_{i,k}$ is the number of platelet tetrahedra shared by vertices $\boldsymbol{v}_i$ and $\boldsymbol{v}_k$ and J_j is the Jacobian associated with the jth platelet tetrahedron. An example of a hierarchy of trivariate best linear spline approximations is shown in Fig. 5.

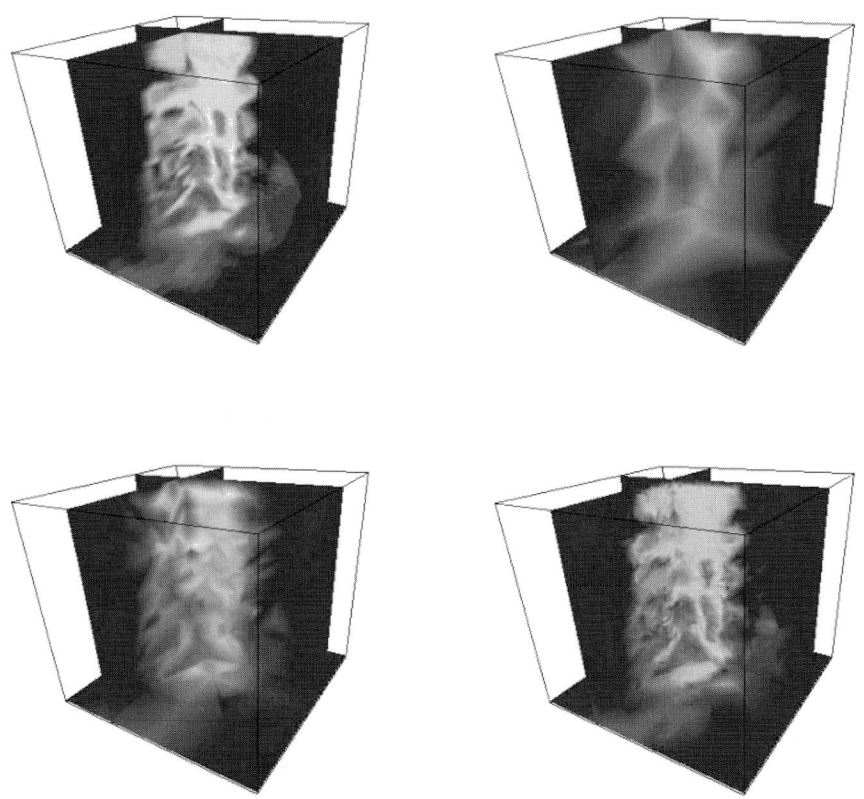

Fig. 5. Three approximations of flame data; original data shown in the upper-left corner (208000 sites). Number of knots for approximations: 68, 514, and 5145; corresponding global error estimates: 0.12, 0.07, and 0.03

5 Finite Element Enhancement

Often, the function being approximated is known only at a finite number of locations. An enhancement to the method we have discussed in previous sections is to take advantage of finitely specified, "discrete" data to save space when storing a hierarchy of approximations. A sequence of knot insertions could potentially reference original data sites rather than explicitly specifying the exact location of each knot when performing edge bisection.

We incorporate the idea of utilizing only originally given data site into the approximation method by modifying the initial input mesh and the bisection step. The vertices in the initial input mesh must be a subset of the original data sites. The convex hull of the data sites suffices for the construction of an initial mesh. Regarding the subdivision steps, in the univariate case, rather than selecting the exact midpoint of an interval subdivision, we select the original data site nearest to the midpoint (while respecting the interval extents) as the new knot. For bivariate and trivariate subdivision, we select the nearest original data site (inside the simplex selected for subdivision) to the midpoint of the longest edge. If the nearest original data site is not unique, we choose one at random. A simplex is not subdivided if there are no original data sites inside it. Care must by taken to ensure a valid mesh since the "snapping" to original data sites can produce inside-out simplices, see Fig. 6.

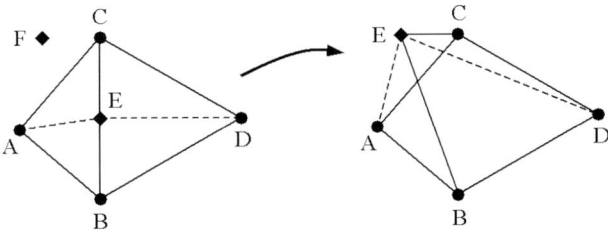

Fig. 6. Inside-out triangles are produced when snapping bisection vertex E to original data site F

6 Incorporating First-derivative Information

When approximating discontinuous data, we have noticed problems with our generated approximations, namely, "over-" and "under-shoots," see Fig. 7 for an example.

To solve this problem we include first-derivative information for spline construction. In addition, we must then also include first-derivative information in the error computations to help focus the refinement near discontinuous

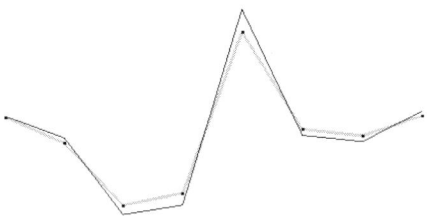

Fig. 7. Demonstration of effect of over- and under-shoots. The lighter polygon shows the original function; the darker polygon shows the approximation

regions and regions with high derivatives/gradients. No algorithmic changes are made otherwise to the original bisection method. The refinement process proceeds by starting with a coarse initial triangulation and iteratively refining the simplices with maximal error.

6.1 The Univariate Case

To incorporate first-derivative information into the computations, a few changes must be made. We re-define (1), the univariate scalar product $\langle f, g \rangle$ of two functions $f(x)$ and $g(x)$, defined over the interval $[a, b]$, as

$$\langle f, g \rangle = \int_a^b w_0 f(x) g(x) + w_1 f'(x) g'(x) dx, \tag{24}$$

where the "weights" w_0 and w_1 are non-negative and sum to one. This new scalar product defines a "*Sobolev*-like" L_2 norm – denoted by $\| \|_{Sob}$ – for a function $f(x)$, which replaces (2):

$$\|f\|_{Sob} = \langle f, f \rangle^{1/2} = \left(\int_a^b w_0 \left(f(x) \right)^2 + w_1 \left(f'(x) \right)^2 dx \right)^{1/2}. \tag{25}$$

This norm allows us to measure the quality of an approximation $f(x)$ of a given function $F(x)$ by considering the norm of $D(x) = F(x) - f(x)$:

$$\|D\|_{Sob} = \langle D, D \rangle^{1/2}. \tag{26}$$

We use this new measure to compute interval-specific error estimates. The univariate global error (7) becomes

$$E = \int_a^b w_0 \left(F(x) - f(x) \right)^2 + w_1 \left(F'(x) - f'(x) \right)^2 dx, \tag{27}$$

where F is the function to be approximated on the interval $[a, b]$ by function f. We want to minimize E, which is equivalent to minimizing

$$\begin{aligned} E = & \int_a^b w_0 \left(f(x) \right)^2 + w_1 \left(f'(x) \right)^2 \\ & - 2 \left(w_0 F(x) f(x) + w_1 F'(x) f'(x) \right) dx \\ & + \int_a^b w_0 \left(F(x) \right)^2 + w_1 \left(F'(x) \right)^2, \end{aligned} \tag{28}$$

which, in matrix notation, reduces to

$$\begin{aligned} E = & \int_a^b \frac{1}{2} [f(x) \ f'(x)] \begin{bmatrix} 2w_0 & 0 \\ 0 & 2w_1 \end{bmatrix} \begin{bmatrix} f(x) \\ f'(x) \end{bmatrix} \\ & - [f(x) \ f'(x)] \begin{bmatrix} 2w_0 F(x) \\ 2w_1 F'(x) \end{bmatrix} dx \\ = & \int_a^b \frac{1}{2} \boldsymbol{f}^T Q \boldsymbol{f} - \boldsymbol{f}^T \boldsymbol{l} dx. \end{aligned} \tag{29}$$

Substituting (3) into (29) yields

$$E = \frac{1}{2}[c_0 \; c_1 \; \ldots \; c_{n-1}] \begin{bmatrix} \int q(f_0,f_0) & \cdots & \int q(f_0,f_{n-1}) \\ \vdots & & \vdots \\ \int q(f_{n-1},f_0) & \cdots & \int q(f_{n-1},f_{n-1}) \end{bmatrix} \begin{bmatrix} c_0 \\ \vdots \\ c_{n-1} \end{bmatrix}$$

$$- [c_0 \; c_1 \; \ldots \; c_{n-1}] \begin{bmatrix} \int l(f_0) \\ \vdots \\ \int l(f_{n-1}) \end{bmatrix} dx$$

$$= \frac{1}{2}\mathbf{c}^T \mathbf{A}\mathbf{c} - \mathbf{c}^T \mathbf{l}, \qquad (30)$$

where $q(f_i, f_j)$ is quadratic and $l(f_i)$ is linear in f_i, f_j, and their derivatives. The "energy" E is minimal for the set of coefficients c_i resulting from the *Ritz equations*, i.e., the linear system

$$\mathbf{A}\mathbf{c} = \mathbf{l}, \qquad (31)$$

see [1]. The elements $a_{i,j}$ of the symmetric, positive definite matrix A are given by

$$a_{i,j} = w_0 \int_a^b f_i(x) f_j(x) dx + w_1 \int_a^b f_i'(x) f_j'(x) dx,$$
$$i, j = 0, \ldots, n-1, \qquad (32)$$

and the elements l_i of the column vector $\mathbf{l}$ are given by

$$l_i = w_0 \int_a^b F(x) f_i(x) dx + w_1 \int_a^b F'(x) f_i'(x) dx,$$
$$i = 0, \ldots, n-1. \qquad (33)$$

Integral values required to compute the matrix elements $a_{i,j}$ are

$$\int_{x_0}^{x_1} (f_0(x))^2 \, dx = \frac{1}{3}\Delta_0, \qquad (34\text{a})$$

$$\int_{x_{n-1}}^{x_n} (f_n(x))^2 \, dx = \frac{1}{3}\Delta_{n-1}, \qquad (34\text{b})$$

$$\int_{x_{i-1}}^{x_{i+1}} (f_i(x))^2 \, dx = \frac{1}{3}(\Delta_{i-1} + \Delta_i), \; i = 1, \ldots, n-1, \text{ and} \qquad (34\text{c})$$

$$\int_{x_i}^{x_{i+1}} f_i(x) f_{i+1}(x) dx = \frac{1}{6}\Delta_i, \; i = 0, \ldots, n-1. \qquad (34\text{d})$$

Terms involving the first derivative are

$$\int_{x_0}^{x_1} (f_0'(x))^2 \, dx = \frac{1}{\Delta_0}, \qquad (35\text{a})$$

$$\int_{x_{n-1}}^{x_n} (f'_n(x))^2\, dx = \frac{1}{\Delta_{n-1}}, \tag{35b}$$

$$\int_{x_{i-1}}^{x_{i+1}} (f'_i(x))^2\, dx = \frac{1}{\Delta_{i-1}} + \frac{1}{\Delta_i},\ i = 1, \ldots, n-1,\ \text{and} \tag{35c}$$

$$\int_{x_i}^{x_{i+1}} f'_i(x) f'_{i+1}(x)\, dx = -\frac{1}{\Delta_i},\ i = 0, \ldots, n-1. \tag{35d}$$

Thus, the matrix $\boldsymbol{A}$ is a tridiagonal matrix, given as a "weighted sum" of two tridiagonal matrices, $\boldsymbol{A}_0$ and $\boldsymbol{A}_1$:

$$\boldsymbol{A} = \frac{1}{6}(w_0 \boldsymbol{A}_0 + w_1 \boldsymbol{A}_1), \tag{36}$$

where the matrices $\boldsymbol{A}_0$ and $\boldsymbol{A}_1$ are given by

$$\boldsymbol{A}_0 = \begin{bmatrix} 2\Delta_0 & \Delta_0 & & & \\ \Delta_0 & 2(\Delta_0 + \Delta_1) & \Delta_1 & & \\ & \ddots & \ddots & \ddots & \\ & & \Delta_{n-2} & 2(\Delta_{n-2} + \Delta_{n-1}) & \Delta_{n-1} \\ & & & \Delta_{n-1} & 2\Delta_{n-1} \end{bmatrix}, \text{ and} \tag{37a}$$

$$\boldsymbol{A}_1 = 6 \begin{bmatrix} \frac{1}{\Delta_0} & \frac{1}{\Delta_0} & & & \\ \frac{-1}{\Delta_0} & \frac{\Delta_0 + \Delta_1}{\Delta_0 \Delta_1} & \frac{-1}{\Delta_1} & & \\ & \ddots & \ddots & \ddots & \\ & & \frac{-1}{\Delta_{n-2}} & \frac{\Delta_{n-2} + \Delta_{n-1}}{\Delta_{n-2}\Delta_{n-1}} & \frac{-1}{\Delta_{n-1}} \\ & & & \frac{-1}{\Delta_{n-1}} & \frac{1}{\Delta_{n-1}} \end{bmatrix}. \tag{37b}$$

An example of a univariate approximation using first-derivative information is shown in Fig. 8. In this example, the function $F(x) = \sin(4\pi x^2)$ was sampled at eight uniformly spaced locations in the interval $[0,1]$, defining a piecewise linear spline $F(x)$, for which the approximations were constructed. It is apparent how the first derivative affects the over- and under-shoots. As more relative weight is assigned to the first derivative, the approximation becomes much better.

6.2 The Bivariate Case

Using a generalization of (26), the error functional we want to minimize in the bivariate case is

$$E = \int_T w_{0,0}\, (F(x,y) - f(x,y))^2 + w_{1,0}\, (F_x(x,y) - f_x(x,y))^2 \\ + w_{0,1}\, (F_y(x,y) - f_y(x,y))^2\, dx dy, \tag{38}$$

where F_x denotes the partial derivative of F with respect to x, f_x denotes the partial derivative of f with respect to x, etc. We want to minimize E,

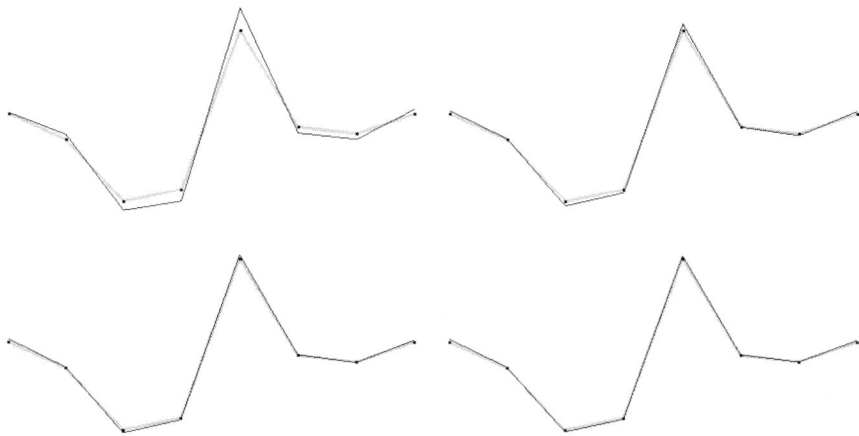

Fig. 8. Four approximations of $F(x) = \sin(4\pi x^2)$ using eight uniformly spaced knots with varying weights (w_0, w_1) given by $(1.0, 0.0)$, $(0.99, 0.01)$, $(0.98, 0.02)$, and $(0.97, 0.03)$ (from upper-left to lower-right corner); the lighter polygon shows the original function (polygon); the darker polygon is the approximation

which corresponds, in matrix notation, to minimizing

$$E = \int_T \frac{1}{2} [f(x,y) \; f_x(x,y) \; f_y(x,y)] \begin{bmatrix} 2w_{0,0} & 0 & 0 \\ 0 & 2w_{1,0} & 0 \\ 0 & 0 & 2w_{0,1} \end{bmatrix} \begin{bmatrix} f(x,y) \\ f_x(x,y) \\ f_y(x,y) \end{bmatrix}$$
$$- [f(x,y) \; f_x(x,y) \; f_y(x,y)] \begin{bmatrix} 2w_{0,0} F(x,y) \\ 2w_{1,0} F_x(x,y) \\ 2w_{0,1} F_y(x,y) \end{bmatrix} dxdy . \quad (39)$$

Substituting (3) into (39) yields, formally, the same equation one obtains in the univariate case. Regarding (31), the elements $a_{i,j}$ of the symmetric, positive definite matrix $\boldsymbol{A}$ are given by

$$a_{i,j} = w_{0,0} \int_T f_i(x,y) f_j(x,y) dxdy + w_{1,0} \int_T f_{ix}(x,y) f_{jx}(x,y) dxdy ,$$
$$+ w_{0,1} \int_T f_{iy}(x,y) f_{jy}(x) dxdy , \quad i,j = 0, \ldots, n-1, \quad (40)$$

where f_{ix} denotes the partial derivative of basis function f_i with respect to x, etc. The elements l_i of the column vector $\boldsymbol{l}$ are given by

$$l_i = w_{0,0} \int_T F(x,y) f_i(x,y) dxdy + w_{1,0} \int_T F_x(x,y) f_{ix}(x,y) dxdy ,$$
$$+ w_{0,1} \int_T F_y(x,y) f_{iy}(x,y) dxdy , \quad i = 0, \ldots, n-1. \quad (41)$$

Integral values required to compute the matrix elements $a_{i,j}$ are

$$\int_{T_i} (f_i(x,y))^2 \, dxdy = \frac{1}{12} \sum_{j=0}^{n_i-1} |J_i|, \qquad (42)$$

where n_i is the number of platelet triangles, T_i, associated with vertex $\boldsymbol{v}_i$ and J_i is the Jacobian, given by (14), associated with the jth platelet triangle. Two basis functions $f_i(x,y)$ and $f_j(x,y)$ whose associated vertices $\boldsymbol{v}_i$ and $\boldsymbol{v}_j$ are connected by an edge imply the non-zero integral value

$$\int_{T_{i,j}} f_i(x,y) f_j(x,y) dxdy = \frac{1}{24} \sum_{k=0}^{n_{i,j}-1} |J_k|, \qquad (43)$$

where $T_{i,j}$ is the set of triangles in common between the platelets of $\boldsymbol{v}_i$ and $\boldsymbol{v}_j$. The linear polynomial interpolating the values one, zero, and zero at the vertices $[x_0, y_0]^T$, $[x_1, y_1]^T$, and $[x_2, y_2]^T$, respectively, has the partial derivatives

$$f_x(x,y) = -\frac{1}{J} \det \begin{bmatrix} 1 & y_1 \\ 1 & y_2 \end{bmatrix} = \frac{y_1 - y_2}{J} \text{ and} \qquad (44a)$$

$$f_y(x,y) = -\frac{1}{J} \det \begin{bmatrix} x_1 & 1 \\ x_2 & 1 \end{bmatrix} = \frac{x_2 - x_1}{J}. \qquad (44b)$$

Integrals involving these partial derivatives are

$$\int_{T_i} (f_{ix}(x,y))^2 \, dxdy = \frac{1}{2} \sum_{j=0}^{n_i-1} \frac{1}{|J_j|} \det{}^2 \begin{bmatrix} 1 & y_{j,1} \\ 1 & y_{j,2} \end{bmatrix} \text{ and} \qquad (45a)$$

$$\int_{T_i} (f_{iy}(x,y))^2 \, dxdy = \frac{1}{2} \sum_{j=0}^{n_i-1} \frac{1}{|J_j|} \det{}^2 \begin{bmatrix} x_{j,1} & 1 \\ x_{j,2} & 1 \end{bmatrix}, \qquad (45b)$$

where $[x_{j,0}, y_{j,0}]^T$, $[x_{j,1}, y_{j,1}]^T$, and $[x_{j,2}, y_{j,2}]^T$ are the counterclockwise-ordered vertices of the n_i platelet triangles associated with vertex $\boldsymbol{v}_i$, see Fig. 9.

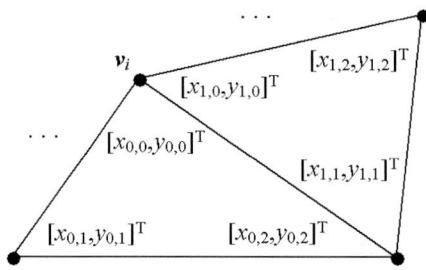

Fig. 9. Indexing scheme for platelet vertices relative to $\boldsymbol{v}_i$ in bivariate case (neighboring triangles oriented counterclockwise)

Other required values are

$$\int_{T_{i,j}} f_{ix}(x,y) f_{jx}(x,y) dx dy \qquad (46a)$$

$$= \frac{1}{2} \sum_{k=0}^{n_{i,j}-1} \frac{1}{|J_k|} \det \begin{bmatrix} 1 & y_{k,1} \\ 1 & y_{k,2} \end{bmatrix} \det \begin{bmatrix} 1 & y_{k,0} \\ 1 & y_{k,1} \end{bmatrix} \text{ and} \qquad (46b)$$

$$\int_{T_{i,j}} f_{iy}(x,y) f_{jy}(x,y) dx dy \qquad (46c)$$

$$= \frac{1}{2} \sum_{k=0}^{n_{i,j}-1} \frac{1}{|J_k|} \det \begin{bmatrix} x_{k,1} & 1 \\ x_{k,2} & 1 \end{bmatrix} \det \begin{bmatrix} x_{k,0} & 1 \\ x_{k,1} & 1 \end{bmatrix}, \qquad (46d)$$

where $n_{i,j}$ is the number of common platelet triangles and $[x_{k,0}, y_{k,0}]^T$, $[x_{k,1}, y_{k,1}]^T$, and $[x_{k,2}, y_{k,2}]^T$ are the vertices of a common triangle of the platelets of v_i and v_j, see Fig. 10.

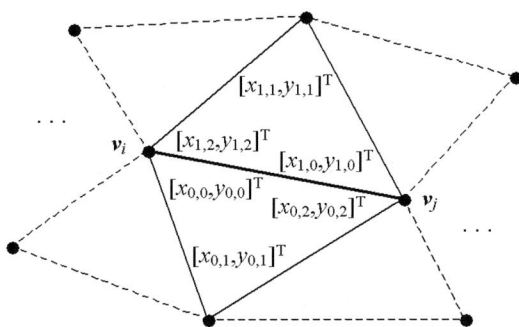

Fig. 10. Indexing scheme for platelet vertices relative to v_i and v_j in bivariate case (neighboring triangles oriented counterclockwise)

An example of a bivariate approximation using first-derivative information is shown in Fig. 11. A checkerboard function was digitized to a 100×100 grid to which a linear spline was fit. The approximations were computed for this spline using the first-derivative information and the finite-element approach described in Sect. 5. It is obvious in this example that the first derivative affects the over- and under-shoots significantly. As more weight is added to the first-derivative information, the better the approximation becomes.

6.3 The Trivariate Case

The error functional we minimize in the trivariate setting is given by

$$E = \int_T \sum_{\substack{i,j,k \geq 0 \\ i+j+k \leq 1}} \left(\frac{\partial^{i+j+k}}{\partial x^i \partial y^j \partial z^k} (F(x,y,z) - f(x,y,z)) \right)^2 dx dy dz, \qquad (47)$$

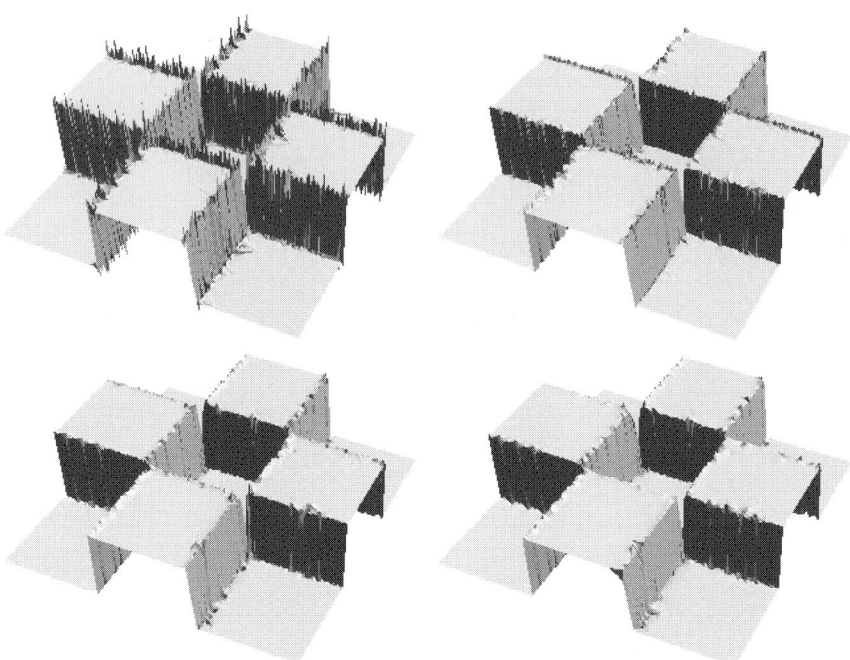

Fig. 11. Four approximations of bivariate checkerboard function with varying weights $(w_{0,0}, w_{1,0}, w_{0,1})$: $(1,0,0)$, $\left(\frac{3}{4}, \frac{1}{8}, \frac{1}{8}\right)$, $\left(\frac{1}{2}, \frac{1}{4}, \frac{1}{4}\right)$, and $\left(\frac{1}{4}, \frac{3}{8}, \frac{3}{8}\right)$, from upper-left to lower-right corner; number of knots varies between 5000 and 6000

which, in matrix notation, corresponds to minimizing

$$E = \int_T \frac{1}{2} [f(x,y,z)\ f_x(x,y,z)\ f_y(x,y,z)\ f_z(x,y,z)]$$
$$\begin{bmatrix} 2w_{0,0,0} & 0 & 0 & 0 \\ 0 & 2w_{1,0,0} & 0 & 0 \\ 0 & 0 & 2w_{0,1,0} & 0 \\ 0 & 0 & 0 & 2w_{0,0,1} \end{bmatrix} \begin{bmatrix} f(x,y,z) \\ f_x(x,y,z) \\ f_y(x,y,z) \\ f_z(x,y,z) \end{bmatrix}$$
$$- [f(x,y,z)\ f_x(x,y,z)\ f_y(x,y,z)\ f_z(x,y,z)]$$
$$\begin{bmatrix} 2w_{0,0,0}F(x,y,z) \\ 2w_{1,0,0}F_x(x,y,z) \\ 2w_{0,1,0}F_y(x,y,z) \\ 2w_{0,0,1}F_z(x,y,z) \end{bmatrix} dxdydz . \tag{48}$$

Substituting (3) into (48) yields, formally, the same equations one obtains for the univariate and bivariate cases. Regarding (31), the elements $a_{i,j}$ of the symmetric, positive definite matrix $\boldsymbol{A}$ are given by

$$a_{i,j} = w_{0,0,0} \int_T f_i(x,y,z) f_j(x,y,z) dxdydz$$

$$+ w_{1,0,0} \int_T f_{ix}(x,y,z) f_{jx}(x,y,z) dxdydz$$

$$+ w_{0,1,0} \int_T f_{iy}(x,y,z) f_{jy}(x,y,z) dxdydz$$

$$+ w_{0,0,1} \int_T f_{iz}(x,y,z) f_{jz}(x,y,z) dxdydz ,$$

$$i,j = 0, \ldots, n-1 . \tag{49}$$

The elements l_i of the column vector $\boldsymbol{l}$ are given by

$$l_i = w_{0,0,0} \int_T F(x,y,z) f_j(x,y,z) dxdydz$$

$$+ w_{1,0,0} \int_T F_x(x,y,z) f_{jx}(x,y,z) dxdydz$$

$$+ w_{0,1,0} \int_T F_y(x,y,z) f_{jy}(x,y,z) dxdydz$$

$$+ w_{0,0,1} \int_T F_z(x,y,z) f_{jz}(x,y,z) dxdydz ,$$

$$i,j = 0, \ldots, n-1 . \tag{50}$$

Integral values required to compute the matrix elements $a_{i,j}$ are

$$\int_{T_i} (f_i(x,y,z))^2 \, dxdydz = \frac{1}{60} \sum_{j=0}^{n_i-1} |J_i| , \tag{51}$$

where n_i is the number of platelet tetrahedra, T_i, associated with vertex $\boldsymbol{v}_i$ and J_i is the Jacobian, given by (20), associated with the jth platelet tetrahedron. Two basis functions $f_i(x,y)$ and $f_j(x,y)$ whose associated vertices $\boldsymbol{v}_i$ and $\boldsymbol{v}_j$ are connected by an edge imply the non-zero integral value

$$\int_{T_{i,j}} f_i(x,y,z) f_j(x,y,z) dxdydz = \frac{1}{120} \sum_{k=0}^{n_{i,j}-1} |J_k| , \tag{52}$$

where $T_{i,j}$ is the set of tetrahedra in common between the platelets of $\boldsymbol{v}_i$ and $\boldsymbol{v}_j$. The linear polynomial interpolating the values one, zero, zero, and zero at the vertices $[x_0, y_0, z_0]^T$, $[x_1, y_1, z_1]^T$, $[x_2, y_2, z_2]^T$, and $[x_3, y_3, z_3]^T$, respectively, has the partial derivatives

$$f_x(x,y,z) = -\frac{1}{J} \det \begin{bmatrix} 1 & y_1 & z_1 \\ 1 & y_2 & z_2 \\ 1 & y_3 & z_3 \end{bmatrix} , \tag{53a}$$

$$f_y(x,y,z) = -\frac{1}{J} \det \begin{bmatrix} x_1 & 1 & z_1 \\ x_2 & 1 & z_2 \\ x_3 & 1 & z_3 \end{bmatrix} , \text{ and} \tag{53b}$$

$$f_z(x,y,z) = -\frac{1}{J} \det \begin{bmatrix} x_1 & y_1 & 1 \\ x_2 & y_2 & 1 \\ x_3 & y_3 & 1 \end{bmatrix} . \tag{53c}$$

Integrals involving these partial derivatives are

$$\int_{T_i} (f_{ix}(x,y,z))^2 \, dxdydz = \frac{1}{6} \sum_{j=0}^{n_i-1} \frac{1}{|J_j|} \det^2 \begin{bmatrix} 1 & y_{j,1} & z_{j,1} \\ 1 & y_{j,2} & z_{j,2} \\ 1 & y_{j,3} & z_{j,3} \end{bmatrix}, \quad (54a)$$

$$\int_{T_i} (f_{iy}(x,y,z))^2 \, dxdydz = \frac{1}{6} \sum_{j=0}^{n_i-1} \frac{1}{|J_j|} \det^2 \begin{bmatrix} x_{j,1} & 1 & z_{j,1} \\ x_{j,2} & 1 & z_{j,2} \\ x_{j,3} & 1 & z_{j,3} \end{bmatrix}, \text{ and } (54b)$$

$$\int_{T_i} (f_{iz}(x,y,z))^2 \, dxdydz = \frac{1}{6} \sum_{j=0}^{n_i-1} \frac{1}{|J_j|} \det^2 \begin{bmatrix} x_{j,1} & y_{j,1} & 1 \\ x_{j,2} & y_{j,2} & 1 \\ x_{j,3} & y_{j,3} & 1 \end{bmatrix}, \quad (54c)$$

where the vertices $[x_{j,1}, y_{j,1}, z_{j,1}]^T$, $[x_{j,2}, y_{j,2}, z_{j,2}]^T$, and $[x_{j,3}, y_{j,3}, z_{j,3}]^T$ denote the boundary vertices of the faces of the platelet tetrahedra associated with vertex $[x_i, y_i, z_i]^T$. Other required values are

$$\int_{T_{i,j}} f_{ix}(x,y,z) f_{jx}(x,y,z) dxdydz$$

$$= \frac{1}{6} \sum_{k=0}^{n_{i,j}-1} \frac{1}{|J_k|} \det \begin{bmatrix} 1 & y_{k,1} & z_{k,1} \\ 1 & y_{k,2} & z_{k,2} \\ 1 & y_{k,3} & z_{k,3} \end{bmatrix} \det \begin{bmatrix} 1 & y_{k,0} & z_{k,0} \\ 1 & y_{k,3} & z_{k,3} \\ 1 & y_{k,2} & z_{k,2} \end{bmatrix}, \quad (55a)$$

$$\int_{T_{i,j}} f_{iy}(x,y,z) f_{jy}(x,y,z) dxdydz$$

$$= \frac{1}{6} \sum_{k=0}^{n_{i,j}-1} \frac{1}{|J_k|} \det \begin{bmatrix} x_{k,1} & 1 & z_{k,1} \\ x_{k,2} & 1 & z_{k,2} \\ x_{k,3} & 1 & z_{k,3} \end{bmatrix} \det \begin{bmatrix} x_{k,0} & 1 & z_{k,0} \\ x_{k,3} & 1 & z_{k,3} \\ x_{k,2} & 1 & z_{k,2} \end{bmatrix}, \text{ and } (55b)$$

$$\int_{T_{i,j}} f_{iz}(x,y,z) f_{jz}(x,y,z) dxdydz$$

$$= \frac{1}{6} \sum_{k=0}^{n_{i,j}-1} \frac{1}{|J_k|} \det \begin{bmatrix} x_{k,1} & y_{k,1} & 1 \\ x_{k,2} & y_{k,2} & 1 \\ x_{k,3} & y_{k,3} & 1 \end{bmatrix} \det \begin{bmatrix} x_{k,0} & y_{k,0} & 1 \\ x_{k,3} & y_{k,3} & 1 \\ x_{k,2} & y_{k,2} & 1 \end{bmatrix}, \quad (55c)$$

where $n_{i,j}$ is the number of common platelet tetrahedra and $[x_{k,0}, y_{k,0}, z_{k,0}]^T$, $[x_{k,1}, y_{k,1}, z_{k,1}]^T$, $[x_{k,2}, y_{k,2}, z_{k,2}]^T$, and $[x_{k,3}, y_{k,3}, z_{k,3}]^T$ are the vertices of a common tetrahedron of the platelets of $\boldsymbol{v}_i$ and $\boldsymbol{v}_j$.

An example of a trivariate approximation using first-derivative information is shown in Fig. 12. A checkerboard function (the trivariate generalization of the bivariate function) was digitized to a $100 \times 100 \times 100$ grid to which a linear spline was fit. The approximations were computed for this spline. Again, the use of first derivative affects approximation quality substantially. Fig. 12 shows a flat-shaded slice through the approximations. As more weight is added to the first derivative, the discontinuities (abrupt changes from zero to one and vice versa) are captured much better.

Fig. 12. Four approximations of trivariate checkerboard function with varying weights $(w_{0,0,0}, w_{1,0,0}, w_{0,1,0}, w_{0,0,1})$: $(1, 0, 0, 0)$, $\left(\frac{3}{4}, \frac{1}{12}, \frac{1}{12}, \frac{1}{12}\right)$, $\left(\frac{1}{2}, \frac{1}{6}, \frac{1}{6}, \frac{1}{6}\right)$, and $\left(\frac{1}{4}, \frac{3}{12}, \frac{3}{12}, \frac{3}{12}\right)$, from upper-left to lower-right corner; number of knots varies between 1400 and 3000

7 Quadratic Simplices

As an alternative to using linear simplices, we show how to use *curved simplices* to better approximate data. Most scientific data sets contain discontinuities – such as a car body geometry or a pressure field discontinuity – that can often be represented much better with curved elements. Discontinuous data sets can often be approximated better by dividing their domains into several smaller domains with curved boundaries. A combination of geometry/boundary and dependent field variable discontinuities can be treated in an integrated fashion. The discussion in this section only treats the bivariate and trivariate cases.

The algorithms that we have described in the previous sections still apply: we begin with a coarse initial triangulation, which may now contain curved simplices, and repeatedly refine this mesh until a global error tolerance is met.

7.1 Mapping the standard simplex

In the bivariate case, we map the standard triangle, see Sect. 3, to a curved triangular region in physical space by mapping the six knots $\boldsymbol{u}_i = [u_{i,j}, v_{i,j}]^T = \left(\frac{i}{2}, \frac{j}{2}\right)$, $i, j \geq 0$, $i + j \leq 2$ (abbreviated in multi-index notation as $|\boldsymbol{i}| = 2$) in parameter space to six corresponding vertices $\boldsymbol{x}_i = [x_{i,j}, y_{i,j}]^T$ in physical space, using a quadratic mapping. The quadratic mapping, using Bernstein-Bézier polynomials $B_i^2(\boldsymbol{u})$ as basis functions, see [8] and [26], is given by

$$\boldsymbol{x}(\boldsymbol{u}) = \begin{bmatrix} x(u,v) \\ y(u,v) \end{bmatrix} = \sum_{|\boldsymbol{i}|=2} \boldsymbol{b}_i B_i^2(\boldsymbol{u}) = \begin{bmatrix} \sum_{|\boldsymbol{i}|=2} c_{i,j} B_{i,j}^2(u,v) \\ \sum_{|\boldsymbol{i}|=2} d_{i,j} B_{i,j}^2(u,v) \end{bmatrix} \quad (56)$$

In the same manner, we define the mapping of the standard tetrahedron, described in Sect. 4, to a curved tetrahedron in physical space, mapping ten knots $\boldsymbol{u}_i = [u_{i,j,k}, v_{i,j,k}, w_{i,j,k}]^T = \left(\frac{i}{2}, \frac{j}{2}, \frac{k}{2}\right)$, $|\boldsymbol{i}| = 2$ in parameter space, to ten corresponding vertices $\boldsymbol{x}_i = [x_{i,j,k}, y_{i,j,k}, z_{i,j,k}]^T$ in physical space. The quadratic mapping in the trivariate case is given by

$$\boldsymbol{x}(\boldsymbol{u}) = \begin{bmatrix} x(u,v,w) \\ y(u,v,w) \\ z(u,v,w) \end{bmatrix} = \sum_{|\boldsymbol{i}|=2} \boldsymbol{b}_i B_i^2(\boldsymbol{u})$$
$$= \begin{bmatrix} \sum_{|\boldsymbol{i}|=2} c_{i,j,k} B_{i,j,k}^2(u,v,w) \\ \sum_{|\boldsymbol{i}|=2} d_{i,j,k} B_{i,j,k}^2(u,v,w) \\ \sum_{|\boldsymbol{i}|=2} e_{i,j,k} B_{i,j,k}^2(u,v,w) \end{bmatrix} . \quad (57)$$

7.2 Initial Simplicial Decomposition

The original grid, its boundaries, and possibly known locations of field discontinuities (in the dependent field variables) influence how we construct an initial simplicial decomposition. We consider domain boundaries and known discontinuities of interest to define a set of vertices and a set of edges (possibly curved) connecting these vertices. (We do not discuss in this paper how to obtain these vertices or edges.) With this information, one can define the unique fields and sub-regions in the overall domain that are bounded by these edges. If the edges specified are linear spline curves, a quadratic curve fitting step must take place to approximate the linear segments by quadratic curves. Each of the sub-domains, bounded by curved edges, can then be triangulated to form an initial mesh, see Fig. 13. Approximation is then performed for each sub-domain independently. It is possible that (quadratic) simplices share knot/vertex locations on field boundaries. In this case, since an approximation is computed independently for each sub-domain, there exist two separate coefficients for the same location.

Bisection of curved edges of mesh simplices is performed in a manner similar to the linear case: we bisect at the midpoint of the arc and insert a knot at this location. The finite-element approach based on a Sobolev-like norm can be applied, as described in Sect. 5.

7.3 Best Approximation for Quadratic Simplices

We denote basis functions associated with simplex vertex $\boldsymbol{v}_i$ by $f_i(\boldsymbol{x})$. The basis function corresponding to edge e_j of a simplex is denoted by $g_j(\boldsymbol{x})$.

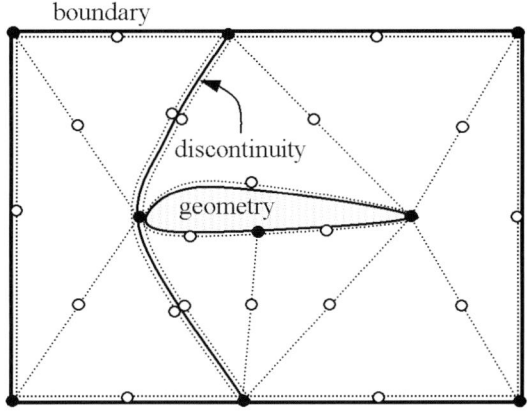

Fig. 13. Decomposition of domain around a wing using bivariate curved simplices. There are two distinct fields defined by the vertices and edges in this example: the field left of the discontinuity and the field right of the discontinuity

The best approximation is denoted by $f(\boldsymbol{x})$, and we write it as a linear combination of the basis functions associated with the simplex corners and edges. Assuming that there are n distinct corner vertices and m distinct edges, we can write the best approximation as

$$f(\boldsymbol{x}) = \sum_{i=0}^{n-1} c_i f_i(\boldsymbol{x}) + \sum_{j=0}^{m-1} d_j g_j(\boldsymbol{x}). \tag{58}$$

In matrix form, the normal equations are

$$\begin{bmatrix} \langle f_0, f_0 \rangle & \cdots & \langle f_0, f_{n-1} \rangle & \langle f_0, g_0 \rangle & \cdots & \langle f_0, g_{m-1} \rangle \\ \vdots & & \vdots & \vdots & & \vdots \\ \langle f_{n-1}, f_0 \rangle & \cdots & \langle f_{n-1}, f_{n-1} \rangle & \langle f_{n-1}, g_0 \rangle & \cdots & \langle f_{n-1}, g_{m-1} \rangle \\ \langle g_0, f_0 \rangle & \cdots & \langle f_0, f_{n-1} \rangle & \langle g_0, g_0 \rangle & \cdots & \langle g_0, g_{m-1} \rangle \\ \vdots & & \vdots & \vdots & & \vdots \\ \langle g_{m-1}, f_0 \rangle & \cdots & \langle g_{m-1}, f_{n-1} \rangle & \langle g_{m-1}, g_0 \rangle & \cdots & \langle g_{m-1}, g_{m-1} \rangle \end{bmatrix} \begin{bmatrix} c_0 \\ \vdots \\ c_{n-1} \\ d_0 \\ \vdots \\ d_{m-1} \end{bmatrix}$$

$$= \begin{bmatrix} \langle F, f_0 \rangle \\ \vdots \\ \langle F, f_{n-1} \rangle \\ \langle F, g_0 \rangle \\ \vdots \\ \langle F, g_{m-1} \rangle \end{bmatrix}. \tag{59}$$

The Bernstein-Bézier quadratic basis functions are defined as

$$B_{\boldsymbol{i}}^2(\boldsymbol{u}) = \frac{2!}{(2-i-j)!i!j!}(1-u-v)^{2-i-j}u^i v^j \quad \text{and} \tag{60a}$$

$$B_{\boldsymbol{i}}^2(\boldsymbol{u}) = \frac{2!}{(2-i-j-k)!i!j!k!}(1-u-v-w)^{2-i-j-k}u^i v^j w^k \tag{60b}$$

in the bivariate and trivariate cases, respectively. Again, we can use of the change-of-variables theorem and implement error estimates as described in Sects. 3 and 4 to compute a best approximation. The needed inner products, defined over the standard simplex in parameter space, are given by

$$\langle B_{i,j}, B_{k,l}\rangle = \frac{1}{180}\begin{bmatrix} 6 & 3 & 1 & 3 & 1 & 1 \\ 3 & 4 & 3 & 2 & 2 & 1 \\ 1 & 3 & 6 & 1 & 3 & 1 \\ 3 & 2 & 1 & 4 & 2 & 3 \\ 1 & 2 & 3 & 2 & 4 & 3 \\ 1 & 1 & 1 & 3 & 3 & 6 \end{bmatrix} \tag{61}$$

in the bivariate case, and

$$\langle B_{i,j,k}, B_{l,m,n}\rangle = \frac{1}{1260}\begin{bmatrix} 6 & 3 & 1 & 3 & 1 & 1 & 3 & 1 & 1 & 1 \\ 3 & 4 & 3 & 2 & 2 & 1 & 2 & 2 & 1 & 1 \\ 1 & 3 & 6 & 1 & 3 & 1 & 1 & 3 & 1 & 1 \\ 3 & 2 & 1 & 4 & 2 & 3 & 2 & 1 & 2 & 1 \\ 1 & 2 & 3 & 2 & 4 & 3 & 1 & 2 & 2 & 1 \\ 1 & 1 & 1 & 3 & 3 & 6 & 1 & 1 & 3 & 1 \\ 3 & 2 & 1 & 2 & 1 & 1 & 4 & 2 & 2 & 3 \\ 1 & 2 & 3 & 1 & 2 & 1 & 2 & 4 & 2 & 3 \\ 1 & 1 & 1 & 2 & 2 & 3 & 2 & 2 & 4 & 3 \\ 1 & 1 & 1 & 1 & 1 & 1 & 3 & 3 & 3 & 6 \end{bmatrix} \tag{62}$$

in the trivariate case.

8 Conclusions and Future Work

We have discussed a best linear spline approximation scheme and several enhancements to improve and generalize. To adapt our method to multi-valued data, one can, at each iteration, approximate each dependent variable separately.

In terms of computational efficiency, large linear systems are produced when dealing with vary large data sets. This is manageable, since the resulting systems are sparse and can easily be treated with sparse-system solvers.

Investigating the effectiveness of curved elements for approximating data is a topic of our current research. With the use of curved elements growing, we believe that using curved elements in approximations, and supporting them directly during rendering, will become an important visualization research area.

9 Acknowledgements

This work was performed under the auspices of the U.S. Department of Energy by University of California, Lawrence Livermore National Laboratory under Contract W-7405-Eng-48. This work was also supported by the National Science Foundation under contract ACI 9624034 (CAREER Award), through the Large Scientific and Software Data Set Visualization (LSSDSV) program under contract ACI 9982251, and through the National Partnership for Advanced Computational Infrastructure (NPACI); the Office of Naval Research under contract N00014-97-1-0222; the Army Research Office under contract ARO 36598-MA-RIP; the NASA Ames Research Center through an NRA award under contract NAG2-1216; the Lawrence Livermore National Laboratory under ASCI ASAP Level-2 Memorandum Agreement B347878 and under Memorandum Agreement B503159; and the North Atlantic Treaty Organization (NATO) under contract CRG.971628 awarded to the University of California, Davis. We also acknowledge the support of ALSTOM Schilling Robotics and SGI. We thank the members of the Visualization and Graphics Research Group at the Center for Image Processing and Integrated Computing (CIPIC) at the University of California, Davis.

References

1. Boehm, W. and Prautzsch, H. (1993), Numerical Methods, A K Peters, Ltd., Wellesley, MA
2. Bonneau, G. P., Hahmann, S. and Nielson, G. M. (1996), BLAC-wavelets: A multiresolution analysis with non-nested spaces. In: Yagel, R. and Nielson, G. M. (Eds.) Visualization '96, IEEE Computer Society Press, Los Alamitos, CA, 43–48
3. Cignoni, P., De Floriani, L., Montani, C., Puppo, E. and Scopigno, R. (1994), Multiresolution modeling and visualization of volume data based on simplicial complexes. In: Kaufman, A. E. and Krüger, W. (Eds.) 1994 Symposium on Volume Visualization, IEEE Computer Society Press, Los Alamitos, CA, 19–26
4. Cignoni, P., Montani, C., Puppo, E., and Scopigno, R. (1997), Multiresolution representation and visualization of volume data, IEEE Transactions of Visualization and Computer Graphics **3(4)**, 352–369
5. Davis, P. J. (1975) Interpolation and Approximation. New York: Dover
6. Duff, I. S., Erisman, A. M., and Reid, J. K. (1986), Direct Methods for Sparse Matrices, Clarendon Press, Oxford, England, 239–251
7. Dyn, N., Floater, M. S., and Iske, A. (2000), Adaptive thinning for bivariate scattered data, Technische Universität München, Fakultät für Mathematik, München, Germany, Report TUM M0006, 2000.
8. Farin, G. (2001), Curves and Surfaces for Computer Aided Geometric Design, fifth edition, Academic Press, San Diego, CA
9. Floater, M. S. and Reimers, M. (2001), Meshless parameterization and surface reconstruction, Computer Aided Geometric Design **18**, 77–92
10. Franke, R. (1982), Scattered data interpolation: Tests of some methods, Math. Comp. **38**, 181–200
11. Gieng, T. S., Hamann, B., Joy, K. I., Schussman, G. L. and Trotts, I. J. (1997), Smooth hierarchical surface triangulations. In: Yagel, R. and Hagen, H. (Eds.) Visualization '97, IEEE Computer Society Press, Los Alamitos, CA, 379–386

12. Gieng, T. S., Hamann, B., Joy, K. I., Schussman, G. L. and Trotts, I. J. (1998), Constructing hierarchies for triangle meshes, IEEE Transactions on Visualization and Computer Graphics **4(2)**, 145–161
13. Gross, M. H., Gatti, R. and Staadt, O. (1995), Fast multiresolution surface meshing. In: Nielson, G. M. and Silver, D. (Eds.) Visualization '95, IEEE Computer Society Press, Los Alamitos, CA, 135–142
14. Grosso, R., Lürig, C. and Ertl, T. (1997), The multilevel finite element method for adaptive mesh optimization and visualization of volume data. In: Yagel, R. and Hagen, H. (Eds.) Visualization '97, IEEE Computer Society Press, Los Alamitos, CA, 387–394
15. Hagen, H., Müller, H. and Nielson, G. M. (Eds.) (1993), Focus on Scientific Visualization, Springer-Verlag, New York, NY
16. Hamann, B. (1994), A data reduction scheme for triangulated surfaces, Computer Aided Geometric Design **11(2)**, Elsevier, 197–214
17. Hamann, B., Jordan, B. W. and Wiley, D. F. (1999), On a construction of a hierarchy of best linear spline approximations using repeated bisection, IEEE Transactions on Visualization and Computer Graphics **5(1)**, 30–46, and **5(2)**, 190 (errata)
18. Hardy, R. L. (1971), Multiquadric equations of topography and other irregular surfaces, Journal of Geophysical Research **76**, 1905–1915
19. Heckbert, P. S. and Garland, M. (1997), Survey of polygonal surface simplification algorithms, Technical Report, Computer Science Department, Carnegie Mellon University, Pittsburg, Pennsylvania, to appear
20. Hoppe, H. (1996), Progressive meshes. In: Rushmeier, H. (Ed.) Proceedings of SIGGRAPH 1996, ACM Press, New York, NY, 99–108
21. Hoppe, H. (1997), View-dependent refinement of progressive meshes. In: Whitted, T. (Ed.) Proceedings of SIGGRAPH 1997, ACM Press, New York, NY, 189–198
22. Kaufman, A. E. (Ed.) (1991), Volume Visualization, IEEE Computer Society Press, Los Alamitos, CA
23. Nielson, G. M. (1993), Scattered data modeling, IEEE Computer Graphics and Applications **13(1)**, 60–70
24. Nielson, G. M., Müller, H. and Hagen, H. (Eds.) (1997), Scientific Visualization: Overviews, Methodologies, and Techniques, IEEE Computer Society Press, Los Alamitos, CA
25. Nielson, G. M. and Shriver B. D. (Eds.) (1990), Visualization in Scientific Computing, IEEE Computer Society Press, Los Alamitos, CA
26. Piegl, L. A. and Tiller, W. (1996), The NURBS Book, second edition, Springer-Verlag, New York, NY
27. Rippa, S. (1992), Long and thin triangles can be good for linear interpolation, SIAM J. Numer. Anal. **29(1)**, 257–270
28. Rosenblum, L. J, Earnshaw, R. A., Encarnação, J. L., Hagen, H., Kaufman, A. E., Klimenko, S., Nielson, G. M., Post, F. and Thalmann, D. (Eds.) (1994), Scientific Visualization – Advances and Challenges, IEEE Computer Society Press, Los Alamitos, CA
29. Staadt, O. G., Gross, M. H. and Weber, R. (1997), Multiresolution compression and reconstruction. In: Yagel, R. and Hagen, H. (Eds.) Visualization '97, IEEE Computer Society Press, Los Alamitos, CA, 337–346

Terrain Modeling Using Voronoi Hierarchies

Martin Bertram[1], Shirley E. Konkle[2], Hans Hagen[1], Bernd Hamann[3], and Kenneth I. Joy[3]

[1] University of Kaiserslautern, Department of Computer Science, P.O. Box 3049, D-67653 Kaiserslautern, Germany.
[2] Pixar Animation Studios, 1200 Park Ave., Emeryville, CA 94608, U.S.A.
[3] Center for Image Processing and Integrated Computing (CIPIC), Department of Computer Science, University of California, Davis, CA 95616-8562, U.S.A.

Abstract. We present a new algorithm for terrain modeling based on Voronoi diagrams and Sibson's interpolant. Starting with a set of scattered sites in the plane with associated function values defining a height field, our algorithm constructs a top-down hierarchy of smooth approximations. We use the convex hull of the given sites as domain for our hierarchical representation. Sibson's interpolant is used to approximate the underlying height field based on the associated function values of selected subsets of the sites. Therefore, our algorithm constructs a hierarchy of Voronoi diagrams for nested subsets of the given sites. The quality of approximations obtained with our method compares favorably to results obtained from other multiresolution algorithms like wavelet transforms. For every level of resolution, our approximations are C^1 continuous, except at the selected sites, where only C^0 continuity is satisfied. Considering n sites, the expected time complexity of our algorithm is $O(n \: log \: n)$. In addition to a hierarchy of smooth approximations, our method provides a cluster hierarchy based on convex cells and an importance ranking for sites.

1 Introduction

Clustering techniques [7] can be used to generate a data-dependent partitioning of space representing inherent topological and geometric structures of scattered data. Adaptive clustering methods recursively refine a partitioning resulting in a multiresolution representation that is of advantage for applications like progressive transmission, compression, view-dependent rendering, and topology reconstruction. For example, topological structures of two-manifold surfaces can be reconstructed from scattered points in three-dimensional space using adaptive clustering methods [6]. In contrast to mesh-simplification algorithms, adaptive clustering methods do not require a grid structure connecting data points. A cluster hierarchy is built in a "top-down" approach, so that coarse levels of resolution require less computation times than finer levels.

We present a Voronoi-based adaptive clustering method for terrain modeling. Arbitrary samples taken from large-scale terrain models are recursively selected according to their relevance. Continuous approximations of the terrain model are constructed based on the individual sets of selected sites using

Sibson's interpolant [10]. We have implemented this algorithm using a Delaunay triangulation, i.e., the dual of a Voronoi diagram, as underlying data structure. Constructing a Delaunay triangulation is simpler than constructing the corresponding Voronoi diagram, since a large number of special cases (where Voronoi vertices have valences greater than three) can be ignored. A major drawback of Delaunay triangulations is that they are not unique, in general. This becomes evident when the selected sites are sampled from regular, rectilinear grids such that either diagonal of a quadrilateral can be used, resulting in random choices affecting the approximation. The corresponding Voronoi diagram, however, is uniquely defined and can be derived directly from a Delaunay triangulation. Sibson's interpolant is also efficiently computed from a Delaunay triangulation. The advantage of our method when compared to Delaunay-based multiresolution methods [4] is that our approximations are unique and C^1 continuous almost everywhere.

2 Adaptive Clustering Approach

Adaptive clustering schemes construct a hierarchy of tesselations, each of which is associated with a simplified representation of the data. We assume that a data set is represented at its finest level of resolution by a set P of n points in the plane with associated function values:

$$P = \{(\mathbf{p}_i, f_i) \mid \mathbf{p}_i \in \mathbb{R}^2,\ f_i \in \mathbb{R},\ i = 1, \ldots, n\}.$$

This set can be considered as a sampled version of a continuous function $f : D \to \mathbb{R}$, where $D \subset \mathbb{R}^2$ is a compact domain containing all points $\mathbf{p}_i$. The points $\mathbf{p}_i$ define the associated parameter values for the samples f_i. We do not assume any kind of "connectivity" or grid structure for the points $\mathbf{p}_i$. For applications more general than terrain modeling, the points p_i can have s dimensions with t-dimensional function values f_i, see Figure 1.

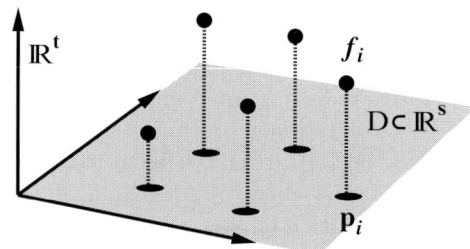

Fig. 1. Scattered sites with associated function values.

The output of an adaptive clustering scheme consists of a number of levels L_j, $j = 0, 1, \ldots$, defined as

$$L_j = \{(\tau_k^j, \tilde{f}_k^j, \varepsilon_k^j) \mid k = 1, \ldots, n_j\},$$

where, for every level with index j, the tiles (or regions) $\{\tau_k^j \subseteq D \mid k = 1,\ldots,n_j\}$ form a partitioning of the domain D, the functions $\tilde{f}_k^j : \tau_k^j \to \mathbb{R}$ approximate the function values of points in the tiles τ_k^j, i.e.,

$$\tilde{f}_k^j(\mathbf{p}_i) \approx f_i \quad \forall \mathbf{p}_i \in \tau_k^j,$$

and the residuals $\varepsilon_k^j \in \mathbb{R} \geq 0$ estimate the approximation error. In principle, any error norm can be chosen to compute the residuals ε_k^j. We note that the error norm has a high impact on the efficiency and quality of the clustering algorithm, since it defines an optimization criterion for the approximations at every level of resolution. We suggest to use the following norm:

$$\varepsilon_k^j = \left(\frac{1}{n_k^j} \sum_{\mathbf{p}_i \in \tau_k^j} \left| \tilde{f}_k^j(\mathbf{p}_i) - f_i \right|^p \right)^{\frac{1}{p}}, \quad p \in [1, \infty], \tag{1}$$

where $n_k^j = |\{\mathbf{p}_i \in \tau_k^j\}|$ is the number of points located in tile τ_k^j. In the case of $p = \infty$, the residual is simply the maximal error considering all sites in the corresponding tile. This error norm can be easily adapted to higher-dimensional function values $f_i \in \mathbb{R}^t$ by using the Euclidean norm of the individual differences, $\|\tilde{f}_k^j(\mathbf{p}_i) - f_i\|$.

A global error ε^j with respect to this norm can be computed efficiently for every level of resolution from the residuals ε_i^j as

$$\begin{aligned}\varepsilon^j &= \left(\frac{1}{n} \sum_{k=1}^{n_j} \sum_{\mathbf{p}_i \in \tau_k^j} \left| \tilde{f}_k^j(\mathbf{p}_i) - f_i \right|^p \right)^{\frac{1}{p}} \\ &= \left(\frac{1}{n} \sum_{k=1}^{n_j} n_k^j \left(\varepsilon_k^j \right)^p \right)^{\frac{1}{p}}.\end{aligned} \tag{2}$$

Starting with a coarse approximation L_0, an adaptive clustering algorithm computes finer levels L_{j+1} from L_j until a maximal number of clusters is reached or a prescribed error bound is satisfied. To keep the clustering algorithm simple and efficient, the approximation L_{j+1} should differ from L_j only in cluster regions with large residuals in L_j. As the clustering is refined, it should eventually converge to a space partitioning, where every tile contains exactly one data point or where the number of points in every tile is sufficiently low leading to zero residuals.

3 Constructing Voronoi Hierarchies

In the following, we describe our adaptive clustering approach for multiresolution representation of scattered data: a hierarchy of Voronoi diagrams [1,9] constructed from nested subsets of the original point set.

The *Voronoi diagram* of a set of points $\mathbf{p}_i$, $i = 1, \ldots, n$ in the plane is a space partitioning consisting of n tiles τ_i. Every tile τ_i is defined as a subset of $\mathbb{R}^2$ containing all points that are closer to $\mathbf{p}_i$ than to any $\mathbf{p}_j$, $j \neq i$, with respect to the Euclidean norm.

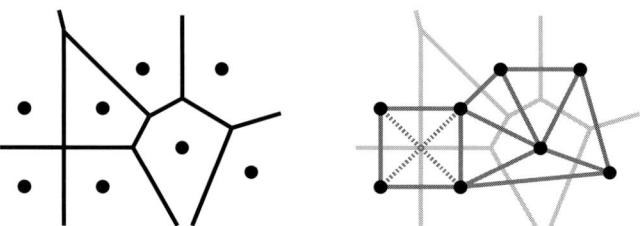

Fig. 2. Planar Voronoi diagram and its dual, the (not uniquely defined) Delaunay triangulation.

A Voronoi diagram can be derived from its dual, the *Delaunay triangulation* [2,5,3,4], see Figure 2. The circumscribed circle of every triangle in a Delaunay triangulation does not contain any other data points. If more than three points are located on such a circle, the Delaunay triangulation is not unique. The Voronoi vertices are located at the centers of circumscribed circles of Delaunay triangles, which can be exploited for constructing a Voronoi diagram. The Voronoi diagram is unique, in contrast to the Delaunay triangulation.

Our method constructs a Delaunay triangulation incrementally by successive insertion of selected points. For every point inserted, all triangles whose circumscribed circles contain the new point are erased. The points belonging to erased triangles are then connected to the new point, defining new triangles that satisfy the Delaunay property, see Figure 3.

When inserting a point located inside a prescribed Voronoi tile, the corresponding tile center is incident to one or more Delaunay triangles to be removed. Since all triangles to be removed define a connected region, point insertion is a local operation of expected constant time complexity (and of $O(n)$ complexity in the extremely rare worst case).

For applications in s-dimensional spaces ($s > 2$), the Delaunay triangulation consists of s-simplices whose circumscribed s-dimensional hyperspheres contain no other point. Our algorithm remains valid for applications using data defined on higher-dimensional domains.

Our adaptive clustering algorithm uses *Sibson's interpolant* [10] for constructing the functions $\tilde{f}_k^j$. Sibson's interpolant is based on blending function values f_i associated with the points $\mathbf{p}_i$ defining the Voronoi diagram. The blending weights for Sibson's interpolant at a point $\mathbf{p} \in \mathbb{R}^2$ are computed by inserting $\mathbf{p}$ temporarily into the Voronoi diagram and by computing the areas a_i that are "cut away" from Voronoi tiles τ_i, see Figure 4. The value of

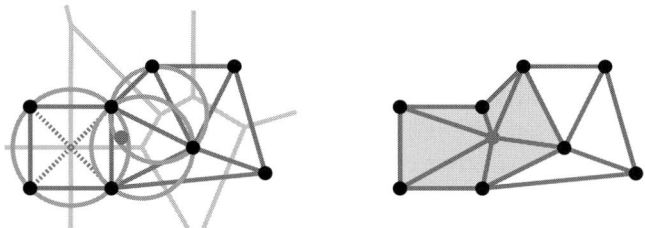

Fig. 3. Construction of Delaunay triangulation by point insertion. Every triangle whose circumscribed circle contains the inserted point is erased. The points belonging to removed triangles are connected to the new point.

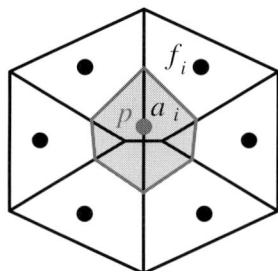

Fig. 4. Computing Sibson's interpolant at point p by inserting p into a Voronoi diagram and using the areas cut away from every tile as blending weights.

Sibson's interpolant at **p** is defined as

$$f(\mathbf{p}) = \frac{\sum_i a_i f_i}{\sum_i a_i}.$$

Sibson's interpolant is C^1 continuous everywhere, except at the points $\mathbf{p}_i$. To avoid infinite areas a_i, we clip the Voronoi diagram against the boundary of the compact domain D. A natural choice for the domain D is the convex hull of the points $\mathbf{p}_i$.

In the following, we provide the individual steps of our clustering algorithm:

(i) Construct the Voronoi diagram for the minimal point set defining the convex hull of all points $\mathbf{p}_i$, $i = 1, \ldots, n$. The tiles of this Voronoi diagram define the cluster regions τ_k^0, $k = 1, \ldots, n_0$, of level L_0.

(ii) From the functions $\tilde{f}_k^j$, defined by Sibson's interpolant and from error norm (1) ($p = 2$), compute all residuals ε_k^0. To avoid square root computations, $(\varepsilon_k^0)^2$ is stored.

(iii) Refinement: $L_j \to L_{j+1}$. Let m denote the index of a maximal residual in L_j, i.e., $\varepsilon_m^j \geq \varepsilon_k^j \ \forall k = 1, \ldots, n_j$. Among all $\mathbf{p}_i \in \tau_m^j$, identify a data point $\mathbf{p}_{max}$ with maximal error given by $max_{\mathbf{p}_i \in \tau_m^j} \{|\tilde{f}_m^j(\mathbf{p}_i) - f_i|\}$. Insert $\mathbf{p}_{max}$ into the Voronoi diagram, resulting in a new tile denoted as $\tau_{n_{j+1}}^{j+1}$, where $n_{j+1} = n_j + 1$.

(iv) Update $\varepsilon^{j+1}_{n_{j+1}}$ and all residuals associated with tiles that have been modified, i.e., all tiles that are adjacent to the new tile $\tau^{j+1}_{n_{j+1}}$ with center $\mathbf{p}_{max}$. (All other tiles remain unchanged, i.e., $\tau^{j+1}_i = \tau^j_i$, $\tilde{f}^{j+1}_i = \tilde{f}^j_i$, and $\varepsilon^{j+1}_i = \varepsilon^j_i$.)
(v) Compute the global approximation error ε^j using the error norm given by equation (2). Terminate the process when a prescribed global error bound is satisfied or when a maximal number of points has been inserted. Otherwise, increment j and continue with step (iii).

We briefly analyze the complexity of our algorithm. As stated above, point insertion into a Delaunay triangulation is a local operation, provided that the Voronoi tile containing the inserted point is known. Analogously, evaluating Sibson's interpolant at a point inside a certain Voronoi tile is a local, expected constant-time operation. If we had an oracle providing the order of insertion and the indices of the Voronoi tiles containing every inserted point, our algorithm would perform in expected linear time (and in $O(n^2)$ time in the worst case, for example when all points are nearly co-circular).

The overall computational cost of our method is determined by the cost for computing residuals, selecting points for insertion, and keeping track of the Voronoi tiles containing these points. When starting with a sufficiently even distribution of original points, we can assume that every point is relocated into a different tile on average $O(\log n)$ times. For computing the residuals, Sibson's interpolant is evaluated at every original point also expectedly $O(\log n)$ times. After locally updating the residuals, a tile with greatest residual can be determined in expected constant time by a comparison-free sorting algorithm like hashing. Thus, the overall expected time complexity of our method is $O(n \log n)$.

For comparison, an algorithm constructing Delaunay triangulations and convex hulls from the scratch (without providing a hierarchy) in expected linear time is described by Maus [8]. A divide-and-conquer method provides a worst-case solution with $O(n \log n)$ time complexity.

4 Numerical Results

We have applied our Voronoi-based clustering approach to approximate the "Crater Lake" terrain data set, courtesy of U.S. Geological Survey. This data set consists of 159272 samples at full resolution. Approximation results for multiple levels of resolution are shown in Figure 5 and summarized in Table 1.

The quality of approximations obtained with our method compares favorably to results obtained from other multiresolution algorithms like wavelet transforms. A standard compression method, for example, uses a wavelet transform followed by quantization and arithmetic coding of the resulting coefficients. Using the Haar-wavelet transform for compression of the Crater-Lake data set (re-sampled on a regular grid at approximately the same resolution) results in approximation errors (for $p = 2$) of 0.89 percent for a 1:10

Table 1. Approximation errors in percent of amplitude for Crater Lake. Figure 5 shows the different levels of resolution.

No. Voronoi Tiles	Error ($p = \infty$) [%]	Error ($p = 2$) [%]
100	31.6	3.13
200	17.1	1.96
300	16.3	1.55
400	13.9	1.33
500	11.9	1.21
1000	10.8	0.80

compression and 4.01 percent for a 1:100 compression [1]. For a Voronoi-based compression locations of the samples need to be encoded as well.

In addition to a hierarchy of smooth approximations, our method provides a cluster hierarchy consisting of convex cells and an importance ranking for sites. Future work will be directed at the explicit representation of discontinuities and sharp features.

5 Acknowledgements

We thank Mark Duchaineau, Daniel Laney, Eric LaMar, and Nelson Max for their helpful ideas and contributions to many discussions. This work was supported by the National Science Foundation under contract ACI 9624034 (CAREER Award), through the Large Scientific and Software Data Set Visualization (LSSDSV) program under contract ACI 9982251, and through the National Partnership for Advanced Computational Infrastructure (NPACI); the Office of Naval Research under contract N00014-97-1-0222; the Army Research Office under contract ARO 36598-MA-RIP; the NASA Ames Research Center through an NRA award under contract NAG2-1216; the Lawrence Livermore National Laboratory under ASCI ASAP Level-2 Memorandum Agreement B347878 and under Memorandum Agreement B503159; the Lawrence Berkeley National Laboratory; the Los Alamos National Laboratory; and the North Atlantic Treaty Organization (NATO) under contract CRG.971628. We also acknowledge the support of ALSTOM Schilling Robotics and SGI. We thank the members of the Visualization and Graphics Research Group at the Center for Image Processing and Integrated Computing (CIPIC) at the University of California, Davis.

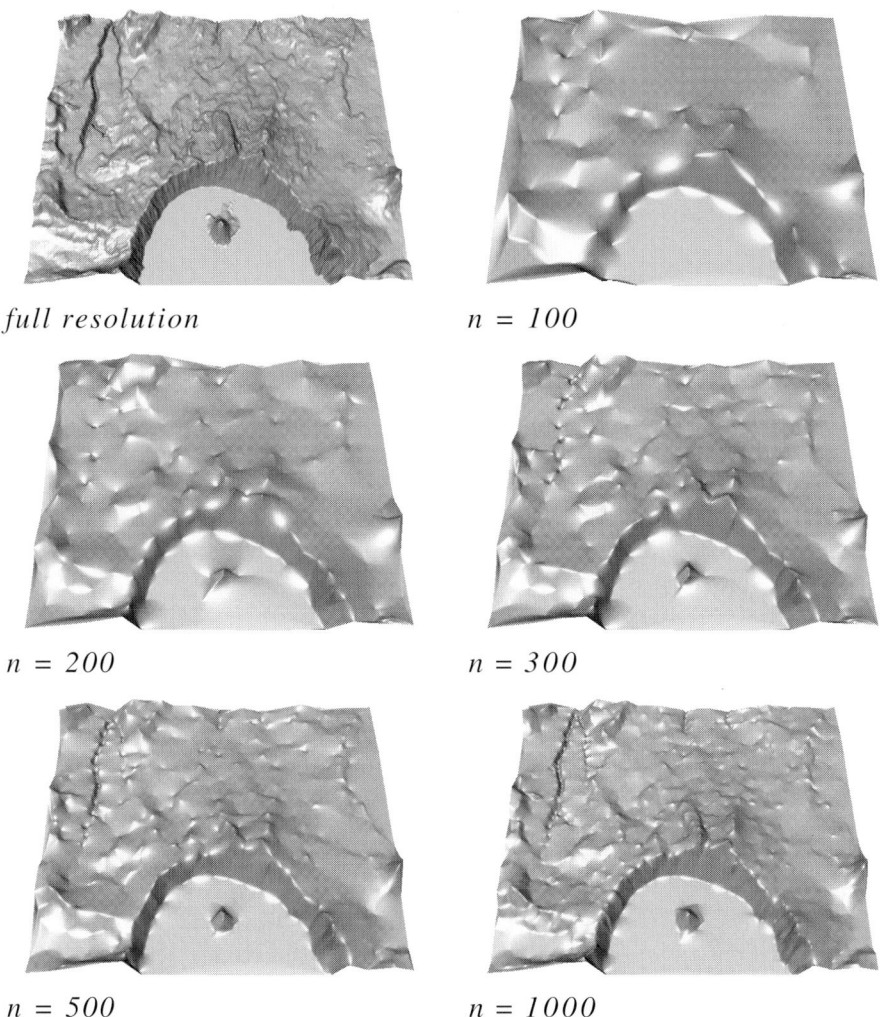

Fig. 5. Crater-Lake terrain data set at different levels of resolution (using $p = 2$). The full-resolution data set consists of 159272 points, courtesy of U.S. Geological Survey. (See Color Plate 6 on page 352.)

References

1. M. Bertram, *Multiresolution Modeling for Scientific Visualization*, Ph.D. Thesis, Department of Computer Science, University of California, Davis, California, 2000. http://daddi.informatik.uni-kl.de/~bertram
2. M. de Berg, M. van Kreveld, M. Overmars, and O. Schwarzkopf, *Computational Geometry: Algorithms and Applications*, Springer, Berlin, Germany, 1997.
3. L. De Floriani, B. Falcidieno, and C. Pienovi, *A Delaunay-based method for surface approximation*, Proceedings of Eurographics '83, Amsterdam, Netherlands, 1983, pp. 333–350, 401.
4. L. De Floriani and E. Puppo, *Constrained Delaunay triangulation for multiresolution surface description*, Proceedings Ninth IEEE International Conference on Pattern Recognition, IEEE, 1988, pp. 566–569.
5. G. Farin, *Surfaces over Dirichlet tesselations*, Computer Aided Geometric Design, Vol. 7, No. 1–4, 1990, pp 281–292.
6. B. Heckel, A.E. Uva, B. Hamann, and K.I. Joy, *Surface reconstruction using adaptive clustering methods*, Proceedings of the Dagstuhl Seminar on Scientific Visualization, Dagstuhl, Germany, May 21–26, 2000 (to appear).
7. B.F.J. Manly, *Multivariate Statistical Methods, A Primer*, second edition, Chapman & Hall, New York, 1994.
8. A. Maus, *Delaunay triangulation and convex hull of n points in expected linear time*, BIT, Vol. 24, No. 2, pp. 151–163, 1984.
9. S.E. Schussman, M. Bertram, B. Hamann and K.I. Joy, *Hierarchical data representations based on planar Voronoi diagrams*, W.C. de Leeuw and R. van Liere (eds.), Data Visualization 2000, Proceedings of the Joint Eurographics and IEEE TCVG Symposium on Visualization (VisSym), Springer Vienna/New York, 2000, pp. 63–72.
10. R. Sibson, *Locally equiangular triangulation*. The Computer Journal, Vol. 21, No. 2, 1992, pp. 65–70.

Multiresolution Representation of Datasets with Material Interfaces

Benjamin F. Gregorski[1], David E. Sigeti[3], John Ambrosiano[3], Gerald Graham[3], Murray Wolinsky[3], Mark A. Duchaineau[2], and Bernd Hamann[1] Kenneth I. Joy[1]

[1] Center for Image Processing and Integrated Computing (CIPIC)
  Department of Computer Science,
  University of California,
  Davis, CA 95616-8562, USA
[2] Center for Applied Scientific Computing (CASC) Lawrence Livermore National Laboratory, P.O. Box 808, L-561, Livermore, CA 94551, USA
  Lawrence Livermore National Laboratory
[3] Los Alamos National Laboratory
  Los Alamos, New Mexico 87545

Abstract. We present a new method for constructing multiresolution representations of data sets that contain material interfaces. Material interfaces embedded in the meshes of computational data sets are often a source of error for simplification algorithms because they represent discontinuities in the scalar or vector field over mesh elements. By representing material interfaces explicitly, we are able to provide separate field representations for each material over a single cell. Multiresolution representations utilizing separate field representations can accurately approximate datasets that contain discontinuities without placing a large percentage of cells around the discontinuous regions. Our algorithm uses a multiresolution tetrahedral mesh supporting fast coarsening and refinement capabilities; error bounds for feature preservation; explicit representation of discontinuities within cells; and separate field representations for each material within a cell.

1 Introduction

Computational physics simulations are generating larger and larger amounts of data. They operate on a wide variety of input meshes, for example rectilinear meshes, adaptively refined meshes for Eulerian hydrodynamics, unstructured meshes for Lagrangian hydrodynamics and arbitrary Lagrange-Eulerian meshes. Often, these data sets contain special physical features such as material interfaces, physical boundaries, or thin slices of material that must be preserved when the field is simplified. In order to ensure that these features are preserved, the simplified version of the data set needs to be constructed using strict L^∞ error bounds that prevent small yet important features from being eliminated.

Data sets of this type require a simplification algorithm to approximate data sets with respect to several simplification criteria. The cells in the approximation must satisfy error bounds with respect to the dependent field variables over each mesh cell, and to the representation of the discontinuities within each cell. In addition, the simplification algorithm must be able to deal with a wide range of possible input meshes.

We present an algorithm for generating an approximation of a computational data set that can be used in place of the original high-resolution data set generated by the simulation. Our approximation is a resampling of the original data set that preserves user-specified as well as characteristic features in the data set and approximates the dependent field values to within a specified tolerance.

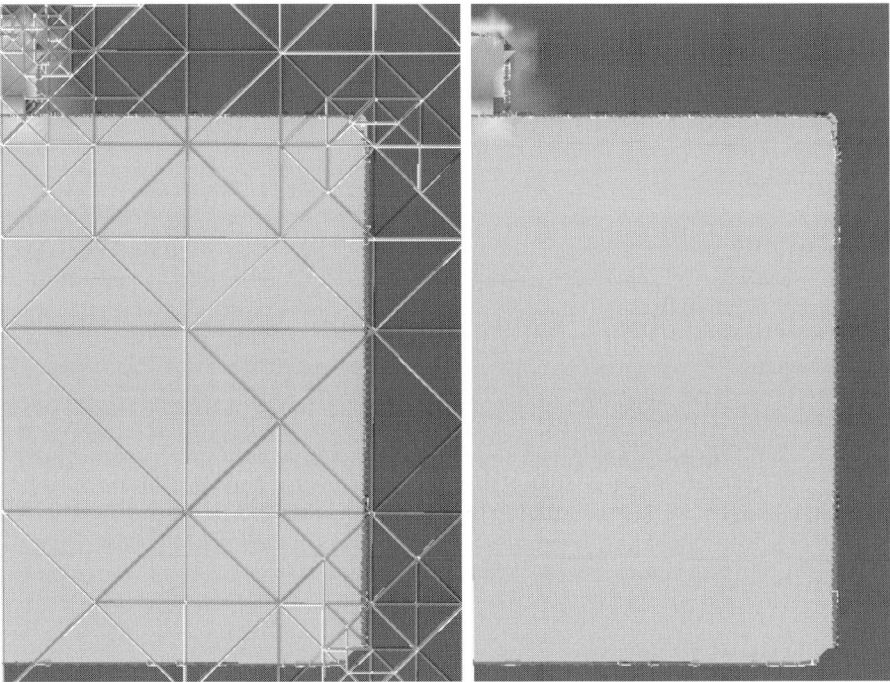

Fig. 1. Cross section of a density field approximated using explicit interface representations and separate field representations. The left picture shows the field along with the approximating tetrahedral mesh. (interface error = 0.15). (See Color Plate 7 on page 353.)

2 Related Work

Hierarchical approximation techniques for triangle meshes, scattered data, and tetrahedral meshes have matured substantially over recent years. The approximation of material interfaces is similar to the approximation or simplification of large polygonal meshes. The approximating mesh represents the large mesh to within a certain error tolerance using a substantially smaller number of triangles. Mesh approximation and simplification algorithms can be divided into decimation techniques and remeshing techniques.

Decimation based techniques attempt to simply the existing geometry by removing vertices, edges or faces and evaluating an error function that determines the fidelity of the new mesh. a large amount of research has been done in the field of surface simplification. In [9] an iterative triangle mesh decimation method is introduced. Triangles in nearly linear regions are identified and collapsed to a point that is comoputed in a locally optimal fashion. In [11], Hoppe describes a progressive mesh simplification method for triangle meshes. An arbitrary mesh is simplified through a series of edge collapse operations to yield a very simply base mesh. Heckbert and Garland present a comprehensive survey of these techniques in [10]. In more recent work, Heckbert and Garland [7] use a quadric error metric for surface simplification. They use vertex pair collapse operations to simplify triangle meshes and they use *quadric matrices* that define a quadric object at each vertex to control the error of the simplified surfaces. In [6], they use the same technique to simplify meshes that have associated colors and textures. Hoppe [12] uses a modified quadric metric for simplifying polygonal meshes while preserving surface attributes such as normal vectors, creases, and colors. Lindstrom [16] developed out-of-core techniques for simplifying large meshes that do not fit in main memory.

Decimation based techniques start with the existing high resolution geometry and remove features until a specified error tolerance has been reached. Another approach to mesh approximation is to generate a completely new mesh through a remeshing or fitting process. The remeshing process starts with a coarse mesh and adds in geometry until the surface approximates the original model to within a specified error tolerance. Lee et al. [15] use a mesh parameterization method for surface remeshing. A simplified base mesh is constructed and a subdivision surface is fitted to the original model. Kobbelt et al. [14] use a shrink wrapping approach to remesh polygonal meshes with subdivison surfaces. Multiresolution methods for reconstruction and simplification have also been explored using subdivision techniques and wavelets [5] and [4].

Our approximation of material interfaces falls into the category of remeshing techniques. As described in Section 5, the material interfaces are given as triangle meshes. Within each of our cells we construct a piecewise linear approximation of the material interfaces to within a specified error tolerance based on the distance between the original mesh and our approximation.

Simplification of tetrahedral meshes has been discussed in [20], [2], [19], [17], and [18]. Zhou et al. [20] present the multiresolution tetrahedral framework that is the basis of our simplification algorithm. This structure is also used by Gerstner and Pajarola [8] for multiresolution iso-surface visualization that preserves topological genus. (This is further discussed in Section

3). Cignoni at al. [2] describe a multiresolution tetrahedral mesh simplification technique built on scattered vertices obtained from the initial dataset. Their algorithm supports interactive level-of-detail selection for rendering purposes. Trotts et al. [19] simplify tetrahedral meshes through edge-collapse operations. They start with an initial high-resolution mesh that defines a linear spline function and simplify it until a specified tolerance is reached. Staadt and Gross [18] describe progressive tetrahedralizations as an extension of Hoppe's work. They simplify a tetrahedral mesh through a sequence of edge collapse operations. They also describe error measurements and cost functions for preserving consistency with respect to volume and gradient calculations and techniques for ensuring that the simplification process does not introduce artifacts such as intersecting tetrahedra. Different error metrics for measuring the accuracy of simplified tetrahedral meshes have been proposed. Lein et al. [13] present a simplification algorithm for triangle meshes using the Hausdorff distance as an error measurement. They develop a symmetric adaption of the Hausdorff distance that is an intuitive measure for surface accuracy.

3 Multiresolution Tetrahedral Mesh

The first basis for our simplification algorithm is the subdivision of a tetrahedral mesh as presented by Zhou et al. [20]. This subdivision scheme has an important advantage over other multiresolution spatial data structures such as an octree as it makes it easy to avoid introducing spurious discontinuities into representations of fields. The way we perform the binary subdivision ensures that the tetrahedral mesh will always be a conformant mesh, i.e., a mesh where all edges end at the endpoints of other edges and not in the interior of edges. The simplest representation for a field within a tetrahedral cell is given by the unique linear function that interpolates field values specified at the cell's vertices. In the case of a conformant mesh, this natural field representation will be continuous across cell boundaries, resulting in a globally C^0 representation.

We have generalized the implementation presented by Zhou et al. by removing the restriction that the input data needs to be given on a regular rectilinear mesh consisting of $(2^N + 1) \times (2^N + 1) \times (2^N + 1)$ cells. A variety of input meshes can be supported by interpolating field values to the vertices of the multiresolution tetrahedral mesh. In general, any interpolation procedure may be used. In some cases, the procedure may be deduced from the physics models underlying the simulation that produced the data set. In other cases, a general-purpose interpolation algorithm will be appropriate.

We construct our data structure as a binary tree in a top-down fashion. Data from the input data set, including grid points and interface polygons, are assigned to child cells when their parent is split.

The second basis for our algorithm is the ROAM system, described in [3]. ROAM uses priority queue-driven split and merge operations to provide optimal real-time display of triangle meshes for terrain rendering applications. The tetrahedral mesh structure used in our framework can be regarded as an extension to tetrahedral meshes of the original ROAM data structure for triangle meshes.

Since our data structure is defined recursively as a binary tree, a representation of the original data can be pre-computed. We can utilize the methods developed in ROAM to efficiently select a representation that satisfies an error bound or a desired cell count. This makes the framework ideal for interactive display. Given a data set and polygonal representations for the material interfaces, our algorithm constructs an approximation as follows:

1. Our algorithm starts with a base mesh of six tetrahedra and associates with each one the interface polygons that intersect it.
2. The initial tetrahedral mesh is first subdivided so that the polygonal surface meshes describing the material interfaces are approximated within a certain tolerance. At each subdivision, the material interface polygons lying partially or entirely in a cell are associated with the cell's children; approximations for the polygons in each child cell are constructed, and interface approximation errors are computed for the two new child cells.
3. The mesh is further refined to approximate the field of interest, for example density or pressure, within a specified tolerance.

For the cells containing material interfaces, our algorithm computes a field representation for each material. This is done by extrapolating *ghost* values for each material at the vertices of the tetrahedron where the material does not exist. A ghost value is an educated guess of a field value at a point where the field does not exist. When material interfaces are present, field values for a given material do not exist at all of the tetrahedron's vertices. Since tri-linear approximation over a tetrahedron requires four field values at the vertices, extra field values are needed to perform the interpolation. Thus for a given field and material, the ghost values and the existing values are used to form the tri-linear approximation within the tetrahedron.

This is illustrated in Figure 2 for a field sampled over a triangular domain containing two materials. For the field sampled at V_0 and V_1, a ghost value at vertex V_2 is needed to compute a linear approximation of the field over the triangle. The approximation is used only for those sample points that belong to this material. Given a function sampled over a particular domain, the ghost value computation extrapolates a function value at a point outside of this domain. When the field approximation error for the cell is computed, the separate field representations, built using these ghost values, are used to calculate an error for each distinct material in the cell. The decomposition process of a cell that contains multiple materials consists of these steps:

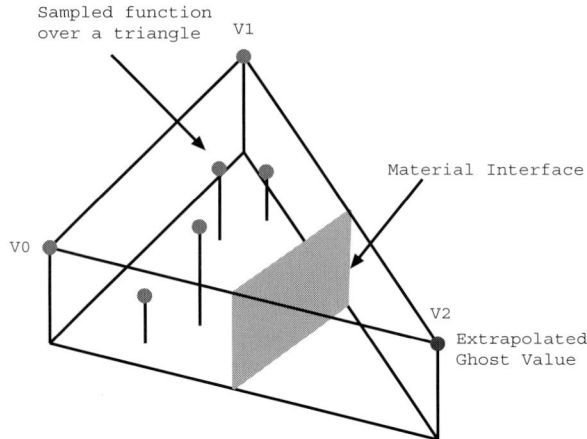

Fig. 2. 2D Ghost Value Computation

1. The signed distance values and ghost values for the new vertex are computed when the vertex is created during a split or subdivision operation. This is done by examining those cells that share the edge being split.
2. The interface representations, i.e., triangle meshes, are associated with the child cells, and the approximating representations of interfaces and their associated errors are computed.
3. The field error for each of the materials is computed, and the maximum value of these errors is defined as the overall error associated with a cell containing multiple materials.

4 General Multiresolution Framework

A multiresolution framework for approximating numerically simulated data needs to be a robust and extensible. The framework must be capable of supporting the wide range of possible input structures used in the simulations and the wide range of output data generated by these simulations. The following properties and operations are desirable for such a framework:

1. **Interactive transition between levels of detail**. The ability to quickly move between different levels of detail allows a user to select a desired approximation at which to perform a calculation (for a visualization application).
2. **Strict L^∞ error bounds**. Strict error bounds prevent small yet important features from being averaged or smoothed out by the approximation process.
3. **Local and adaptive mesh refinement and local error computations**. Local mesh refinement allows the representation to be refined only

in the areas of interest while keeping areas of little interest at relatively lower resolutions. This is essential for maintaining interactivity and strict cell count on computers with limited resources. The error calculations for datasets consisting of millions or billions of cells should not involve a large amount of original data.

4. **Accommodating different mesh types**. Computational simulations are done on a large variety of mesh structures. and it is cumbersome to write a multiresolution algorithm for each specific structure. In order for a framework to be useful it should be easily adaptable to a broad class of input meshes.

5. **Explicit representation of field and/or material discontinuities**. Discontinuities are very important in scientific datasets and very often need to be preserved when the datasets are approximated. A multiresolution framework should support the explicit representation and approximation of these discontinuities.

Our multiresolution recursive tetrahedral framework satisfies these design criteria. Tetrahedral cells allow us to use linear basis functions to approximate the material interfaces and the dependent field variables in a cell. A representation of the original data can be computed in a pre-processing step, and we can utilize methods developed for the ROAM system [3] to efficiently select a representation that satisfies an error bound or a desired cell count. This makes the framework ideal for interactive display. Strict L^∞ error bounds are incorporated into the refinement process.

The framework supports various input meshes by resampling them at the vertices of the tetrahedral mesh. The resampling error of the tetrahedral mesh is a user specified variable. This error defines the error between our field approximation using the tetrahedral mesh and the field defined by the input dataset and its interpolation method. The resampled mesh can be refined to approximate the underlying field to within a specified tolerance or until the mesh contains a specific number of tetrahedra. The resampled field is a linear approximation of the input field. The boundaries of the input mesh are represented in the same manner as the surfaces of discontinuity. The volume of space outside of the mesh boundary is considered as a separate material with constant field values. This region of *empty* space is easy to evaluate and approximate; no ghost values and no field approximations need to be computed. This allows a non-rectilinear input mesh to be embeded into the larger rectilinear mesh generated by the refinement of the tetrahedral grid. Discontinuities are supported at the cell level allowing local refinement of the representations of surfaces of discontinuity in geometrically complex areas. The convergence of the approximation depends upon the complexity of the input field and the complexity of the input mesh. For meshes with complex geometry and interfaces but simple fields, the majority of the work is done approximating the input geometry and material interfaces. For meshes with simple geometry and interfaces but complex fields, the majority of work is done approximating the field values.

The framework has several advantages over other multiresolution spatial data structures such as an octree. The refinement method ensures that the tetrahedral mesh will always be free of cracks and *T-intersections*. This makes it easy to guarantee that representations of fields and surfaces of discontinuity are continuous across cell boundaries.

Our resampling and error bounding algorithms require that an original data set allow the extraction of the values of the field variables at any point and, for a given field, the maximum difference between the representation over one of our cells and the representation in the original dataset over the same volume.

5 Material Interfaces

A material interface defines the boundary between two distinct materials. Figure 3 shows an example of two triangles crossed by an single interface (smooth curve). This interface specifies where the different materials exist within a cell. Field representations across a material interface are often discontinuous. Thus, an interface can introduce a large amount of error to cells that cross it. Instead of refining an approximation substantially in the neighborhood of an interface, the discontinuity in the field is better represented by explicitly representing the surface of discontinuity in each cell. Once the discontinuity is represented, two separate functions are used to describe the dependent field variables on either side of the discontinuity. By representing the surface of discontinuity exactly, our simplification algorithm does not need to refine regions in the spatial domain with a large number of tetrahedra.

5.1 Extraction and Approximation

In the class of input datasets with which we are working, material interfaces are represented as triangle meshes. In the case that these triangle meshes are not known, they are extracted from volume fraction data by a material interface reconstruction technique, see [1]. (The volume fractions resulting from numerical simulations indicate what percentages of which materials are present in each cell.) Such an interface reconstruction technique produces a set of crack-free triangle meshes and normal vector information that can be used to determine on which side and in which material a point lies.

Within one of our tetrahedra, an approximate material interface is represented as the zero set of a signed distance function. Each vertex of a tetrahedron is assigned a signed distance value for each of the material interfaces in the tetrahedron. The signed distance from a vertex **V** to an interface mesh **I** is determined by first finding a triangle mesh vertex $\mathbf{V}_i$ in the triangle mesh describing **I** that has minimal distance to **V**. The sign of the distance is determined by considering the normal vector $\mathbf{N}_i$ at $\mathbf{V}_i$. If $\mathbf{N}_i$ points towards

V, then **V** is considered to be on the *positive side* of the interface; otherwise it is considered to be on the *negative side* of the interface. The complexity of this computation is proportional to the complexity of the material interfaces within a particular tetrahedra. In general a coarse cell in the mesh will contain a large number of interface polygons, and a fine cell in the mesh will contain a small number of interface polygons. The signed distance values are computed as the mesh is subdivided. When a new vertex is introduced via the mesh refinement, the computation of the signed distance for that vertex only needs to look at the interfaces that exist in the tetrahedra around the split edge. If those tetrahedra do not contain any interfaces, no signed distance value needs to be computed.

In Figure 3, the true material interface is given by the smooth curve and its approximation is given by the piecewise linear curve. The minimum distances from the vertices of the triangles to the interface are shown as dotted lines. The distances for vertices on one side of the interface (say, above the interface) are assigned positive values and those on the other side are assigned negative values. These signed distance values at the vertices determine linear functions in each of the triangles, and the approximated interface will be the zero set of these linear functions. Because the mesh is conformant, the linear functions in the two triangles will agree on their common side, and the zero set is continuous across the boundary. The situation in three dimensions is analogous.

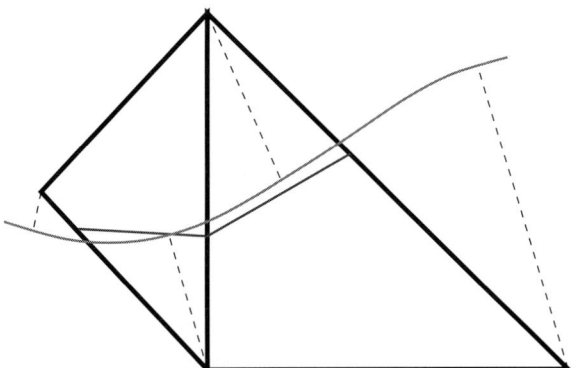

Fig. 3. True and approximated interfaces.

Figure 4 shows a two-dimensional example of a triangle with several material interfaces and their approximations. In this figure, the thin, jagged lines are the original boundaries and the thick, straight lines are the approximations derived from using the signed distance values. For the interface between materials A and B, the thin, dashed lines from vertices A, B, and C indicate the points on the interface used to compute the signed distance values. The

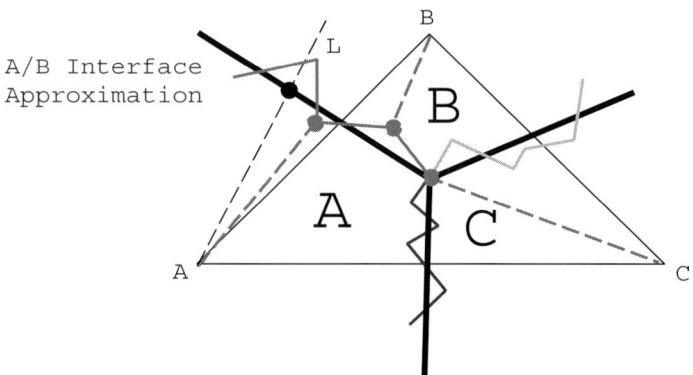

Fig. 4. Triangle with three materials (A, B, and C) and three interfaces.

dashed line L demonstrates that the projection of a point onto an approximation does not always lie inside the cell. The signed distance function is assumed to vary linearly in the cell, i.e., a tetrahedron. The distance function is a linear function $f(x, y, z) = Ax + By + Cz + D$. The coefficients for the linear function defining a boundary representation are found by solving a 4×4 system of equations, considering the requirement that the signed distance function over the tetrahedron must interpolate the signed distance values at the four vertices.

The three-dimensional example in Figure 5 shows a tetrahedron, a material interface approximation, and the signed distance values d_i for each vertex V_i. The approximation is shown as a plane cutting through the tetrahedron. The normal vector N indicates the positive side of the material boundary approximation. Thus, the distance to V_0 is positive and the distances for V_1, V_2, and V_3 are negative.

We note that a vertex has at most one signed distance value for each interface. This ensures that the interface representation is continuous across cell boundaries. If a cell does not contain a particular interface, the signed distance value for that interface is meaningless for that cell. Given a point **P** in an interface polygon and its associated approximation B_r, the error associated with **P** is the absolute value of the distance between **P** and B_r. The material interface approximation error associated with a cell is the maximum of these distances, considering all the interfaces within the cell.

6 Discontinuous Field Representations

Cells that contain material interfaces typically also have discontinuities in the fields defined over them. For example, the density field over a cell that contains both steel and nickel is discontinuous exactly where the two materials

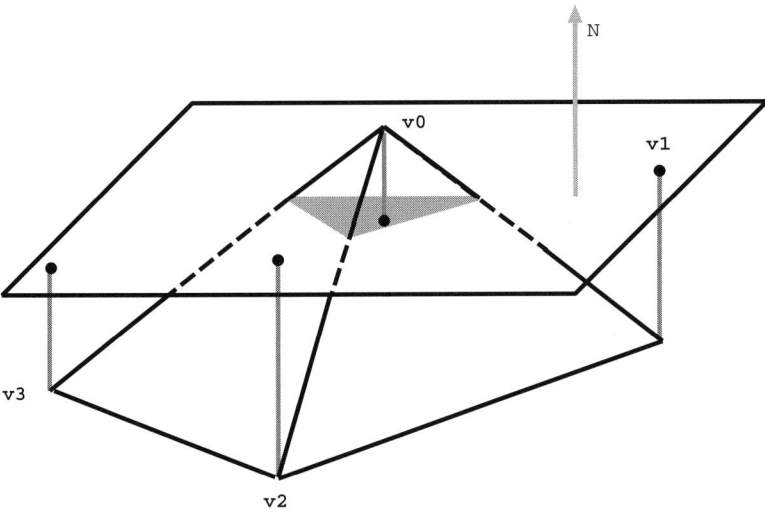

Fig. 5. Tetrahedron showing signed distance values and the corresponding boundary approximation.

meet. In these situations, it is better to represent the density field over the cell as two separate fields, one field for the region containing only the first material and one for the second material. One way to accomplish field separation is to divide the cell into two distinct cells at the material interface. In Figure 6, the triangle would be divided into a quadrilateral for material A and a triangle for material B. The disadvantage of this method is that it introduces new cell types into the mesh which makes it harder to have continuous field representations across cells. Furthermore if new cell types are introduced, we loose the multiresolution structure and adaptive refinement capabilities of the tetrahedral mesh. Our algorithm represents the discontinuity by constructing a field representation for each material in the cell. Each of the vertices in a cell must have distinct field values for each material in the cell.

For a vertex **V** that does not reside in material **M**, we compute a ghost value for the field associated with material **M** at vertex **V**. This ghost value is an extrapolation of the field value for **M** at **V**. The process is illustrated in Figure 6. The known field values are indicated by the solid circles. A_0 and A_1 represent the known field values for material A, and B_0 represents the known field value for material B. Vertices A_0 and A_1 are in material A, and thus ghost values for material B must be calculated at their positions. Vertex B_0 lies in material B, and thus a ghost value for material A must be calculated at its position. As described in Section 3, the ghost value computation is performed when the vertex is created during the tetrahedral refinement process.

6.1 Computation of Ghost Values

The ghost values for a vertex **V** are computed as follows:

1. For each material interface present in the cells that share the vertex, find a vertex V_{min} in a triangle mesh representing an interface with minimal distance to **V**. (In Figure 6, these vertices are indicated by the dashed lines from A_0, A_1, and B_2 to the indicated points on the interface.)
2. Evaluate the data set on the far side of the interface at V_{min} and use this as the ghost value at **V** for the material on the opposite side of the interface.

Only one ghost value exists for a given vertex, field and material. This ensures that the field representations are C^0 continuous across cell boundaries. For example, consider vertex V_0 of the triangle in Figure 4. The vertex V_0 lies in material A, and therefore we must compute ghost values for materials B and C at vertex A_0. The algorithm will examine the three material boundaries and determine the points from materials B and C that are closest to A_0. The fields for materials B and C are evaluated at these points, and these values are used as the ghost values for A_0. These points are exactly the points that were used to determine the distance map that defines the approximation to the interface. This computation assumes that the field remains constant on the other side of the interface. Alternatively, a linear or higher order function can be used to approximate the existing data points within the cell and to extrapolate the ghost value.

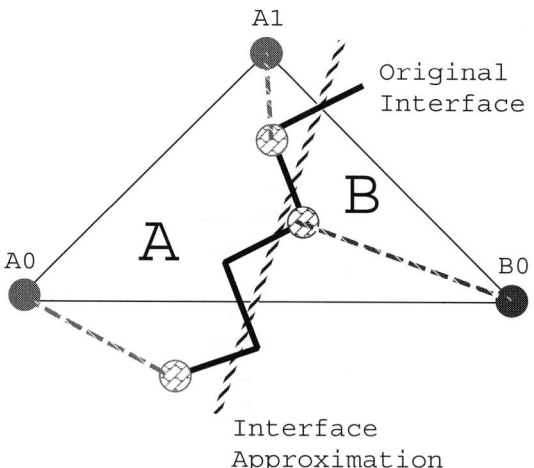

Fig. 6. Ghost value computation for a triangle containing two materials.

7 Error Metrics

The error metrics employed in our framework are similar to the nested error bounds used in the ROAM system. Each cell has two associated error values: a field error and a material interface error. In order to calculate the field errors for a leaf cell in our tetrahedral mesh hierarchy, we assume that the original data set can be divided into *native data elements*. Each of these is presumed to have a well defined spatial extent and a well defined representation for each field of interest over its spatial domain. The simplest example of a native data element is just a grid point that holds field values. Other possibilities are blocks of grid points treated as a unit, cells with a non-zero volume and a field representation defined over the entire cell, or blocks of such cells. For a given field, we assume that it is possible to bound the difference between the representation over one of our leaf cells and the representation over each of the native data elements with which the given cell intersects. The error for the given field in the given cell is the maximum of the errors associated with each of the intersecting native data elements. Currently, we are dealing with native data elements that are grid points of zero volume.

The field error e_T for a non-leaf cell is computed from the errors associated with its two children according to:

$$e_T = \max\{e_{T_0}, e_{T_1}\} + |z(v_c) - z_T(v_c)|, \tag{1}$$

where e_{T_0} and e_{T_1} are the errors of the children; v_c is the vertex that splits the parent into its children; $z(v_c)$ is the field value assigned to v_c; and $z_T(v_c)$ is the field value that the parent assigns to the spatial location of v_c. The approximated value at v_c, $z_T(v_c)$, is calculated as:

$$z_T(v_c) = \frac{1}{2}(z(v_0) + z(v_1)), \tag{2}$$

where v_0 and v_1 are the vertices of the parent's split edge. This error is still a genuine bound on the difference between our representation and the representation of the original data set. However, it is looser than the bound computed directly from the data. The error computed from the children has the advantage that the error associated with a cell bounds not only the deviation from the original representation but also the deviation from the representation at any intermediate resolution level. Consequently, this error is *nested* or monotonic in the sense that the error of a child is guaranteed not to be greater than the error of the parent. Once the errors of the leaf cells are computed, the nested bound for all cells higher in the tree can be computed in time proportional to K, where K is the number of leaf cells in the tree. This can be accomplished by traversing the tree in a bottom-up fashion.

The material interface error associated with a leaf node is the maximum value of the errors associated with each of the material interfaces in the node. For each material interface, the error is the maximum value of the errors associated with the vertices constituting the triangle mesh defining the interface and being inside the cell. The error of a vertex is the absolute value of the distance between the vertex and the interface approximation. The material interface error **E** for a cell guarantees that no point in the original interface polygon mesh is further from its associated approximation than a distance of **E**. This error metric is an upper bound on the deviation of the original interfaces from our approximated interfaces. A cell that does not contain a material interface is considered to have an interface error of zero.

8 Results

We have tested our algorithm on a data set resulting from a simulation of a hypersonic impact between a projectile and a metal block. The simulation was based on a logically rectilinear mesh of dimensions $32 \times 32 \times 52$. For each cell, the average density and pressure values are available, as well as the per-material densities and volume fractions. The physical dimensions in x, y, and z directions are [0,12] [0,12] and [−16,4.8].

There are three materials in the simulation: the projectile, the block, and *empty* space. The interface between the projectile and the block consists of 38 polygons, the interface between the projectile and empty space consists of 118 polygons and the interface between empty space and the block consists of

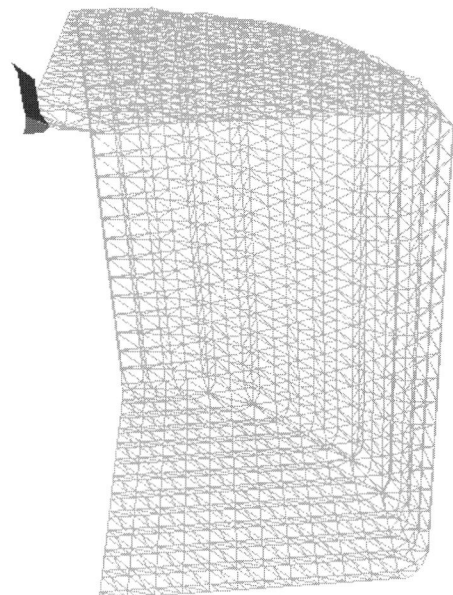

Fig. 7. Original triangular meshes representing material interfaces. (See Color Plate 8 on page 353.)

17574 polygons. Figure 7 shows the original interface meshes determined from the volume fraction information. The largest mesh is the interface between the metal block and empty space; the next largest mesh in the top, left, front corner is the interface between the projectile and empty space; the smallest mesh is the interface between the projectile and the block.

Figure 8 shows a cross-section view of the mesh created by a cutting plane through the tetrahedral mesh. The darker lines are the original interface polygons, and the lighter lines are the approximation to the interface. The interface approximation error is 0.15. (An error of 0.15 means that all of the vertices in the original material interface meshes are no more that a physical distance of 0.15 from their associated interface approximation. This is equivalent to an error of (0.5–1.5)% when considered against the physical dimensions.) A total of 3174 tetrahedra were required to approximate the interface within an error of 0.15. The overall mesh contained a total of 5390 tetrahedra. A total of 11990 tetrahedra were required to approximate the interface to an error of 0.15 and the density field within an error of 3. The maximum field approximation error in the cells containing material interfaces is 2.84, and the average field error for these cells is 0.007. These error measurements indicate that separate field representations for the materials on either side of a discontinuity can accurately reconstruct the field.

Figures 8 and 1 compare the density fields generated using linear interpolation of the density values and explicit field representations on either side of the discontinuity. These images are generated by intersecting the cutting plane with the tetrahedra and evaluating the density field at the intersection points. A polygon is drawn through the intersection points to visualize the density field. In the cells where material interfaces are present, the cutting plane is also intersected with the interface representation to generate data points on the cutting plane that are also on the interface. These data points are used to draw a polygon for each material that the cutting plane intersects.

Figure 1 shows that using explicit field representations in the presence of discontinuities can improve the quality of the field approximation. This can be seen in the flat horizontal and vertical sections of the block where the cells approximate a region that contains the block and empty space. In these cells, the use of explicit representations of the discontinuities leads to an exact representation of the density field. The corresponding field representations using linear interpolation, shown in Figure 8, capture the discontinuities poorly. Furthermore, Figure 1 captures more of the dynamics in the area where the projectile is entering the block (upper-left corner). The linear interpolation of the density values in the region where the projectile is impacting the block smooths out the density field, and does not capture the distinct interface between the block and the projectile.

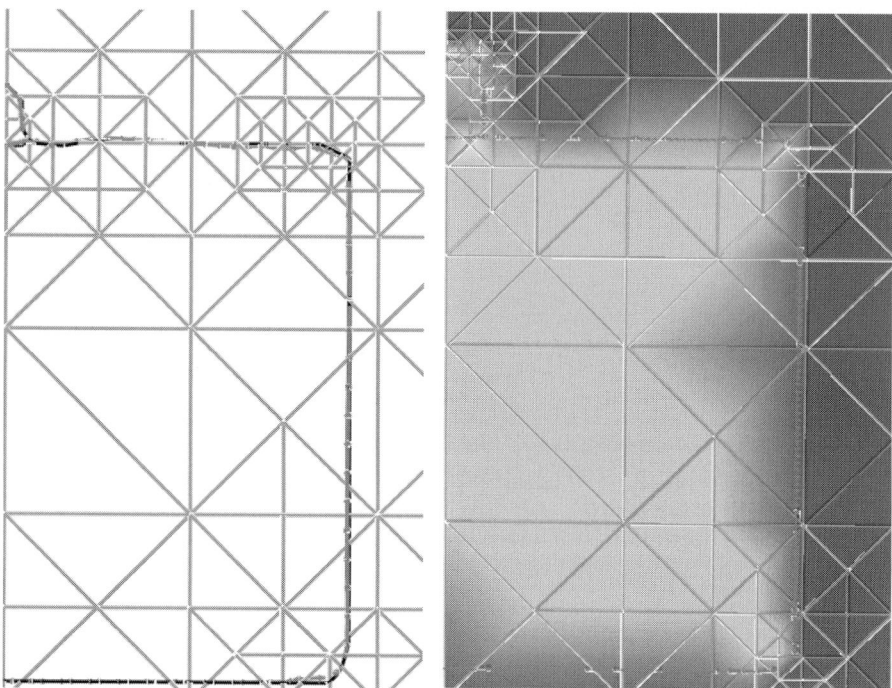

Fig. 8. Cross section of the tetrahedral mesh. The left picture shows the original interfaces and their approximations. The picture on the right shows the density field using linear interpolation. (See Color Plate 9 on page 354.)

9 Conclusions and Future Work

We have presented a method for constructing multiresolution representations of scientific datasets that explicitly represents material interfaces. Our algorithm constructs an approximation that can be used in place of the original data for visualization purposes. Explicitly representing material and implicit field discontinuities allows us to use multiple field representations to better approximate the field within each cell. The use of the tetrahedral subdivision allows us to generalize our algorithm to a wide variety of data sets and to support interactive level-of-detail exploration and view-dependent simplification.

Future work will extend our error calculations to support complex native data element types such as tetrahedra and curvilinear hexahedra. Our current ghost value computation assumes that the field is constant on the other side of the interface. Higher-order extrapolation methods will be investigated for ghost value computation to determine if an improved field approximation can be obtained. Similarly, material interfaces are defined by approximations based on linear functions. The tradeoff between cell count and higher-order approximation methods should be investigated to determine if a better approximation can be obtained without a great increase in computational complexity. We also plan to apply our algorithm to more complex unstructured data sets.

10 Acknowledgments

This work was supported by, Los Alamos National Laboratory and by the National Science Foundation under contract ACI 9624034 (CAREER Award), through the Large Scientific and Software Data Set Visualization (LSSDSV) program under contract ACI 9982251, and through the National Partnership for Advanced Computational Infrastructure (NPACI); the Lawrence Livermore National Laboratory under ASCI ASAP Level-2 Memorandum Agreement B347878 awarded to the University of California, Davis. We thank the members of the Visualization Group at the Center for Image Processing and Integrated Computing (CIPIC) at the University of California, Davis and the members of the Advanced Computing Laboratory at Los Alamos National Laboratory.

References

1. Kathleen S. Bonnell, Daniel R. Schikore, Kenneth I. Joy, Mark Duchaineau, and Bernd Hamann. Constructing material interfaces from data sets with volume-fraction information. In *Proceedings Visualization 2000*, pages 367–372. IEEE Computer Society Technical Committee on Computer Graphics, 2000.
2. Paolo Cignoni, Enrico Puppo, and Roberto Scopigno. Multiresolution Representation and Visualization of Volume Data. *IEEE Transactions on Visualization and Computer Graphics*, 3:352–369, October 1997.
3. M. A. Duchaineau, M. Wolinsky, D. E. Sigeti, M. C. Miller, C. Aldrich, and M. B. Mineev-Weinstein. ROAMing terrain: Real-time optimally adapting meshes. In R. Yagel and H. Hagen, editors, *IEEE Visualization '97*, pages 81–88. IEEE Computer Society Press, 1997.
4. Mark A. Duchaineau, Martin Bertram, Serban Porumbescu, Bernd Hamann, and Kenneth I. Joy. Interactive display of surfaces using subdivision surfaces and wavelets. In T.L. Kunii, editor, *Proceedings of 16th Spring Conference on Computer Graphics, Comenius University, Bratislava, Slovak Republic*, 2001.
5. Mark A. Duchaineau, Serban Porumbescu, Martin Bertram, Bernd Hamann, and Kenneth I. Joy. Dataflow and re-mapping for wavelet compression and view-dependent optimization of billion-triangle isosurfaces. In G. Farin, H. Hagen, and B. Hamann, editors, *Hierarchical Approximation and Geometrical Methods for Scientific Visualization*. Springer-Verlag, Berlin, Germany, 1999.
6. M. Garland and P. S. Heckbert. Simplifying surfaces with color and texture using quadric error metrics. In *IEEE Visualization '98 (VIS '98)*, pages 263–270, Washington - Brussels - Tokyo, October 1998. IEEE.

7. Michael Garland and Paul S. Heckbert. Surface simplification using quadric error metrics. In Turner Whitted, editor, *SIGGRAPH 97 Conference Proceedings*, Annual Conference Series, pages 209–216. ACM SIGGRAPH, Addison Wesley, August 1997.
8. Thomas Gerstner and Renato Pajarola. Topology preserving and controlled topology simplifying multiresolution isosurface extraction. In T. Ertl, B. Hamann, and A. Varshney, editors, *Proceedings Visualization 2000*, pages 259–266. IEEE Computer Society Technical Committee on Computer Graphics, 2000.
9. B. Hamann. A data reduction scheme for triangulated surface. In *Computer Aided Geometric Design*, volume 11, pages 197–214, 1994.
10. Paul S. Heckbert and Michael Garland. Survey of polygonal surface simplification algorithms. Technical report.
11. Hugues Hoppe. Progressive meshes. In Holly Rushmeier, editor, *SIGGRAPH 96 Conference Proceedings*, Annual Conference Series, pages 99–108. ACM SIGGRAPH, Addison Wesley, August 1996. held in New Orleans, Louisiana, 04-09 August 1996.
12. Hugues H. Hoppe. New quadric metric for simplifying meshes with appearance attributes. In David Ebert, Markus Gross, and Bernd Hamann, editors, *IEEE Visualization '99*, pages 59–66, San Francisco, 1999. IEEE.
13. Reinhard Klein, Gunther Liebich, and Wolfgang Straßer. Mesh reduction with error control. In Roni Yagel and Gregory M. Nielson, editors, *Proceedings of the Conference on Visualization*, pages 311–318, Los Alamitos, October 27– November 1 1996. IEEE.
14. Leif P. Kobbelt, Jens Vorsatz, Ulf Labsik, and Hans-Peter Seidel. A shrink wrapping approach to remeshing polygonal surface. In P. Brunet and R. Scopigno, editors, *Computer Graphics Forum (Eurographics '99)*, volume 18(3), pages 119–130. The Eurographics Association and Blackwell Publishers, 1999.
15. Aaron W. F. Lee, Wim Sweldens, Peter Schröder, Lawrence Cowsar, and David Dobkin. MAPS: Multiresolution adaptive parameterization of surfaces. In Michael Cohen, editor, *SIGGRAPH 98 Conference Proceedings*, Annual Conference Series, pages 95–104. ACM SIGGRAPH, Addison Wesley, July 1998. ISBN 0-89791-999-8.
16. Peter Lindstrom. Out-of-Core simplification of large polygonal models. In Sheila Hoffmeyer, editor, *Proceedings of the Computer Graphics Conference 2000 (SIGGRAPH-00)*, pages 259–262, New York, July 23–28 2000. ACM-Press.
17. Anwei Liu and Barry Joe. Quality local refinement of tetrahedral meshes based on bisection. *SIAM Journal on Scientific Computing*, 16(6):1269–1291, November 1995.
18. Oliver G. Staadt and Markus H. Gross. Progressive tetrahedralizations. In David Ebert, Hans Hagen, and Holly Rushmeier, editors, *Proceedings of Visualization 98*, pages 397–402. IEEE Computer Society Press, October 1998.
19. Isaac J. Trotts, Bernd Hamann, Kenneth I. Joy, and David F. Wiley. Simplification of tetrahedral meshes. In Holly Rushmeier David Ebert and Hans Hagen, editors, *Proceedings IEEE Visualization '98*, pages 287–296. IEEE Computer Society Press, October 18–23 1998.

20. Yong Zhou, Baoquan Chen, and Arie E. Kaufman. Multiresolution tetrahedral framework for visualizing regular volume data. In Roni Yagel and Hans Hagen, editors, *IEEE Visualization '97*, pages 135–142. IEEE Computer Society Press, November 1997.

Approaches to Interactive Visualization of Large-scale Dynamic Astrophysical Environments

Andrew J. Hanson and Philip Chi-Wing Fu

Indiana University, Bloomington, Indiana, USA

Abstract. Dynamic astrophysical data require visualization methods that handle dozens of orders of magnitude in space and time. Continuous navigation across large scale ranges presents problems that challenge conventional methods of direct model representation and graphics rendering. In addition, the frequent need to accommodate multiple scales of time evolution, both across multiple spatial scales and within single spatial display scales, compounds the problem because direct time evolution methods may also prove inadequate.

We discuss systematic approaches to building interactive visualization systems that address these issues. The concepts of homogeneous power coordinates, pixel-driven environment-map-to-geometry transitions, and spacetime level-of-detail model hierarchies are suggested to handle large scales in space and time. Families of techniques such as these can then support the construction of a virtual dynamic Universe that is scalable, navigable, dynamic, and extensible. Finally, we describe the design and implementation of a working system based on these principles, along with examples of methods that support the visualization of complex astrophysical phenomena such as causality and the Hubble expansion.

1 Introduction

Supporting the interactive exploration of large-scale time-varying data sets presents a broad spectrum of design challenges. Here we extend and expand upon our introductory work [8] on interacting with large spatial scales, and pay particular attention to combining spatial and temporal scaling and to dealing with the resulting representation issues. Our domain is data sets that may or may not involve a large quantity of data, but which are large in the number of orders of magnitude of spatio-temporal scale required to support interactive rendering. Our ultimate goal is a seamless strategy for interactively navigating through large data sets of this type, particularly astrophysical environments, which form the obvious prototype. In addition, we suggest some novel approaches such as the "Lightcone Clock" to help visualize causality and cosmological features as they apply to the spacetime scales of the entire Universe.

1.1 Background

The nature of the physical Universe has been increasingly better understood in recent years, and cosmological concepts have undergone a rapid evolution (see, e.g., [11], [2], or [5]). Although there are alternate theories, it is generally believed that the large-scale relationships and homogeneities that we see can only be explained by having the universe expand suddenly in a very early "inflationary" period. Subsequent evolution of the Universe is described by the Hubble expansion, the observation that the galaxies are flying away from each other. We can attribute different rates of this expansion to domination of different cosmological processes, beginning with radiation, evolving to matter domination, and, relatively recently, to vacuum domination (the Cosmological Constant term) [4].

We assume throughout that we will be relying as much as possible on observational data, with simulations used only for limited purposes, e.g., the appearance of the Milky Way from nearby intergalactic viewpoints. The visualization of large-scale astronomical data sets using fixed, non-interactive animations has a long history. Several books and films exist, ranging from "Cosmic View: The Universe in Forty Jumps" [3] by Kees Boeke to "Powers of 10" [6,13] by Charles and Ray Eames, and the recent Imax film "Cosmic Voyage" [15]. We have added our own contribution [9], "Cosmic Clock," which is an animation based entirely on the concepts and implementation described in this paper.

In order to go beyond these fixed-sequence films and into the realm of live, interactive virtual reality experiences, we must provide the viewer with substantial additional interactive capabilities. One key idea is to sidestep possible scaling problems in the graphics pipeline by defining all data at unit scale and making transitions among representations using scale tags corresponding to logarithms of the true magnitudes [8]. Multiple representation modes then handle varying spacetime scales.

In this paper, we first discuss the phenomenology of scaling breakdown in typical graphics systems; with this motivation in hand, we then turn to the design issues encountered in building a dynamic astrophysics visualization system handling scale ranges covering huge orders of magnitude; in particular, we include implementation details, concepts, and extensions that were of necessity omitted in the abbreviated description in [8]. Then, we extend our conceptual structure to include homogeneous power coordinates in both space and time, and present our basic approach to transparent rescaling in our virtual model of the Universe. Finally, we give an overview of the system implementation, and present several examples and visualization techniques particularly relevant to cosmology.

1.2 Challenges of Large-scale Dynamic Astrophysics Visualization

To navigate through large-scale dynamic astrophysical data sets, we must overcome difficulties such as the following:

- **Spatial magnitude.** Our data sets sweep through huge orders of magnitude in space. For instance, the size of the Earth is of order 10^7 meters while its orbit around the Sun is of order 10^8 meters. However, the bright stars in the night sky, the stars in the Galaxy near our Solar System, go out to about 10^{15} m, while the visible Universe may stretch out to 10^{27} m. Small but relevant physical scales may go down as far as the Planck length, 10^{-35} m. Designing a unified system to navigate continuously through all these data sets is difficult.
- **Temporal magnitude.** Besides huge orders of magnitude in space, we encounter huge orders of magnitude in time. It takes one day for the Earth to rotate once on its axis, while it takes one year to orbit around the Sun, and Pluto takes almost 248 years. In contrast, our galaxy takes 240 million years to complete one rotation, and the Universe is around 15 billion years old. Thus, handling motion and animation timing properly in astronomical visualization requires special attention.
- **Reference Frames.** There is no absolute reference center for the Universe. We cannot have the same reference frame fixed to the Earth as it moves around the Sun and to the Sun as it moves through the Milky Way. In a dynamic astronomical environment, nothing is at rest. Neither the Ptolemaic Model nor the Copernican model is adequate to simulate the dynamic interaction among data sets because we can place the observer anywhere. Thus our scales, our reference frames, our relative distances, and our navigation from frame to frame must all be carefully represented.

In the next three sections, we discuss techniques for dealing with the above problems: large spatial scale, large time scale, and modeling approaches consistent with navigation in dynamic large-scale environments. Finally, we present our approach to an implementation that addresses these issues.

2 Large Spatial Scales

2.1 Some Experimental Observations

We have observed a variety of scale-related breakdowns in computer graphics systems. Many modern graphics environments can actually be switched from program computation to hardware-assisted computation, where the CPU software implementation may be slower but more accurate. We will implicitly assume that we are dealing with an OpenGL-like hardware-assisted polygon projection environment; so far as can be determined, the limits below are those reflecting native hardware graphics board performance for OpenGL.

The most significant experimental breakdown modes appear to be the following:

1. **Geometry:** Geometry distortions and breakdowns typically result from the limits of floating point precision in matrix manipulation as implemented in the hardware graphics pipeline. Some graphics environments, notably VRML browsers, apparently attempt automatically to adjust such features as the view matrix and depth range, resulting in the destruction of geometry tens of orders of magnitude before the effect appears in bare OpenGL.
2. **Depth Buffer:** Depth buffer breakdowns occur due to limits in the depth buffer precision when several objects have huge scale differences. Using a linear scale of depth in the conventional hardware does not permit the large range of depth differences that can be supported by a logarithmic scale. Yet, if one actually had a log-scaled depth buffer, it could not precisely handle depth *differences* of objects as well as a normal linear scale. In practice, the depth buffer range is rescaled to (0.0, 1.0), which can easily result in artifacts depending on the buffer word size, and perspective projection causes significantly nonlinear behavior with depth.
3. **Lighting:** Lighting formulas break down due to the accuracy limits of matrix inversion combined with lighting calculations. One must somehow compute the normal vector in world space, and, if any scale operation has been permitted, a complete matrix inversion is involved; this process normally breaks down much earlier than the linear matrix multiplications affecting raw geometry.

In Table 1 we summarize the behavior observed using a test geometry consisting of lit unit cubes with bare scale matrices corresponding to each tested scale. In many geometry pipelines, uniform behavior can be seen to break down at $10^{\pm 37}$. Lighting transformations fail, probably due to the matrix inversion process, at around $10^{\pm 13}$. The most dramatic behavior occurs in VRML systems such as CosmoPlayer under both PC/Windows and SGI/IRIX, probably due to automatic view adjustment: here geometry scaling breaks down significantly earlier, around 10^{12} on an SGI, and around 10^{18} on a PC, while lighting fails around 10^{13}; the scaling behavior of a set of nested cubes in VRML is shown in Figure 1.

2.2 Homogeneous Power Coordinates in Space

To handle large spatial scales, we use the homogeneous power coordinate method introduced in [8]. Points and vectors in three-dimensional space, say $P = (X, Y, Z)$, are represented by homogeneous power coordinates, $p = (x, y, z, s)$, such that

$$s = \log_k ||P||$$
$$(x, y, z) = (X/k^s, Y/k^s, Z/k^s)$$

where k is the base. In practice, k is normally 10.

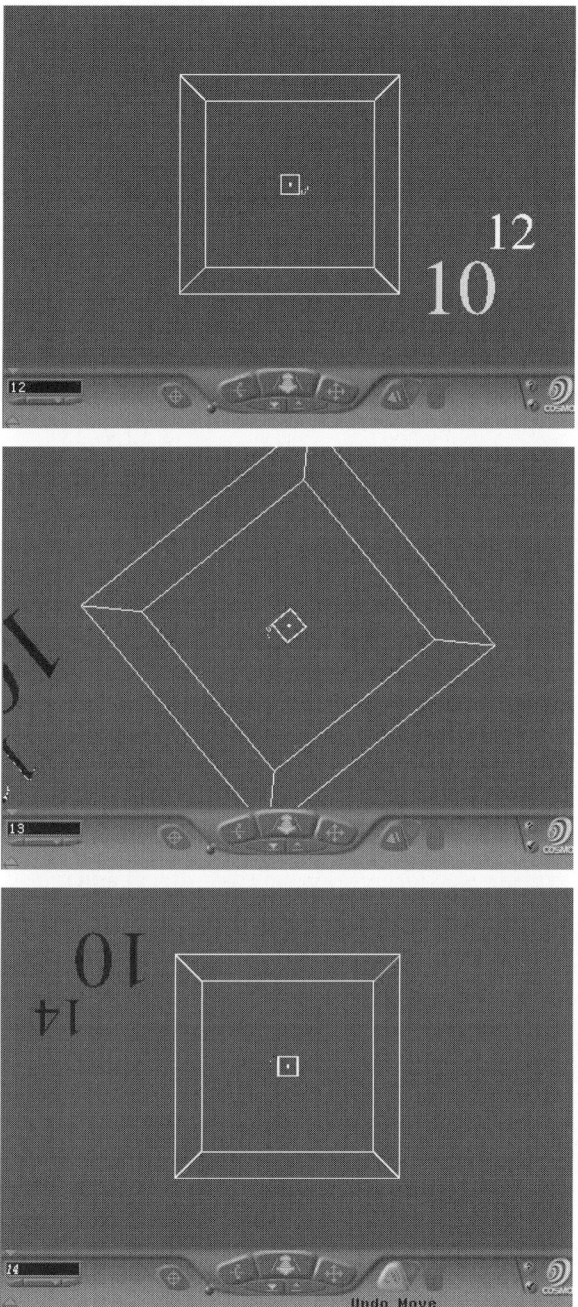

Fig. 1. Breakdown of SGI CosmoPlayer running unit VRML cubes near scale parameter values of 10^{13}. (Top) 10^{12}: Normal behavior. (Center) 10^{13}: Lighting fails (blackened lettering) simultaneously with geometry. (Bottom) 10^{14}: Geometric accuracy is lost from here on. (See Color Plate 10 on page 355.)

Table 1. Observed platform dependence for lighting and geometry breakdowns, including the CosmoPlayer VRML browser and several bare OpenGL implementations. All units are powers of ten, and values should be considered approximate.

Platform	Light	Geometry
PC/Windows VRML/CosmoPlayer		
Micron GEForce3	$13 \Rightarrow 14$	$18 \Rightarrow 19$
	$-20 \Rightarrow -21$	$-20 \Rightarrow -21$
Dell Inspiron laptop	$13 \Rightarrow 14$	$37 \Rightarrow 38$
	$-12 \Rightarrow -13$	$-37 \Rightarrow -38$
SGI/IRIX VRML/CosmoPlayer		
SGI O2:	$12 \Rightarrow 13$	$12 \Rightarrow 13$
	$-12 \Rightarrow -13$	$-12 \Rightarrow -13$
PC/Windows OpenGL		
Micron GEForce3	$12 \Rightarrow 13$	$18 \Rightarrow 19$
	$-15 \Rightarrow -16$	$-19 \Rightarrow -20$
Dell Inspiron laptop	$12 \Rightarrow 13$	$37 \Rightarrow 38$
	$-12 \Rightarrow -13$	$-38 \Rightarrow -39$
SUN/SOLARIS OpenGL		
SUN Creator 3D	$18 \Rightarrow 19$	$37 \Rightarrow 38$ (segfault)
	$-19 \Rightarrow -20$	$-38 \Rightarrow -39$
SUN elite 3D	$18 \Rightarrow 19$	$37 \Rightarrow 38$
	$-19 \Rightarrow -20$	$-38 \Rightarrow -39$
SGI/IRIX OpenGL		
SGI O2:	$12 \Rightarrow 13$	$36 \Rightarrow 37$
	$-19 \Rightarrow -20$	$-36 \Rightarrow -37$
SGI Octane	$12 \Rightarrow 13$	$36 \Rightarrow 37$
	$-12 \Rightarrow -13$	$-36 \Rightarrow -37$
SGI Onyx2	$12 \Rightarrow 13$	$36 \Rightarrow 37$
	$-12 \Rightarrow -13$	$-35 \Rightarrow -36$

In this way, we can separate the original physical representation into its order of magnitude, the log scale (s), and its unit-length position (or direction) (x, y, z) relative to the original representation. This formulation helps us not only to avoid hitting the machine precision limits during calculation, but also gives us a systematic way to model data sets at different levels-of-detail (LOD) and position them in our virtual environment relative to the current viewing scale.

2.3 Navigating Large Spatial Scales

In our system, we define the navigation scale, s_{nav}, as the order of magnitude at which we are navigating through space; that is, one unit in the virtual environment is equivalent to $10^{s_{\text{nav}}}$ m in our real Universe. Consequently, traveling through one unit in the virtual environment gives the sensation of traveling through $10^{s_{\text{nav}}}$ m in the physical Universe. We can thus adapt the step-size to an exploration of the Earth, the Solar System or the whole Galaxy. Note that we typically ignore the distortion effects of special relativity during such navigation, though we may retain the causal effects; conceptually this is consistent with rescaling the speed of light so the current screen viewing scale behaves non-relativistically.

An example of a reasonable value of s_{nav} for exploring something the size of the Earth would be 7 (with base 10 metric units). That is, the size of the Earth is reduced to about 1 unit in the virtual environment. To go beyond the Earth to the galaxy, whose order of magnitude is 10^{21} m, we must increase s_{nav}. Otherwise, with 10^7 m as the step-size, we would need to take billions of steps to go through the Milky Way. Another motivation for environmental rescaling is that we have to "exist" at an appropriate scale to look at a particular thing (the "Alice in Wonderland" effect). Details of the use of s_{nav} to assist navigation are discussed in section 4.

2.4 Level-of-Detail in Space

Besides assisting navigation, we employ s_{nav} to select different levels of detail (LOD's) in space. The use of LOD's to select rendered graphics objects relative to a display pixel has been explored, e.g., in [10,1,12,14]. As suggested in [8], data sets with scales appropriately greater than s_{nav} can be represented as environment maps with minimal perception error, provided the depiction error relating the environment map to its 3D rendering is on the order of one screen pixel. Figure 2 depicts a typical cubic environment map substituting for a very large database of distant stars.

On the other hand, when the scale of a data set is small compared to s_{nav}, we can pick up a smaller scale space LOD and use that to represent the

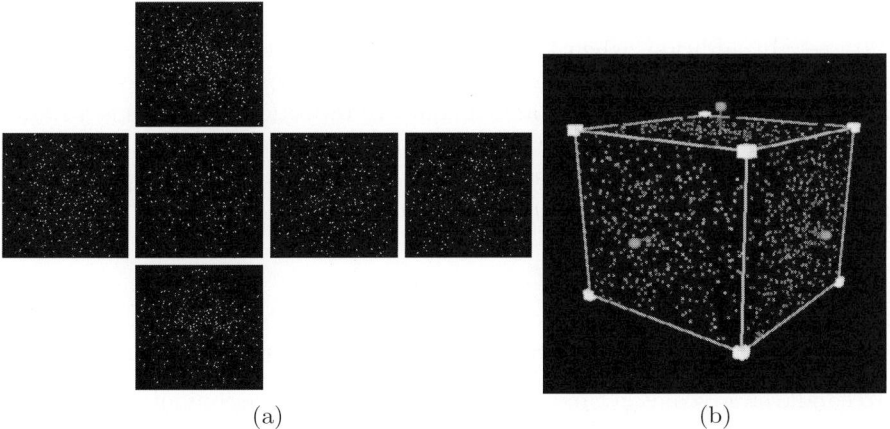

Fig. 2. (a) Pieces of cubic environment map used for the nearby stars when $s_{\text{nav}} <$ 13. (b) Appearance of wrapped map from outside. (See Color Plate 11 on page 356.)

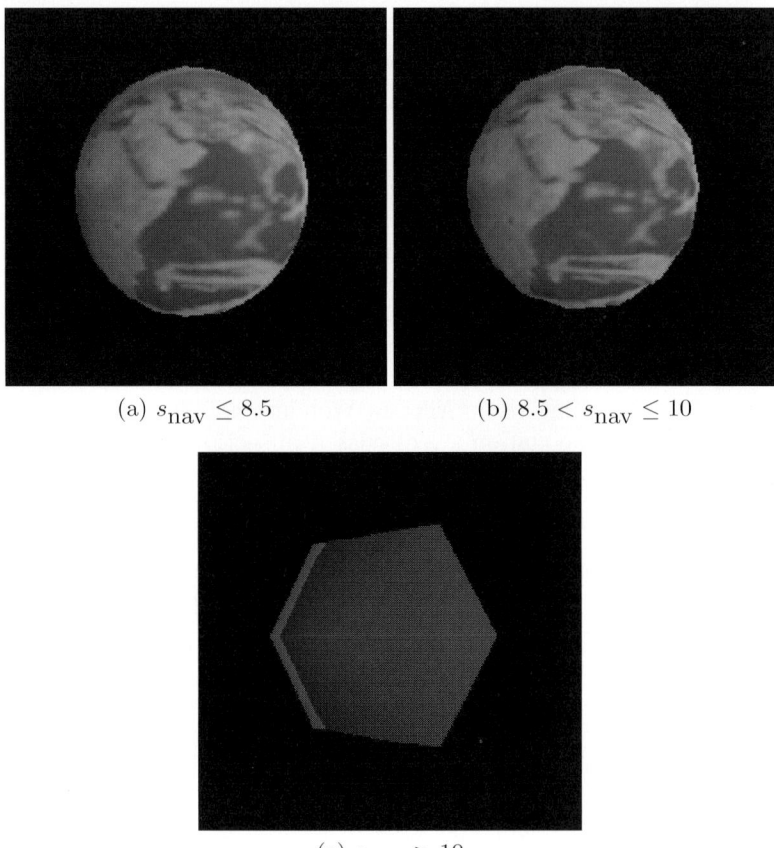

(a) $s_{\text{nav}} \leq 8.5$ (b) $8.5 < s_{\text{nav}} \leq 10$

(c) $s_{\text{nav}} > 10$

Fig. 3. Space-LOD representations of the Earth. (See Color Plate 12 on page 356.)

data set. In addition, when the object size is smaller than one pixel on the screen, we can ignore its rendering unless it is for some reason "flagged" as an important object that warrants an out-of-proportion icon (this is a classic map-making strategy). Figure 3 shows three different space-LOD models of the Earth. Details concerning space-LODs and their relation to s_{nav} are given in the *Rotating Scale algorithm* of section 4.

3 Large Time Scales

3.1 Representing Large Scale Ranges and Steps in Time

To handle large scales in time, we developed a log scale technique based on the same general concepts as our spatial scaling. To specify the period of rotation, an orbital period, or other periodic motion of an astronomical data set, we enter the triple: (*magnitude scale unit*) in the dataset description file, which stands for $magnitude \times 10^{scale} unit$. For example, (3.65 2 *day*) expresses the same information about the Earth's orbital period as (1 0 *year*). The system automatically converts this to seconds, (3.1536×10^7 seconds) and stores the period of motion as $\log_{10}(3.1536 \times 10^7) = 7.4988$. Likewise, we can store this $\log_{10}$ value, denoted as t_{motion}, for every defined motion.

By analogy with the spatial scale s_{nav}, we can define an animation scale, t_{nav}; quantitatively, this time variable specifies the equivalence between one second of observer's screen time (the wall clock) and $10^{t_{nav}}$ seconds in the simulated Universe. That is, if t_{nav} is 7.4988, we could observe the Earth in the virtual Universe going around the sun once each second. Using this rule, we can determine the observable period of motion, t' (in seconds), for a specified t_{nav} with the formula:

$$t' = 10^{t_{motion} - t_{nav}}$$

Consequently, if we want to visualize the rotation of the Earth on its axis, we can reduce t_{nav} from 7.4988 to 4.9365, which corresponds to a time scale of one day. By controlling t_{nav}, we can then manipulate relative simulation time scales to optimize the visualization properties of the simulated motion.

Note that special attention must be paid to the relative scales of t_{motion} and t_{nav} before computing t'. Using a single time LOD representation when t_{motion} differs substantially from t_{nav} will cause unacceptable errors in the time-stepped motion of objects animated with t'. As discussed in the next section, we must therefore add logic to check the ranges of t_{motion} and t_{nav} separately, choose an appropriate time-scaled LOD model, and make sure the chosen steps in t' move the model appropriately in viewer spacetime.

3.2 Levels-of-Detail in Time

For a given value of t_{nav}, the Earth will orbit too slowly around the Sun in viewer time, while the moon will orbit too fast. In the former case, we still want to visualize Earth's *direction* of motion even if there is no apparent movement; in the latter case, we still want to see the subset of space (the orbit) occupied by the moon over time, and we still want to see the other planets move in screen time. To have a richer visualization for these "too slow" and "too fast" situations, we can once again make use of t_{motion} and t_{nav}.

To handle "too fast" motions, we represent the path as a comet tail. Quantitatively, the length of the comet tail denotes the past history of the data set model in the preceding second of viewer-time. Therefore, since t_{motion} is the $\log_{10}$ period of orbital motion, the angle θ swept by the comet tail would be

$$\theta = 2\pi/t' = 2\pi \times 10^{t_{nav} - t_{motion}} \quad (1)$$

If we increase t_{nav}, the length of the comet trail will increase; then, when $\theta \geq 2\pi$ ($t_{nav} \geq t_{motion}$), the a full orbit will take less than one viewer second.

At this point, we remove the moving geometric model from the representation, and show only its circular orbit trail. As illustrated in Figure 9(a,b), we can use the lengths of neighboring comet trails to quantitatively visualize relative motions of each planetary or satellite orbit.

For "too slow" motion, we examine the θ term in Eq. (1). When $\theta \leq \pi/180$, i.e.,

$$t_{\text{nav}} \leq t_{\text{motion}} - \log_{10}(360) \;,$$

as shown in Figure 9(c), we display a little arrow indicating its direction of motion. In the actual implementation, we resize the arrow according to the value of s_{nav} to avoid stroboscopic effects.

Motion Blur Model. Certain common astronomical models such as planets and moons not only move in their orbits, but spin on their axes. When the object's surface is represented by a complex texture, this rotation creates a texture representation problem exactly analogous to the too-slow/too-fast orbit problem. Whenever the textured object reaches a time scale where the texture is moving by a large amount in a unit of screen time, the motion loses smoothness and undesirable effects such as stroboscopic aliasing become apparent. We handle this problem by adopting a texture-based motion blur method (see Figure 10) that gives the object a smoother appearance as it speeds up as well as eliminating stroboscopic effects. For very fast viewer-time rotation speeds, the texture turns into blurred bands exactly analogous to the satellite trails.

4 Modeling our Virtual Universe

Using our scalable representations for space and time variables, we are able to model the full range of physical scales in our virtual Universe. In this section, we first examine the concept of scalable data sets, and then show how we connect scalable models together to form our principal internal data structure, the Interaction Graph. This graph enables us to represent a dynamic virtual Universe that is scalable, navigable, and extensible.

4.1 Scalable Data Sets

Basically, there are two 3D spaces that we must deal with. One is the computational space supported by the graphics hardware environment, which has distinct accuracy limits on the matrix computations transforming the viewpoint and geometry, the combined normal and lighting computations, and the depth buffer. The other is the intrinsic data-representation space. We represent all models, ranging from point-set models of the most distant galaxies to Calabi-Yau spaces suggested by string theory at the Planck length, as unit-sized three-dimensional ideals with multiple level-of-detail (LOD) representations. We will see that this gives us full control of the scaling and the rendering precision.

If we put the ideal models directly into the computational space, say, in meters, we may not have enough computational precision to render the ideal data structures without errors; lighting and the depth buffer in particular can cause major problems even if the matrix accuracy is sufficient for the geometry alone. Adjusting the scale encoding gives us full control over the quality of the rendering process.

To encode the true size of each data set, we define s_{dataset} to be the $\log_{10}$ value of its actual size (in meters). For example, the diameter of the Earth is about $12,756$ km, so its s_{dataset} value becomes $\log_{10}(1.2756 \times 10^7) = 7.1057$. The designer can specify $12,756$ km in the dataset description file of the Earth and the system will compute $s_{\text{dataset}} = 7.1057$ automatically. We can also specify a visible range $[s_{\min}, s_{\max}]$ for each data set. If the current s_{nav} is out of this range, we assume that the data set can be ignored. If s_{nav} is within the range supported by the LOD representations for a model set, we must select a particular space-LOD model. (See Figure 3 for an example.) The system then rescales the selected space-LOD model as shown in Figure 4.

Rotating Scale algorithm. Our scale-control procedure has the unique feature of *reusing* the computational scales near unity over and over. At $s_{\text{nav}} = 1$, we use the LOD appropriate for scale exponent $= 1$, and render with scaling $10^{1-1} = 1$; at $s_{\text{nav}} = 10$, we use the LOD appropriate for scale exponent $= 10$, and render with scaling $10^{10-10} = 1$, thus maintaining complete model accuracy. In effect, the scale "rotates" back to unity over and over as we zoom in or out through the data sets and their matched LOD models. The following

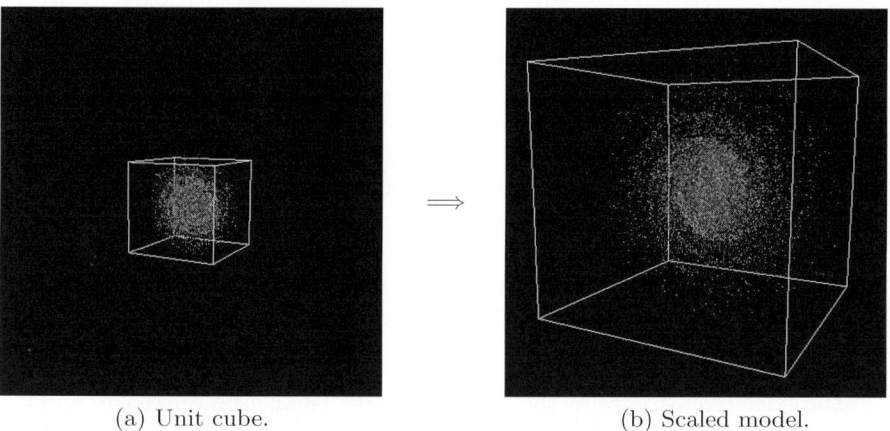

(a) Unit cube. (b) Scaled model.

Fig. 4. Scaling a unit-sized ideal model by 10 to the power $(s_{\text{dataset}} - s_{\text{nav}})$. (See Color Plate 13 on page 357.)

pseudo-code represents the procedure by which we continuously adjust the rendered model scale to the viewer's screen (navigation) scale:

Rotating Scale Algorithm:

```
FOR EACH dataset BEGIN
  IF (s_min <= s_nav <= s_max) BEGIN
    - select the space-LOD matching s_nav
    - if (space-LOD is not an environment map)
-- scale it by 10^(s_dataset-s_nav)
    - attach model to the scene graph
  END-IF
END-FOR-EACH
```

Recall that when s_{dataset} is larger than s_{nav} and the nearest object in the dataset is too far away to move by more than a single pixel in the current navigation context, we can represent that data set as an environment map. A single (distant) environment map can itself be rendered at unit scale provided it is rendered first, with depth buffering disabled, so the depth remains at its maximum value. However, since more than one data set may qualify for this treatment (e.g., nearby stars, Galaxy, local galaxies, etc.), these data must in fact be rendered in order as *distance-sorted transparent environment maps*, and the depth buffer reinitialized before 3D rendering. If particular elements of a data set are too close to satisfy the single-pixel displacement criterion during a particular navigation sequence, they can in principle be treated separately as 3D objects in order to maintain the advantage of rendering the majority of the data set as a fixed environment map. By keeping s_{dataset} judiciously close to s_{nav} for each LOD, we can typically guarantee that no more than four orders of magnitude of hardware transformation scaling are required, and thus the scaling on the unit-sized data models never approaches the experimentally observed limits for the onset of numerical error in any category, for any platform.

4.2 The Interaction Graph

After we select our data models according to the s_{nav} scale key, we can link data sets together in the internal Interaction Graph data structure. In the current implementation, two kinds of interaction are supported, though more are planned for additional applications such as extended gas clouds with known velocity components:

- **Orbiting.** Periodic orbits of one data set around another. Examples are the Earth and the Sun, the Moon and the Earth, the Asteroids and the Sun, Saturn's rings and Saturn.
- **Locking.** One data set maintains a fixed position or orientation with respect to another data set. An example is the use of different observational

coordinate systems such as polar coordinates relative to the Earth (Equatorial Coordinates), relative to the Sun (Ecliptic Coordinates), relative to the galaxy (Galactic Coordinates), or relative to the whole set of galaxies (Super Galactic Coordinates). Since nearby stars are typically expressed in Equatorial coordinates, we may lock this data set to the Earth to maintain the observed position relative to the Earth. When we animate the motion of the Solar system within the galaxy, the coordinate-locking forces the nearby star data set to move correctly relative to the galaxy. This is described in detail as the *cascade animation update algorithm* in section 4.5 below. Similarly, we have to lock the local galaxies, distant galaxies, and the CMB (Cosmic Microwave Background or Cosmic Background Radiation) data sets to the Galaxy if they are represented using Galactic coordinates.

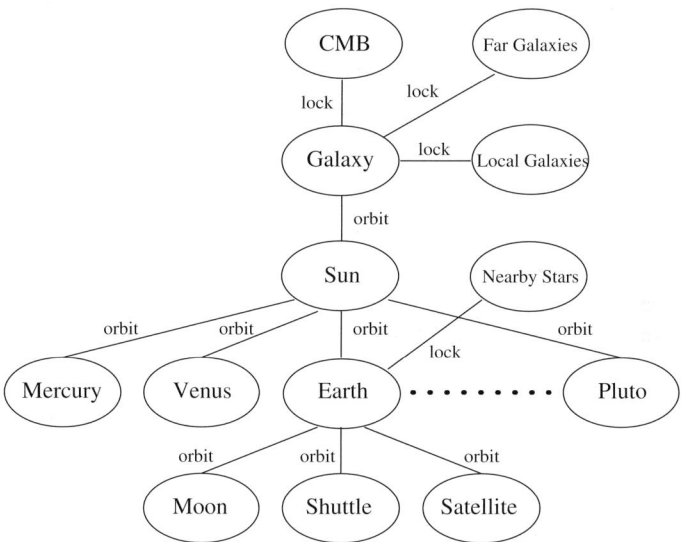

Fig. 5. Interaction Graph for the Universe data set collection.

In the interaction graph, each data set is a node and node-node interactions are represented as edges. Figure 5 depicts a typical interaction graph. Note that this interaction graph differs from the traditional scene graph structure (the tree structure traversed by a rendering system such as Inventor or Performer). No particular data set is singled out as the root node; the interaction graph serves mainly to 1) organize the data sets by their interactions, 2) facilitate scaling, and 3) allow cascaded animation update.

4.3 Scalable Universe

We are now ready to navigate through a scalable Universe. We begin by exploiting homogeneous power coordinates, $p = (x, y, z, s)$, to describe the relative positions between data sets (nodes in the interaction graph). That is, for each edge in the interaction graph, (x, y, z) is the displacement of one data set relative to the other, while (s) is the $\log_{10}$ distance (in meters). Using this representation allows us to rescale the distance between data sets and to determine a relative distance d with respect to s_{nav}:

$$d = 10^{s - s_{\text{nav}}}$$

Note that once s_{nav} is changed, all the distances among data sets are changed accordingly. This keeps the Universe scalable without requiring an absolute center for scale operations.

4.4 Navigable Universe: Centers-of-Navigation

Another use of the interaction graph is to specify a navigation path. The Universe is so vast, and our regions of dense data are so limited, that a completely free-form fly-through makes little sense: human space and time scales are far too limited, and rescaled travel in arbitrary directions is more likely to get the user lost than educated. The interaction graph allows the designer to locate data sets of interest for the users and to carry out smooth traversals of the interesting data in the interaction graph. It also easily supports more sophisticated paradigms such as constrained navigation [7,16].

When we navigate from the Earth, out into the Solar System, the nearby stars, and out beyond the Milky Way galaxy, we must dynamically change the effective center of the virtual environment at each different navigation scale to provide meaningful context. In our system, we can define which data set is to be the effective center (data set of interest) at each different scale of navigation, s_{nav}. Figure 6 depicts a typical example: the center is specified to be the Earth when $s_{\text{nav}} \leq 10$, the Sun when $10.5 \leq s_{\text{nav}} \leq 19.5$, and the galaxy when $s_{\text{nav}} \geq 20$. In effect, the central object of the current s_{nav} is the origin of the virtual environment. For in-between scales such as $10 \leq s_{\text{nav}} \leq 10.5$, the system interpolates between data sets (the Earth and the Sun) to produce a smooth transition.

4.5 Dynamic Universe (Animation Update)

Knowing the center of virtual environment, we can carry out animation updates on each data set by traversing the interaction graph from the data set serving as the central focus. The algorithm for this animation update is the following:

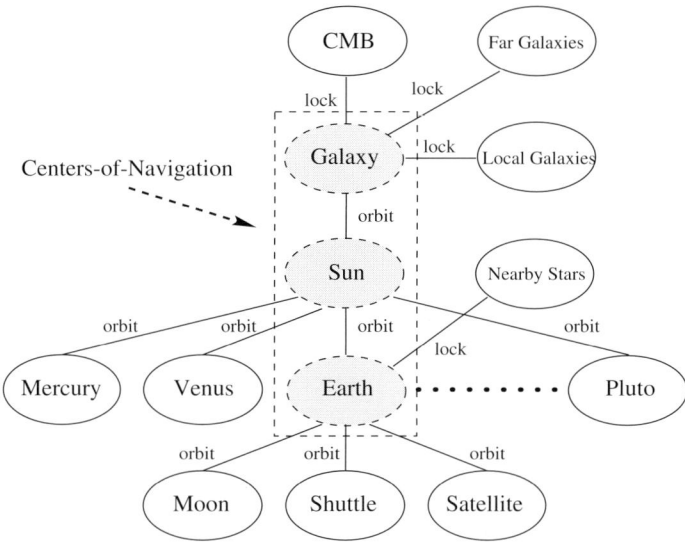

Fig. 6. Centering the Virtual Environment.

Cascade Animation Update Algorithm:

```
- Get time elapsed (dt) since previous update
- Pick up centering data set(s) at s_nav
- Interpolate the centers if necessary
- Position the centering data set(s)
- Traverse the interaction graph from the
  center(s) recursively:
  BEGIN
    - Update rotational motion by dt
    - Update orbital motion by dt
        (relative displacements between objects are in
homogeneous power coordinates)
  END
```

Using this algorithm, we can correctly update relative positions between data sets, as well as placing appropriate data sets of interest at the centers of the virtual environment at different navigation scales. In particular, when we fix the center at the Earth, we can observe the Sun apparently orbiting around the Earth and see the epicycles traced by neighboring planets.

4.6 Expanding Universe: Cosmology

Another astrophysical effect that can be incorporated is the Hubble expansion ([11], [2]). Since Hubble's original discovery that the Universe is expanding, both the observational data and the corresponding theories have been significantly refined. According to our current best understanding ([5],[4]), the

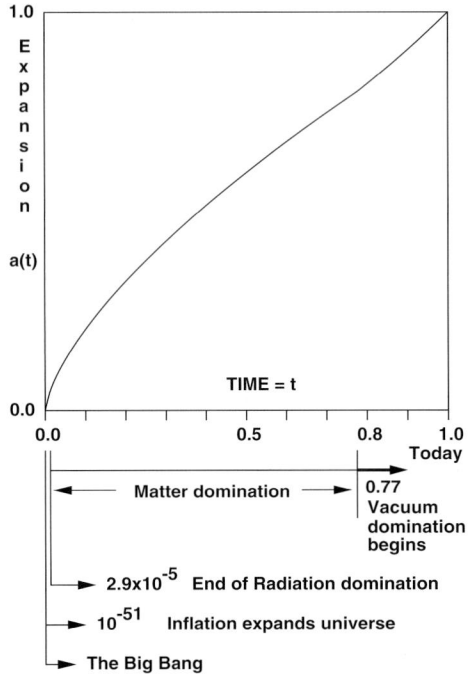

Fig. 7. The expansion of the Universe, $a(t)$. t is in units of the age of the Universe, and $a(t)$ is in units of the current radius of the Universe.

time evolution of the size of the universe is believed to follow the cosmological expansion curve in Figure 7, which plots the effective radius of the Universe $a(t)$ as a function of time t in normalized units.

Since light travels at a finite speed, different data sets with different scales relative to the Earth actually correspond to snapshots taken at significantly different times in cosmological history. The further away a data set is from the Earth, the longer it took the detected light to travel to our instruments. Therefore, when we move our viewpoint out from the Earth into the farther reaches of the Universe, we are looking at historical records of the data sets rather than current states. This phenomenon can be visualized using our "Lightcone Clock" technique described below in section 5.5. We can use the distances to the Earth (when known) to deduce the corresponding "clock" time t for any light-emission event.

If we now wish to know what the Universe looked like *at the time the light was emitted*, we can use t to look up the effective radius of the Universe and factor in a scaling by the $a(t)$ term. In effect, we modify the original equations in subsections 4.1 and 4.3 using

$$\text{model size} = a(t) \times 10^{s_{\text{dataset}} - s_{\text{nav}}}$$

and

$$d = a(t) \times 10^{s - s_{\text{nav}}} .$$

This Hubble-adjusted coordinate system is typically referred to as *Physical Coordinates*. If we ignore the effect of the Hubble expansion during navigation and assume the objects are located as though the Universe has always had the size it has today, we are effectively visualizing the cosmos using a coordinate system referred to as *Comoving Coordinates* in the astronomical literature. The latter system has the advantage of presenting the viewer with a fixed reference system for distances.

5 Implementation and Results

5.1 System Overview

Figure 8 shows the three main components in our system: the Dataset, the Dataset Manager, and the Front-End. The Dataset structurally organizes the astronomical data sets in a three-level hierarchy. The Dataset Manager parses the data sets, the script file, and the Interaction Graph, then provides the information needed to continuously update the animation; it provides a clean stream of properly scaled data models for the Rendering Engine. Finally, the Front-End creates the user visuals based on the stream supplied to it by the Dataset Manager; it constantly tracks user input events, controls the navigation, manages the scene graph, and requests data models from the Dataset Manager.

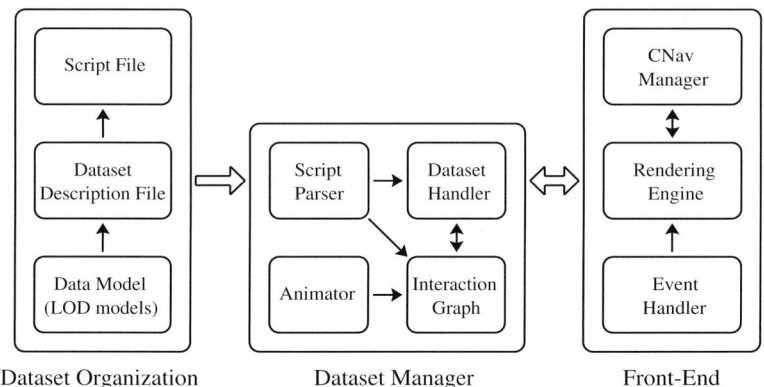

Fig. 8. System Overview.

The entire system is implemented using IRIS Performer and the CAVE™ library in C++. The Dataset Manager interpreting the geometric datasets involves only the Performer library, so that much of the system is portable to non-CAVE^(TM)-based rendering systems. Since we have a well-defined interface between the Dataset Manager and the Front-End, we can replace the existing Front-End (Rendering engine and Event Handler) with a CAVE^(TM)-free Front-End in order to support desktop SGI or Linux workstations. The user interface permits the adjustment of both s_{nav} and t_{nav} at will using the joystick on the wand. Users can also open a menu-bar in the virtual environment and change various options (such as Hubble expansion, lighting, etc) with the wand whenever necessary.

5.2 Dataset Organization

- **Data Model (space-LOD model).** Conceptually, all unit-sized data models (space-LODs) are included in this lowest layer of the three-level hierarchy. These data models can be any IRIS-Performer-readable file such as .iv (Open Inventor), .pfb (Performer binary format), etc. In addition, we provide file templates defining spherical and cubic environment maps, as well as the possibility of $3D$ texture data. Note that all data models must be scaled to unit size (bounded by a unit cube).
- **Dataset Description File.** In this layer, we put the physical information (e.g., size, coordinate system, period of rotation, orbital period, ...). The spatial LOD model information, which specifies the keys for switching scale ranges, is also supplied in the dataset description file for each data set. All the physical quantities for each data set are provided here.
- **Script File.** This layer implements a scripting language describing the astronomical environment and the interactions between data sets. Examples of major keywords and fields in the scripting language are listed in Table 2 to provide the general idea. Altogether, there are four categories of keywords: CNav (Constrained Navigation [7,16]), Modeling, Animation, and Centering. The CNav commands define a set of CNav manifolds that restrict the users' movements in a sensible way. The Modeling commands load in data sets from dataset description files and instantiate the loaded data sets in the virtual environment. The Animation commands define the motions and interactions relating astronomical objects. The Centering command specifies which data set is to be treated as the origin or focus of attention at a particular navigation scale.

5.3 Dataset Manager

Instantiating the Interaction Graph. The Dataset Manager first initializes and then maintains interactive updates of the comprehensive data structure we refer to as the Interaction Graph. The Interaction Graph is a graph-like construction integrating all the information loaded from the dataset description files. At each node, there is a pointer to the Performer *pfDCS* structure in

Table 2. Example keywords in the scripting language.

CNav:	Syntax:
DEFNAV	DEFNAV <cnav grid file> AS <cnav model>
USENAV	USENAV <cnav model>
STARTNAV	STARTNAV <cnav model>
Modeling:	Syntax:
DEF	DEF <dataset desc. file> AS <model name>
USE	USE <model name> AS <instance name>
PLACE	PLACE <instance name>
Animation:	Syntax:
ROTATE	ROTATE <instance name> AXIS <vector vx,vy,vz>
ORBIT	ORBIT <instance name> AROUND <center instance> AXIS <vector vx,vy,vz> <option>
LOCK	LOCK <instance name1> <instance name2>
Centering:	Syntax:
CENTER	CENTER <instance name> ATSCALE <scale>

each dataset handler. Animation updates can thus be reflected immediately in the scene graph structure in the Front-End.

- **Script Parser.** Both the dataset description file and the script file are plain text files describing the spatial data sets and the virtual environment. The Script Parser in the Dataset Manager parses the script file and, whenever required, loads specified dataset description files. During parsing, the Parser instantiates one dataset handler for each data set and progressively constructs the Interaction Graph.
- **Dataset Handler.** Dataset handlers correspond to instances of dataset description files. Each dataset handler stores the information concerning its data sets and maintains the current orientation and positional information in the Performer *pfDCS* dynamic coordinate system structure. Scaling of data models is implemented by the *Rotating Scale algorithm* (section 4), which returns properly scaled data models (*pfDCS*) to the Front-End upon request.
- **Animator.** The Animator is a stand-alone thread that constantly applies the *Cascade Animation Update algorithm* (section 4.5) to the interaction graph in order to update the positional and orientation information in the *pfDCS* structure whenever the system is idle. Time rescaling is handled in this module.

5.4 Front-End

- **Event Handler.** The Event Handler interprets inputs from the trackers and pointers (e.g., head tracking and wand devices of the CAVE^{TM}) as well as the keyboard. It then updates the corresponding information stored in the Rendering Engine.

- **CNav Manager.** Whenever the Rendering Engine encounters navigation-related input, it consults the CNav Manager. The CNav Manager then analyzes the input and returns the interpolated navigation states such as s_{nav}, the current position and orientation, the viewing direction, etc. Note that positional information is stored in the CNav structure as homogeneous power coordinates rather than ordinary (x, y, z) coordinates.
- **Rendering Engine.** When s_{nav} changes, the Rendering Engine executes the *Rotating Scale algorithm* and activates the dataset handlers to renew the *pfDCS* structures attached to the scene graph so that the data models are properly rescaled. The Rendering Engine is also responsible for updating various states (e.g., t_{nav}, ...) in the Dataset Manager so that Data Handler and Animator have up-to-date information.

5.5 Results

Time-Adjusted Representations. Figure 9 in the color plate section shows a series of Solar System models for different values of t_{nav}. Going from Figure 9(a) to Figure 9(c), we can see that the length of comet tails (Pluto, Neptune and Uranus) shorten as t_{nav} goes from 10.0 down to 9.0; finally, when t_{nav} reaches 6.0, these planets move too slowly to have apparent motions in user time, and the comet tails are replaced by arrows representing long-term motion. Note that all images were captured during real-time system operation.

Motion Blur Representations. In Figure 10 we show a family of representations of a rotating textured object, in this case the Earth. For rotation speeds that are commensurate with screen time, the full detailed texture is appropriate. However, when the animation speeds up, the detailed texture would exhibit undesirable stroboscopic effects; progressively motion-blurring the texture creates an intuitively appealing visualization of the animation.

The Lightcone Clock. To support a more flexible and informative visualization of the whole body of cosmological data, we introduce an iconic representation of the Universe, the *Lightcone Clock*. Conceptually, when we look at the night sky, our view cuts through the entire spatial model; we can think of the viewed objects as filling a solid view frustum or, for symmetry, a cone. But now we go one step further: in order to see at one glance the *entire* spherical volume containing all the data models of our Universe, we morph each spherical surface at constant radius from the observer into a much smaller surface at the same radius, thus turning a solid sphere into a solid cone. In a very real technical sense, this is the visible light cone of four-dimensional spacetime reduced to a stack of disks, each corresponding to one particular age band of light-rays. The correspondence between the

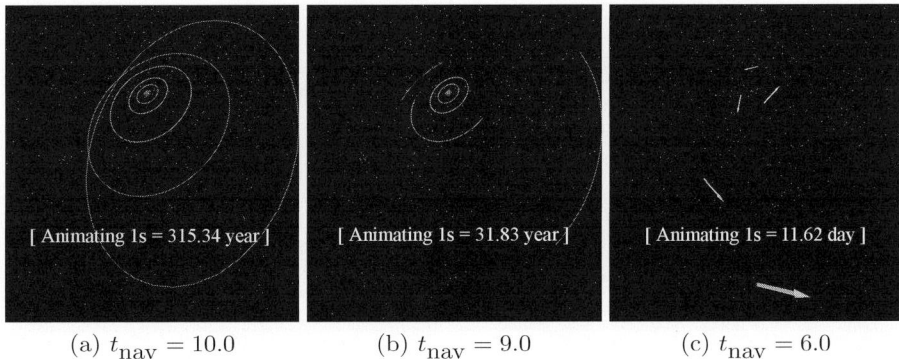

(a) $t_{\mathrm{nav}} = 10.0$ (b) $t_{\mathrm{nav}} = 9.0$ (c) $t_{\mathrm{nav}} = 6.0$

Fig. 9. "Too fast" representations at (a) roughly 300 years per screen second and (b) roughly 30 years per screen second. (c) The "too slow" representation for the motion of planets in our Solar System at a scale of approximately 10 days per screen second. (See Color Plate 14 on page 357.)

(a) $t_{\mathrm{nav}} \leq 4.4$ (b) $4.4 < t_{\mathrm{nav}} \leq 4.9$ (c) $4.9 < t_{\mathrm{nav}}$

Fig. 10. Motion Blur Representation : (a) Normal Earth texture. (b) Texture blur for one rotation in two screen seconds. (c) Texture blur for one rotation in less than one screen second. (See Color Plate 15 on page 358.)

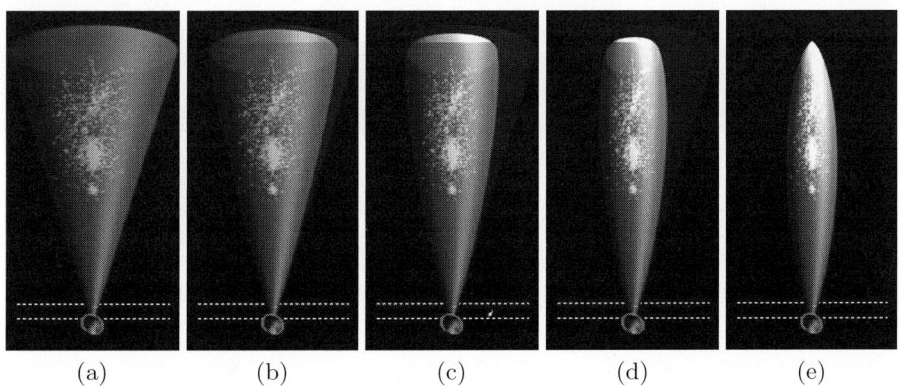

(a)   (b)   (c)   (d)   (e)

Fig. 11. Morphing the Lightcone Clock from Comoving Coordinates to $a(t)$-rescaled Physical Coordinates. (See Color Plate 16 on page 358.)

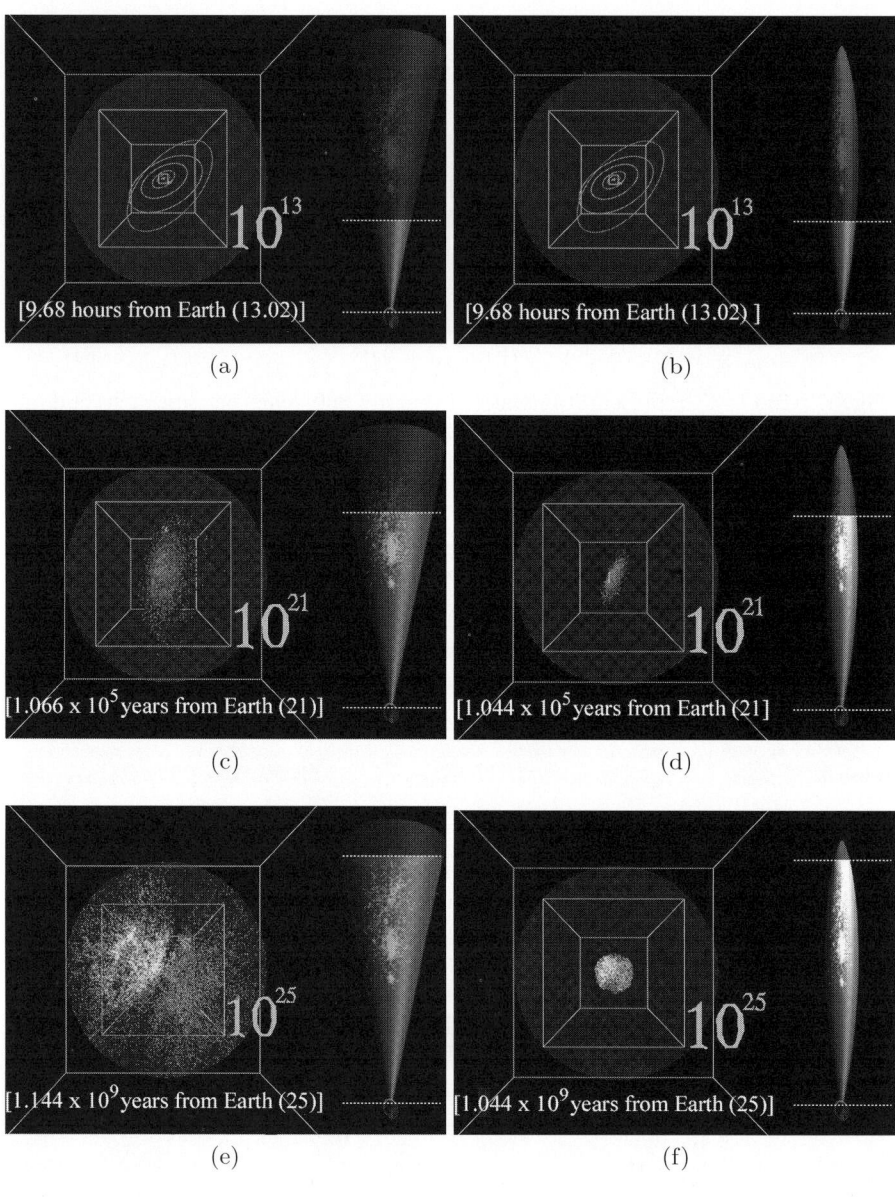

Comoving Coordinates $a(t)$-rescaled Physical Coordinates

Fig. 12. Navigation with Comoving Coordinates and $a(t)$-rescaled Physical Coordinates, with the Comoving Coordinate scales superimposed for absolute reference. (See Color Plate 17 on page 359.)

time when the currently-observed light was emitted from each disk and the emitter's distance from us in comoving coordinates motivates our describing this visualization as a *clock*.

In order to provide a more effective summary viewpoint of the datasets in the Lightcone Clock, we have adopted a logarithmic scale. The canonical viewing origin on the Earth is at the bottom tip of the Lightcone Clock, and the final upper surface is the event horizon for the first visible radiation, the Cosmic Microwave Background. Figure 11(a) shows the shape of the Lightcone when we use comoving coordinates. When we take the Hubble expansion into account by incorporating the effective radius $a(t)$, the Universe in physical coordinates shrinks with the size of $a(t)$; at the upper (CMB) end of the Hubble-corrected Lightcone Clock, the Universe was less than $1/1000$ of its present size. Figure 11(a-e) illustrates the morphing from the comoving-coordinate Lightcone to the $a(t)$-rescaled physical-coordinate Lightcone.

Finally, Figure 12 shows example screen shots captured from our system at a variety of different scales: 10^{13} m, 10^{21} m, and 10^{25} m. The left column corresponds to comoving coordinates, while the right column corresponds to the $a(t)$-rescaled physical coordinates. In each of these figures, we place a corresponding Lightcone Clock on the right to illustrate the current s_{nav} context and the chosen cosmological coordinate system.

6 Conclusion

In this paper, we have presented a the design and implementation of a visualization system that addresses the problem of effective interactive navigation across huge scale ranges of space and time. This was accomplished with a combination of techniques, including log-scale treatments of time and space, a systematic approach to the representation of spacetime scalable data sets, the interaction graph representing our virtual Universe, and, finally, the introduction of the Lightcone Clock, which can be used to indicate the true age of astronomical data due to the finite speed of light as well as to provide an intuitive representation of the size of the Universe.

Acknowledgments

This research was supported by NASA grant number NAG5-8163. This research was also made possible in part by NSF infrastructure grant CDA 93-03189. We are grateful for the extensive help and participation of Eric A. Wernert and the generosity of the Indiana University Advanced Visualization Laboratory. We thank P.C. Frisch, S. Carroll, D. York, D. Eisenstein, and E. Kolb of the University of Chicago for their assistance with astrophysics and cosmology issues.

References

1. P. Astheimer and M.-L. Pöche. Level-of-detail generation and its application to virtual reality. In *Proceedings of the VRST '94 Conference*, pages 299–309, 1994.
2. Jeremey Bernstein. *An Introduction to Cosmology*. Prentice-Hall, 1998.
3. Kees Boeke. *Cosmic View: The Universe in Forty Jumps*. John Day, 1957.
4. Sean Carroll. The cosmological constant. *Living Reviews in Relativity*, 1999. http://www.livingreviews.org: a refereed electronic journal.
5. Sean Carroll. Cosmology for string theorists, 1999. http://pancake.uchicago.edu/ carroll/tasi99.ps.
6. Charles Eames and Ray Eames. Powers of Ten, 1977. 9 1/2 minute film, made for IBM.
7. A. J. Hanson and E. Wernert. Constrained 3D navigation with 2D controllers. In *Proceedings of Visualization '97*, pages 175–182. IEEE Computer Society Press, 1997.
8. A.J. Hanson, Chi-Wing Fu, and E.A. Wernert. Very large scale visualization methods for astrophysical data. In *Data Visualization 2000*, pages 115–124. Springer Verlag, May 29-31, 2000, Amsterdam, the Netherlands 2000.
9. Andrew J. Hanson and Philip C.W. Fu. Cosmic clock, 2000. Siggraph Video Review, vol. 134, scene 5.
10. L.E. Hitchner and M.W. McGreevy. Methods for user-based reduction of model complexity for virtual planetary exploration. *Proceedings of the SPIE – The Int. Soc. for Optical Eng.*, 1913:622–636, 1993.
11. David Layzer. *Constructing the Universe*. Scientific American Books, 1984.
12. P. Maciel and P. Shirley. Visual navigation of large environments using textured clusters. In *1995 Symposium on Interactive 3D Graphics*, pages 95–102, 1995.
13. Philip Morrison and Phylis Morrison. *Powers of Ten*. Scientific American Books, 1982.
14. Martin Reddy. *Perceptually Modulated Level of Detail for Virtual Environments*. Computer science, University of Edinburgh, Edinburgh, Scotland, 1997. PhD Thesis: CST-134-97.
15. Bayley Silleck. Cosmic voyage, 1996. 35 minute film, a presentation of the Smithsonian Institution's National Air and Space Museum and the Motorola Foundation.
16. E.A. Wernert and A.J. Hanson. A framework for assisted exploration with collaboration. In *Proceedings of Visualization '99*, pages 241–248. IEEE Computer Society Press, 1999.

Data Structures for Multiresolution Representation of Unstructured Meshes

Kenneth I. Joy[1], Justin Legakis[2], and Ron MacCracken[3]

[1] Center for Image Processing and Integrated Computing
Computer Science Department
University of California,
Davis, CA 95616 USA
[2] NVIDIA Corporation
2701 San Tomas Expressway
Santa Clara, CA 95050
[3] Centric Software, Inc.
50 Las Colinas Lane
San Jose, CA 95119

Abstract. A major impediment to the implementation of visualization algorithms on very-large unstructured scientific data sets is the suitable internal representation of the data. Not only must we represent the data elements themselves, but we must also represent the connectivity or topological relationships between the data. We present three data structures for unstructured meshes that are designed to fully represent the topological connectivity in the mesh, but also minimize the data storage requirements in representing the mesh. The key idea is to represent the topology of the mesh by the use of a single data item – the lath – which can be used to encapsulate the topological relationships within the mesh. We present and analyze algorithms that query the spatial relations and properties of these data structures, and analyze the data structures of the dual mesh induced by each.

1 Introduction

With the rapid increase in the power of workstations, the development of state-of-the-art imaging systems, and the increasing sophistication of computational simulations, enormous quantities of information relevant to a particular problem area are now being produced. The primary difficulty faced in many decision and planning situations is analyzing these massive data sets and extracting the relevant information that provides solutions to the problems at hand. These *scientific data sets* are usually multi-valued, meaning that multiple dependent variables – *e.g.*, velocity, pressure, temperature, salinity, sound speed, chemical or nuclear contamination, or even entire "matrices" (tensors) – are associated with each data point.

The topological structure underlying the data set may belong to various topological types: it may be *structured*, where the faces or cells are topologically equivalent to a square or a cube, respectively; or it may be *unstructured*, with a face arrangement consisting of triangular or quadrilateral element, cell arrangement consisting of hexahedral or tetrahedral elements, or combinations of various types [19]. The underlying data structure for structured data sets is completely defined by the indices of the data elements, *e.g.*, each node $\mathbf{p}_{i,j}$ in a mesh is connected with the nodes $\mathbf{p}_{i-1,j}$, $\mathbf{p}_{i+1,j}$, $\mathbf{p}_{i,j-1}$, and $\mathbf{p}_{i,j+1}$. This connectivity simplifies the data representation dramatically, and allows efficient algorithms for moving about the data. With the unstructured mesh however, no such elegant data structure exists, as there is no rigid connectivity rules at the vertices. In this paper, we discuss data structures that represent two-dimensional unstructured meshes.

The representation of these very large unstructured data sets impacts a number of problems in computer graphics and visualization:

- The unstructured meshes generated by subdivision algorithms [5,4,8,7] require a structure that can represent faces with an arbitrary number of edges. The subdivision steps in these algorithms require additional vertices to be inserted into the mesh, and a new connectivity structure to be generated.
- The polygonal representations of implicit surfaces [13] requires the storage of large, unstructured meshes and requires methods to easily traverse these structures.
- The mesh-simplification routines for very-large data sets [14,10,12,22,17] require a data structure that can be easily traversed and locally modified as data elements are collapsed.
- Techniques based on Delaunay triangulations [6,18,3] or Voronoi diagrams require a general data structure that can represent an unstructured mesh. The duality between the Delaunay Triangulation and the Voronoi diagram requires a mesh structure that can easily product its dual mesh. In these algorithms the grids are frequently generated or modified by inserting points and performing local mesh modifications.

To service this variety of applications, a simple data structure is necessary that defines the topology of the mesh, is compact, and facilitates the implementation of efficient algorithms that operate on the data.

We present a method of representing data structures on these unstructured meshes that is based upon the used of a single data type – the lath[1]. In a lath representation, this single data type will be used to represent each of the vertices, edges, and faces of the mesh – and give direct links to the data held at each vertex. Traversals, or accesses of the elements of the mesh, can be accomplished easily via a set basic set of traversal operators that take lath elements as input and output, and a set of queries that take lath elements and return sets of vertices, edges or faces in the neighborhood of the lath. These queries can be used to develop in-place algorithms on the mesh.

---

[1] Laths are the thin strips of material fastened together to form the network of elements forming a mesh or grid.

In Section 2 we discuss data structures related to the *winged-edge representation* that have been extensively used in non-manifold geometric modeling. These edge-based data structures are the ones that have inspired this work. In Section 3, we present three data structures for unstructured surface meshes: the *split-edge structure*, the *half-edge structure*, and the *corner structure*. We define five basic traversal operators on the elements of data structure, and analyze the differences between these operators in the three representations. We also examine the nine fundamental data-access operations that allow queries on the data structure and analyze the algorithms that implement them. Implementation of the three structures is straightforward, and is discussed in Section 4. Conclusions and future work are addressed in Section 5.

2 Related work

Boundary representations have become the fundamental representation technique in geometric modeling. In these methods, surfaces, edges and vertices of solid objects are represented explicitly, and the topological information about geometrical relationships between these basic elements are represented in a data structure. Various data structures have been presented in [1,2,20] that both efficiently store these relationships and allow modeling operations to be performed on the solids.

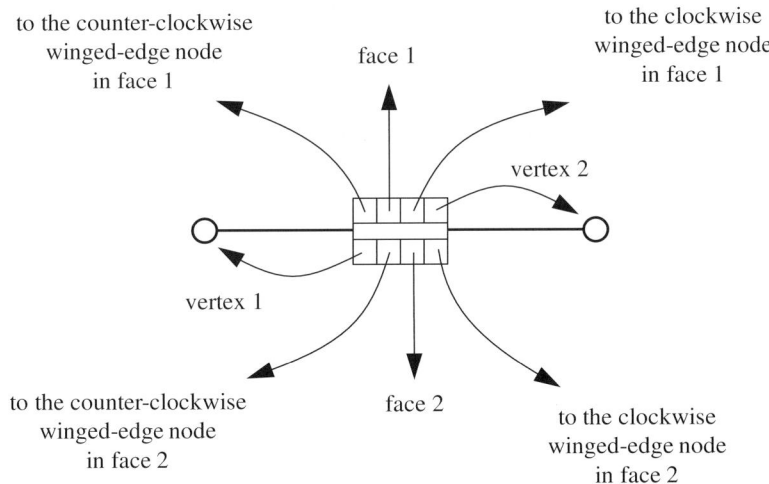

Fig. 1. The winged-edge node: This node has two groups of links that represent the uses of both "sides" of the edge. It contains a link to the vertex, as well as links to both clockwise and counter-clockwise edges and the data structures for the faces adjacent to the edge.

Ideally, the boundary models are manifolds: any neighborhood on the surface of the model is homeomorphic to an open disk in the plane. Unfortunately, the results of many operations commonly used in geometric modeling (specifically the Boolean set operations) produce non-manifold models – *e.g.*, structures containing a single wire edge in space, structures with two surfaces touching at a single point, or two distinct structures sharing a face. The data structures representing these non-manifold boundary models have received considerable study over the past two decades.

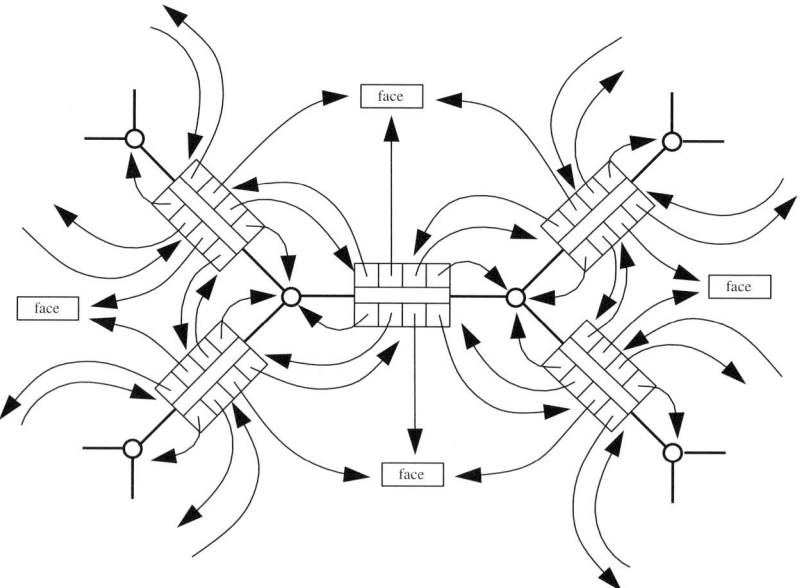

Fig. 2. The winged-edge node when used in a mesh.

Most edge-based data representation schemes are based upon the winged-edge representation of Baumgart [2], who generated a data structure for models that represent the bounding surface of an arbitrary polyhedral structure. Baumgart based his data structure on a single edge record which includes both directional information and information about the faces containing the edge. The winged-edge element, depicted in its most general form, is shown in Figure 1. Each winged-edge node contains links to the two vertices that bound the edge, the two faces that bound the edge, and the winged-edge nodes in the clockwise and counter-clockwise direction around the two adjacent faces, forming a doubly-linked list. The face link is used to access a face data structure, and the vertex link points to a separate vertex data structure. Figure 2 illustrates the use of this element in a mesh.

The primary problem with the classic winged-edge structure is the bundling of the two roles of the edges within one record, and most variations of edge-

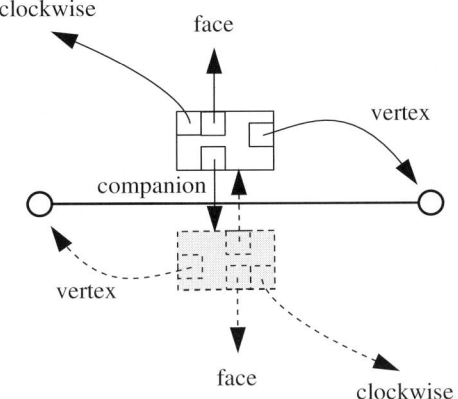

Fig. 3. The split-edge representation of an edge in a mesh.

based data structure are based upon the splitting of the winged-edge element – decoupling the two uses of the edge into two (or more) separate records (see [1]). The first of these, *the split-edge representation*, is commonly credited to Eastman [9]. This representation is achieved by splitting each winged edge into two halves, one for each of the two adjacent faces (see Figure 3). Connectivity is maintained by adding an explicit pointer in each edge record that references the opposite "half". Figure 4 illustrates the use of the split-edge nodes in a mesh.

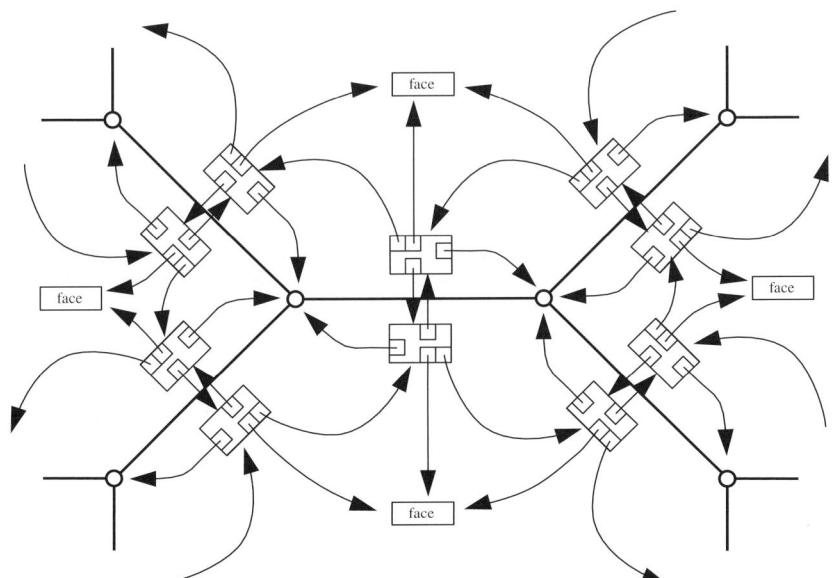

Fig. 4. The split-edge representation of an edge in a mesh.

Other variations on the winged-edge data structure also attempt to give a unique representation to the multiple uses of each edge. The hybrid-edge representation of Kalay [15] and Mantyla [16][2] attempts to use the best characteristics of the winged-edge structure and the split-edge structure. It breaks the representation of an edge into a single edge node that represents the edge itself, and two directional nodes, called segments, that specify the directional uses of the node.

By examining Figure 4, a symmetric representation of the structure can be visualized by noting the seemingly dual role played by faces and vertices. In this case two vertices bound an edge, as well as two faces, and the edge-based structure can be constructed so that the representation of faces and vertices is remarkably similar. This symmetry was first observed by Woo [21].

In mesh-based applications used in multiresolution analysis and visualization, the meshes represent manifolds, and the data is held in the vertices of the mesh. Here the primary operations focus on movement about the mesh and queries that inquire about information in the neighborhood of a vertex or face. These operations are required for the in-place calculations necessary for subdivision algorithms, mesh-reduction algorithms, or mesh-enhancement methods. Thus, we define a *mesh* $\mathcal{M}$ to be a set of vertices $\{\mathbf{p}_0, \mathbf{p}_1, ..., \mathbf{p}_n\}$ and an associated simplicial complex which specifies the connectivity of the vertices. Each edge is defined by two vertices that are connected in the simplicial complex. Each face is defined by a minimal connected loop of vertices. We assume that the mesh is well-connected (*i.e.*, no vertex lies on an edge not containing that vertex), all faces are closed in a mesh, and no two adjoining faces of the mesh intersect.

3 Lath-based data structures

In our representations, a data structure storing a mesh will be based upon a single data type called a *lath*. Each lath will be identified with exactly one vertex, one edge and one face of the mesh and each face-edge pair, face-vertex pair, or edge-vertex pair in the mesh can be associated with a single lath. Thus, each edge of the mesh will have two laths associated with it: one for each face-edge pair (two faces per edge), or for each edge-vertex pair (two vertices per edge). There is no need for edge-based elements or face-based elements in the data structure, as each of these can be specified by specifying a single lath element. Lists of laths can represent lists of edges, lists of faces or lists of vertices.

We assume that vertices contain the geometric information and laths contain the topological, or connectivity, information in the mesh. The lath ele-

---

[2] We note that Mantyla called his representation a *half-edge* data structure – a term that we will use in Section 3. His representation is very close to the hybrid-edge representation of Kalay.

ments can be connected in various ways, and we present three examples of lath-based data structures in the following sections.

3.1 The split-edge representation

In the split-edge data structure the topology of the mesh is carried by a lath element that is similar to the split-edge element given in Section 2.[3] Here

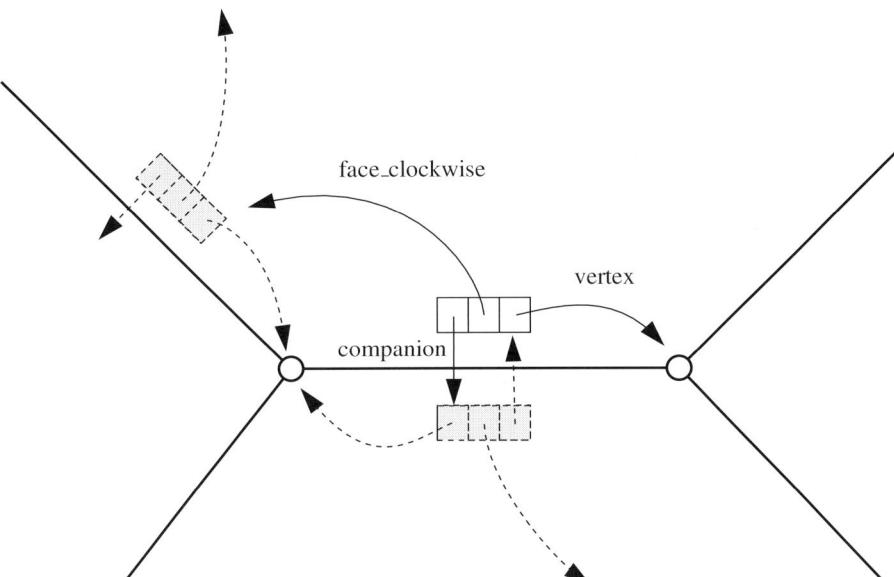

Fig. 5. The lath element for the split-edge representation of a mesh. Three links are explicit in the lath element: a link to the vertex information, a link to the lath that forms the edge companion of the element, and a link to the lath is the next lath in a clockwise traversal of laths in the same face.

a lath element L contains three separate links: First, a link to the vertex information associated with L; second, a link (the *companion* link) to a lath that represents the same edge as L, but the opposite vertex; and third, a link (the *face_clockwise* link) to the lath that follows L in a clockwise traversal of the face that L represents. The split-edge lath can be pictured as in Figure 5 and its use in a mesh is illustrated in Figure 6. Every edge-face pair, edge-vertex pair, and face-vertex pair in the mesh is associated with exactly one lath, and each lath is identified with exactly one vertex, one edge, and one face.

From Figure 6, we can see that the lath structure sets up a contiguous structure in the mesh, and induces two basic loops around the mesh elements:

---

[3] Here we have eliminated the face links for clarity.

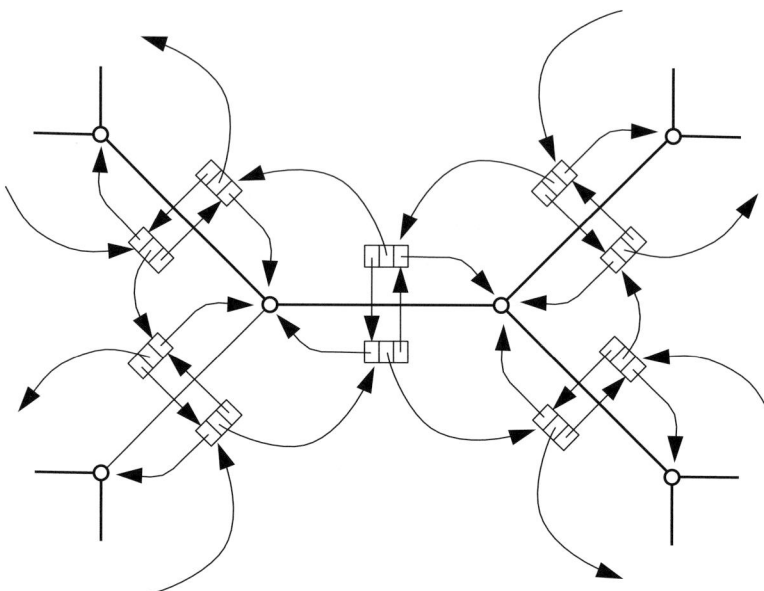

Fig. 6. The split-edge lath elements in a mesh. We note the basic loops induced by the split-edge laths: One in a clockwise direction around the faces of the mesh, and one in a counter-clockwise direction about the vertex.

one in a clockwise direction about the face of the mesh (Figure 7), and one in a counter-clockwise direction about the vertex (Figure 8).

We can traverse this structure in several ways, and for uniformity define the following operators on laths: Given a lath L,

- ec(L) returns the edge companion of L – the lath element that represents the same edge as L, but the opposite vertex and face.
- cf(L) returns the lath that follows L in a clockwise traversal of the face that L represents.
- ccf(L) returns the lath that follows L in a counter-clockwise traversal of the face that L represents.
- cv(L) returns the lath that follows L in a clockwise traversal of laths about the vertex that L represents.
- ccv(L) returns the lath that follows L in a counter-clockwise traversal of laths about the vertex that L represents.

In this case, the operators ec and cf are embodied in the split-edge data structure. The function ccv(L) is quickly identified as returning the lath defined by cf(ec(L)). The function ccf(L) can be obtained in two ways:

(1) by traversing laths representing the face (via cf), until reaching the lath L' where cf(L') = L, or

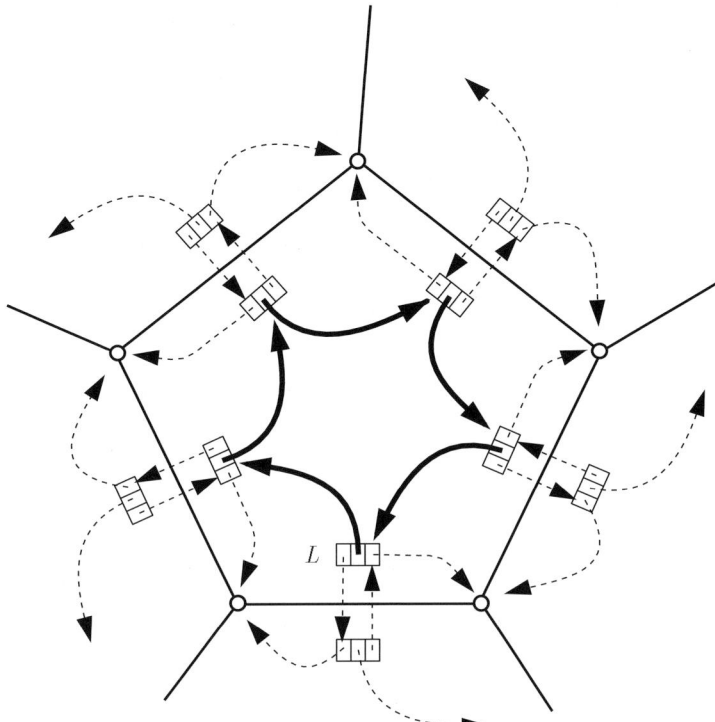

Fig. 7. A face loop in the split-edge structure. Starting with the lath L and following *face_clockwise* links, a sequence of laths is generated that all reference the same face.

(2) by traversing laths surrounding the vertex (using ccv) until we reach the lath L' with $\mathsf{ccv}(L') = L$, then $\mathsf{ccf}(L) = \mathsf{ec}(L')$.

In the first case, this algorithm is linear in the number of edges in the face represented by L. In the second case, this algorithm is linear in the number of edges radiating from the vertex represented by L. The operators depend only on the local complexity in the mesh, not the size of the entire mesh.

The function cv can also be defined in two ways:

(1) as $\mathsf{cv}(L) = \mathsf{ec}(\mathsf{ccf}(L))$,
(2) or by successively traversing laths surrounding the vertex (using ccv). The lath L' is returned by the algorithm, where L' has the property that $\mathsf{ccv}(L') = L$.

These operators allow us to move about the mesh by using the lath elements.

Boundary Considerations Boundaries are easily represented in this structure by storing `Null` values in the companion links (see Figure 9). This implies

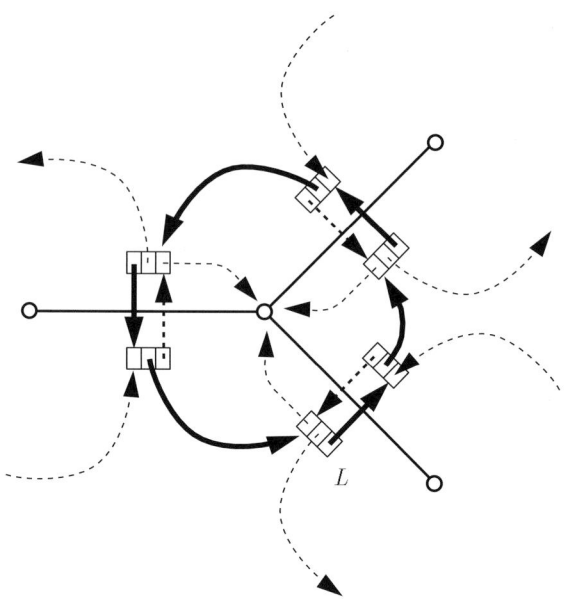

Fig. 8. A vertex loop in the split-edge structure. Starting with the lath L, and following *companion* links and *face_clockwise* links, a sequence of laths is generated that all reference the same vertex.

that the edge-companion function ec(L) may not return meaningful information, and alternate steps must be substituted in the traversal operators. Since cf is always defined in this structure, all operators can be implemented without the ec operation if necessary.

Data access primitives in the split-edge data structure Given a lath representing one of the three elements of the mesh (vertex, edge, or face), the data structure can be queried for all of the neighboring elements of a given type. We can identify 9 separate queries, each returning a list of laths that represent the desired elements in the mesh:

- $\mathcal{Q}_{VV}$: Given a vertex V, find all vertices that share an edge with V.[4]
- $\mathcal{Q}_{VE}$: Given a vertex V, find all edges that radiate from V.
- $\mathcal{Q}_{VF}$: Given a vertex V, find all faces of the mesh that contain this vertex.
- $\mathcal{Q}_{EV}$: Given an edge E, find its two vertices.
- $\mathcal{Q}_{EE}$: Given an edge E, find all edges that share a vertex with E.
- $\mathcal{Q}_{EF}$: Given an edge E, find all faces of the mesh that contain E.
- $\mathcal{Q}_{FV}$: Given a face F, find all its vertices.
- $\mathcal{Q}_{FE}$: Given a face F, find all its edges.

---

[4] The result of this query is commonly called the *one-neighborhood* of the vertex.

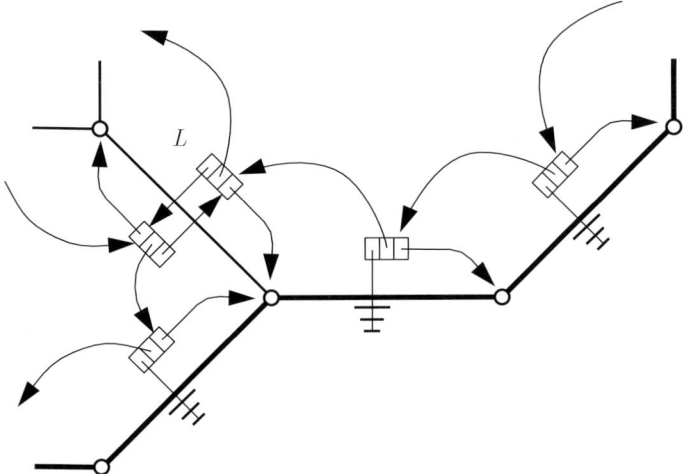

Fig. 9. The boundary of a mesh with an associated split-edge data structure. Here the companion links of the split-edge elements are Null on the boundary.

- $\mathcal{Q}_{FF}$: Given a face F, find all faces of the mesh that share an edge or a vertex with F.[5]

The four basic queries are $\mathcal{Q}_{EF}$, $\mathcal{Q}_{EV}$, $\mathcal{Q}_{FE}$, and $\mathcal{Q}_{VE}$, as they either return information stored explicitly in laths, or they return information stored in the two basic loops introduced in the split-edge structure.

- $\mathcal{Q}_{EF}$– Given a lath L representing an edge-face pair in the mesh, laths representing the two faces bounding the edge are given by L and $\mathsf{ec}(L)$. If the edge represented by L is on the boundary, then this query returns only L, since $\mathsf{ec}(L)$ does not exist.
- $\mathcal{Q}_{EV}$– Given a lath L representing an edge of the mesh, the two laths that represent the vertices that bound the edge are given by L and $\mathsf{cf}(L)$.
- $\mathcal{Q}_{FE}$– Given a lath L representing a face-edge pair in the mesh, laths representing the edges of the face are given by successively applying the cf operator – following the clockwise loop about the face.
- $\mathcal{Q}_{VE}$– Given a lath L representing an edge-vertex pair, laths representing edges that radiate from the vertex are obtained by successively applying the ccv operator – following the counter-clockwise loop about the vertex.

The remaining queries are implemented in terms of these basic queries.

- $\mathcal{Q}_{FV}$– Given a lath L representing a face, the laths representing the vertices of the face are the same as those of $\mathcal{Q}_{FE}$.
- $\mathcal{Q}_{VV}$– Given an lath L, laths representing the vertices of the edges that radiate from the vertex represented by L are given by applying $\mathsf{cf}(L)$ for each lath L in $\mathcal{Q}_{VE}$.

---

[5] The result of this query is commonly called the *stencil* of the face (see [11]).

- $\mathcal{Q}_{\text{VF}}$ is identical with $\mathcal{Q}_{\text{VE}}$, as each lath identified by $\mathcal{Q}_{\text{VE}}$ belongs to a unique face.
- $\mathcal{Q}_{\text{EE}}$– Given a lath L representing an edge E, $\mathcal{Q}_{\text{EE}}$ produces a set of laths that correspond to the edges that radiate from the two vertices of E. This is implemented by taking the union of the results of $\mathcal{Q}_{\text{VE}}$ for both L and $\mathsf{cf}(L)$. To insure that two laths do not exist in the output for the edge represented by L, either L or $\mathsf{ec}(L)$ (which is generated by $\mathcal{Q}_{\text{VE}}(\mathsf{cf}(L))$) is removed before output.
- $\mathcal{Q}_{\text{FF}}$– Given a lath L representing a face-edge pair, all such faces are obtained by applying $\mathcal{Q}_{\text{VE}}$ to each lath returned by $\mathcal{Q}_{\text{FE}}$. Unfortunately, this produces multiple laths in the output for each face. To obtain a single lath for each face in the output, we use a marking strategy that operates as follows:
  - Mark each lath encountered in $\mathcal{Q}_{\text{FE}}$.
  - For each lath L encountered by $\mathcal{Q}_{\text{VE}}$, if L is unmarked, insert it in the output list. Then mark all laths corresponding to edges of the face that L represents – *i.e.*, all laths in $\mathcal{Q}_{\text{FE}}(L)$.

  The output list contains laths returned by this query, one lath per face.

It is possible that $\mathcal{Q}_{\text{VE}}$ (and $\mathcal{Q}_{\text{VF}}$, which is identical in this structure) fails on the boundary. For example, in Figure 9, $\mathcal{Q}_{\text{VE}}(L)$ returns two laths before encountering a Null link on the boundary. To retrieve all laths that represent edges that radiate from the vertex, the query must use cv to traverse around the vertex in the opposite direction.

3.2 The half-edge representation

In the half-edge data structure the topology of the mesh is carried by a lath element that is defined as in Figure 10. Here a lath element L contains three separate links: First, a link to the vertex information associated with L; second, a link (the *companion* link) to a lath that represents the same edge as L, but the opposite vertex; and third, a link (the *vertex_clockwise* link) to the lath that is the next lath in a clockwise vertex traversal of the vertex that L represents.

From Figure 11, we can see that the lath structure sets up a contiguous structure in the mesh and again induces two basic loops around the mesh elements: one in a counter-clockwise direction about the face (Figure 12) and one in a clockwise direction about a vertex (Figure 13). We associate a lath L with the face traversed by this counter-clockwise loop, and with the edge bounded by the vertices given by the vertex pointers of L and L's companion. In this way, every edge-face pair, edge-vertex pair and face-vertex pair in the mesh is associated with exactly one lath, and each lath is identified with exactly one vertex, one edge, and one face.

We can traverse this structure by setting up the same operators as in the split-edge representation. For a given lath L,

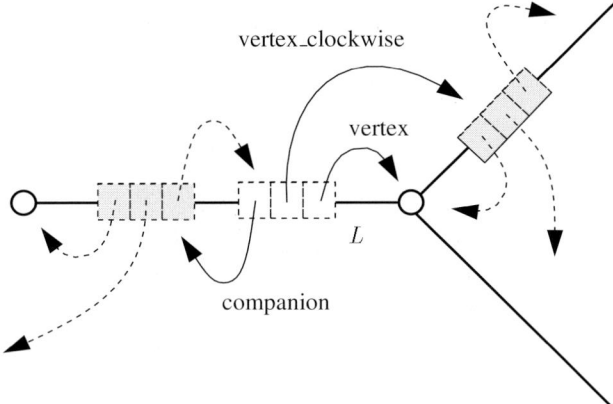

Fig. 10. The lath element for a half-edge representation of a mesh. Three links are explicit in the lath element: a link to the vertex information, a link to the lath that forms the edge companion of the element, and a link to the lath that is the next lath in a clockwise traversal about the vertex defined by L.

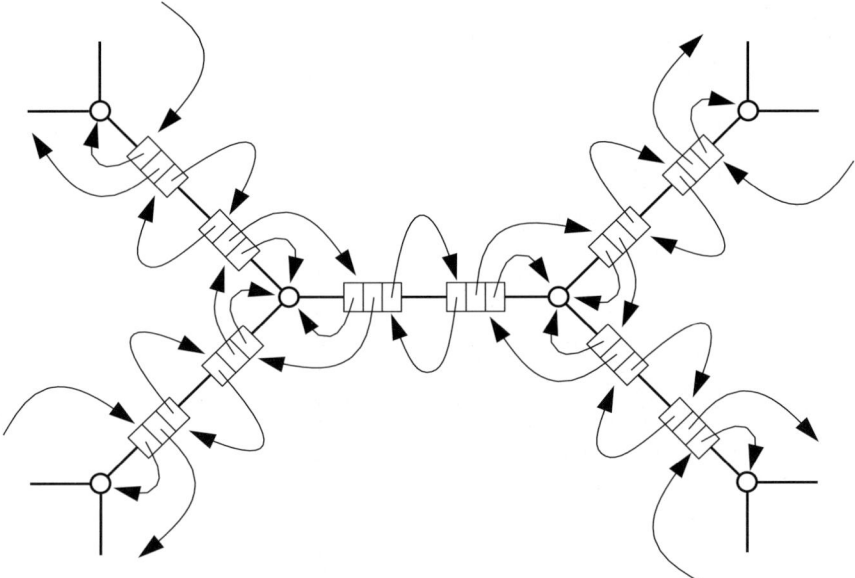

Fig. 11. The half-edge lath elements in a mesh. We note the basic loops induced by the split-edge laths: One in a counter-clockwise direction around the faces of the mesh, and one in a clockwise direction about the vertex.

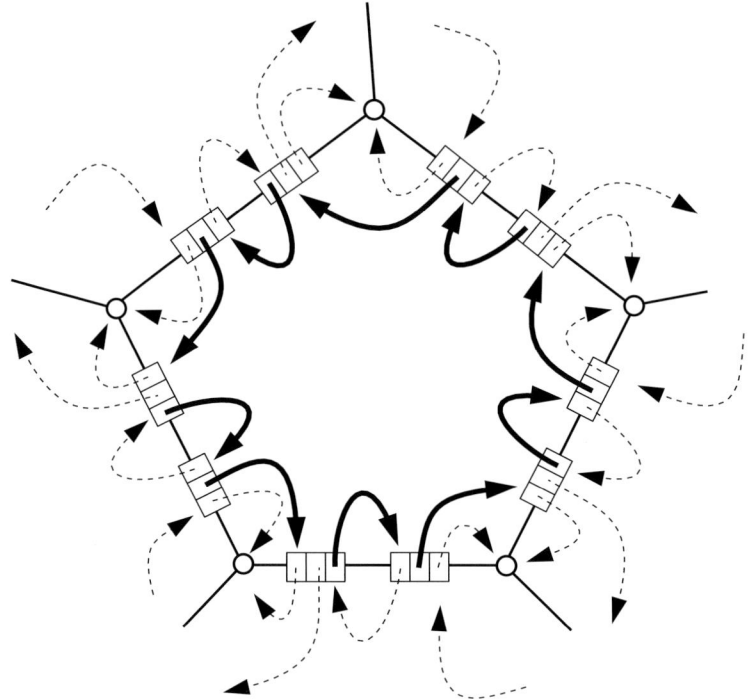

Fig. 12. A face loop in the half-edge structure. Starting with the lath L, and following *vertex_clockwise* links and *companion* links, a sequence of laths is generated that all reference the same face.

- ec(L) returns the edge companion of L – this information is contained in the links of the lath element, and returns the lath that represents the same edge, but opposite vertex and adjacent face that bound the edge.
- cv(L) returns the lath that follows L in a clockwise traversal of laths about the vertex that L represents. This information is contained in the lath through the *vertex_clockwise* link.
- ccf(L) returns the lath that follows L in a counter-clockwise traversal of the face that L represents. It is implemented as ccf(L) = ec(cv(L))
- cf(L) returns the lath that follows L in a clockwise traversal of the face that L represents. This can be obtained in two ways: first by traversing (through ccf) the edges of the face in a counter-clockwise manner until reaching the lath L' that has the property ccf(L') = L; or secondly, by successively applying cv to ec(L) until reaching L'.
- ccv(L) returns the lath that follows L in a counter-clockwise traversal of laths about the vertex that L represents. This can be obtained in two ways: First by traversing (via cv) the clockwise link about the vertex until obtaining a lath L' such that cv(L') = L; or alternatively, as cf(ec(L)).

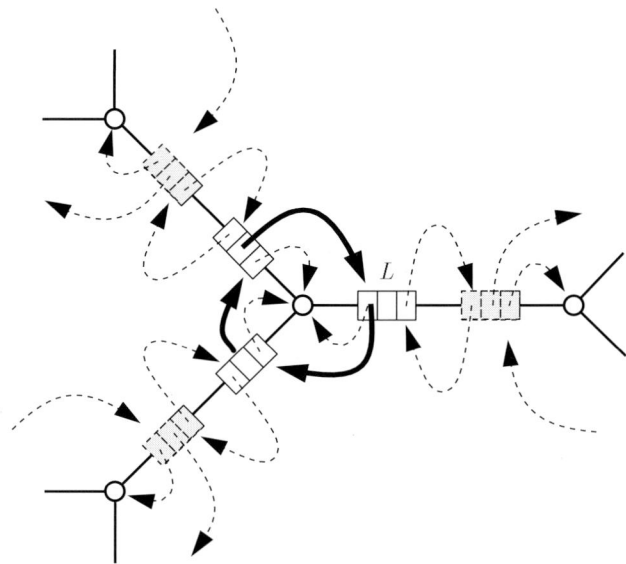

Fig. 13. A vertex loop in the half-edge structure. Starting with the lath L, and following *vertex_clockwise* links, a sequence of laths is generated that all reference the same vertex.

ec, ccf and cv can be implemented in constant time. cf and ccv are linear in the number of edges belonging to a face, or the number of edges radiating from a vertex, depending on the implementation.

Boundary Considerations Boundaries are represented in the half-edge structure by storing Null values in the *vertex_clockwise* links (see Figure 14). In this case, the cv operator may not be defined, and steps must be taken to use alternate operators to move about the mesh in the area of the boundary. However, since there are always two laths that exist on each edge, the ccf operator and the ec operator are always defined.

Data access primitives in the half-edge data structure Given a lath representing one of the three elements of the mesh (vertex, edge, or face), the half-edge data structure can be queried for all of the neighboring elements of a given type.

The four basic queries are Q_{EF}, Q_{EV}, Q_{FE}, and Q_{VE}, as they either return information stored explicitly in laths, or they return information stored in the two basic loops introduced by the data structure. The queries are basically the same here, except that we use ccf to traverse the edges of the face, and cv to traverse the edges around the vertex. The boundaries are handled differently as the ec and ccf operators are always defined, but the cv operator may not be.

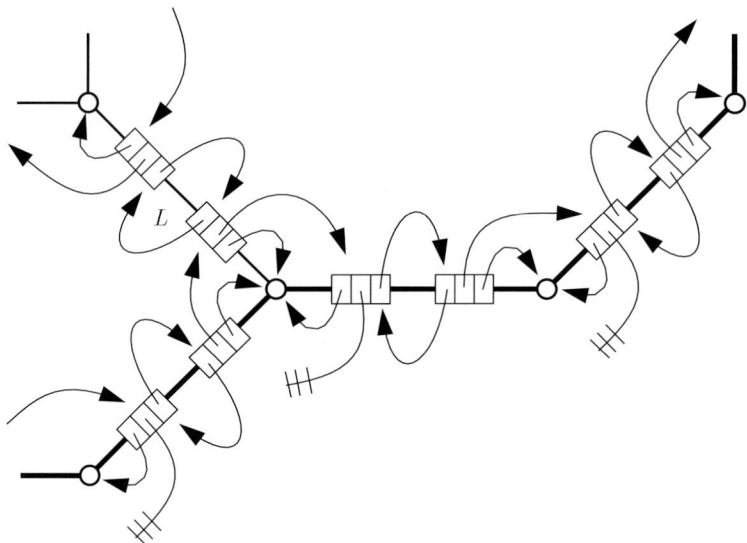

Fig. 14. The boundary of a mesh with an associated half-edge data structure. Here the *vertex_clockwise* links of the half-edge elements are `Null` on the boundary.

- $\mathcal{Q}_{\text{EF}}$– Given a lath L representing an edge in the mesh, laths representing the two faces bounding the edge are given by L and $\text{ec}(L)$. In this case, there are always two laths that represent an edge.
- $\mathcal{Q}_{\text{EV}}$– Given a lath L representing an edge of the mesh, the two laths that represent the vertices that bound the edge are given by L and $\text{ec}(L)$.
- $\mathcal{Q}_{\text{FE}}$– Given a lath L representing a face-edge pair in the mesh, laths representing the edges of the face are found by successively using the ccf operator – following the counter-clockwise loop about the face.
- $\mathcal{Q}_{\text{VE}}$– Given a lath L representing an edge-vertex pair, the laths representing edges that radiate from the vertex are obtained by successively using the cv operator – following the clockwise loop about the vertex.

The remaining queries are implemented in terms of these basic queries.

- $\mathcal{Q}_{\text{FV}}$– Given a lath L representing a face, the laths representing the vertices of the face are the same as those of $\mathcal{Q}_{\text{FE}}$.
- $\mathcal{Q}_{\text{VV}}$– Given an lath L, we can obtain the required laths by taking $\text{ec}(L)$ for each lath L in $\mathcal{Q}_{\text{VE}}$.
- $\mathcal{Q}_{\text{VF}}$is identical with $\mathcal{Q}_{\text{VE}}$, as each lath identified by $\mathcal{Q}_{\text{VE}}$belongs to a unique face.
- $\mathcal{Q}_{\text{EE}}$– Given a lath L representing an edge E, $\mathcal{Q}_{\text{EE}}$produces a set of laths that correspond to the edges that radiate from the two vertices of E. This is implemented by taking the union of the results of $\mathcal{Q}_{\text{VE}}$for both L and its companion $\text{ec}(L)$. Again, we insert only one of L and $\text{ec}(L)$ in order to have only one lath per edge in the output.

- $\mathcal{Q}_{FF}$ is defined exactly as in the split-edge representation with the above modifications to $\mathcal{Q}_{EV}$ and $\mathcal{Q}_{FE}$.

It is possible that $\mathcal{Q}_{VE}$ (and $\mathcal{Q}_{VF}$, which is identical in this structure) encounters Null links on the boundary. For example, in Figure 9, $\mathcal{Q}_{VE}(L)$ returns two laths before encountering a Null link on the boundary. To retrieve all laths that represent edges that radiate from the vertex, the query must use ccv to traverse around the vertex in the opposite direction until another Null link is encountered. Unlike the split-edge structure, one always encounters laths that represent an edge-vertex pair.

3.3 Duality

The dual of a mesh is constructed by swapping the roles of the faces and vertices in the mesh. If, in each lath of the data structure, we replace the vertex link by a link to a similar face structure, we obtain a data structure for the dual mesh. If the mesh is represented by a half-edge structure (see Figure 15), the counter-clockwise vertex loop is transformed into a clockwise loop on the face of each element of the dual, and the counter-clockwise face loop of the half-edge structure is transformed into a clockwise loop about each vertex in the dual. That is, the half-edge structure induces a split-edge data structure on the dual mesh. Conversely, if the original mesh is represented with a split edge structure, a half-edge data structure is induced on the dual mesh (see Figure 16).

3.4 The Corner Data Structure

In the corner data structure the topology of the mesh is carried by a lath element that is defined as in Figure 17. Here an element L contains three separate links: First, a link to the vertex information associated with L; second, a link (the *face_clockwise* link) to a lath that represents the next lath in a clockwise traversal of laths of the face L represents; and third, a link (the *vertex_clockwise* link) to the lath that is the next lath in a clockwise traversal of the vertex that L represents. Here the lath elements are easily identified with each vertex-face pair. The edge identified with each lath L is that edge bounded by the vertex of L and the vertex of the lath identified by the *face_clockwise* link.

From Figure 18, we can see that the lath structure sets up a contiguous structure in the mesh. Again, there are two basic loops that can be identified, one which traverses the clockwise laths about a face (Figure 19), and one which traverses the clockwise laths about a vertex (Figure 20). We can traverse this structure by setting up the same operators as in the split-edge and half-edge representations. Given a lath L,

- cf(L) returns the lath that follows L in a clockwise traversal of the face that L represents. This information is given directly in the *face_clockwise* link of L.

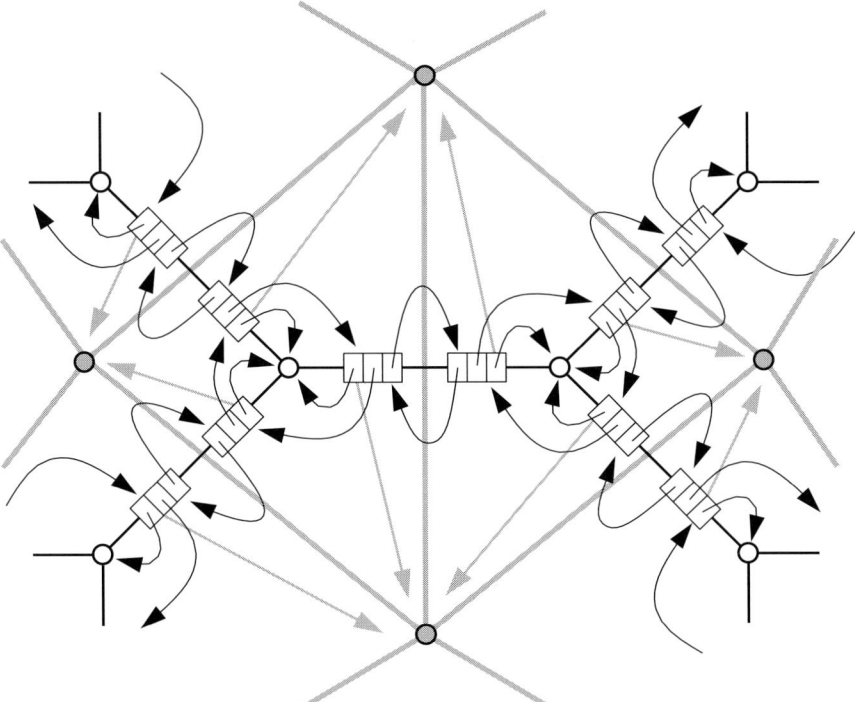

Fig. 15. Representing the dual of a mesh with a half-edge data structure. The original mesh and data structure are shown in black with the dual mesh shown in gray. The gray vertex links are the links to the face nodes in the dual structure. Note the counter-clockwise loops about the face, and the clockwise loops about the vertex in the dual mesh. (See Color Plate 18 on page 360.)

- cv(L) returns the lath that lath follows L in a clockwise traversal of laths about the vertex that L represents. This information is given directly in the lath through the *vertex_clockwise* link of L.
- ec(L) returns the edge companion of L – In the corner representation, this information is not given directly in the lath links, but can be calculated by cv(cf(L)). If L represents a boundary edge of the mesh, the edge companion does not exist.
- ccf(L) returns the lath that follows L in a counter-clockwise traversal of the face that L represents. It is implemented in two ways: as ccf(L) = cv(ec(L)), or by traversing the face clockwise (via cf) until reaching the lath L' where cf(L') = L.
- ccv(L) returns the lath that follows L in a counter-clockwise traversal of laths about the vertex that L represents. This can also be obtained in two ways: as ccv(L) = ec(cf(L)), or by traversing about the vertex clockwise (via cv) until reaching the lath L' where cv(L') = L.

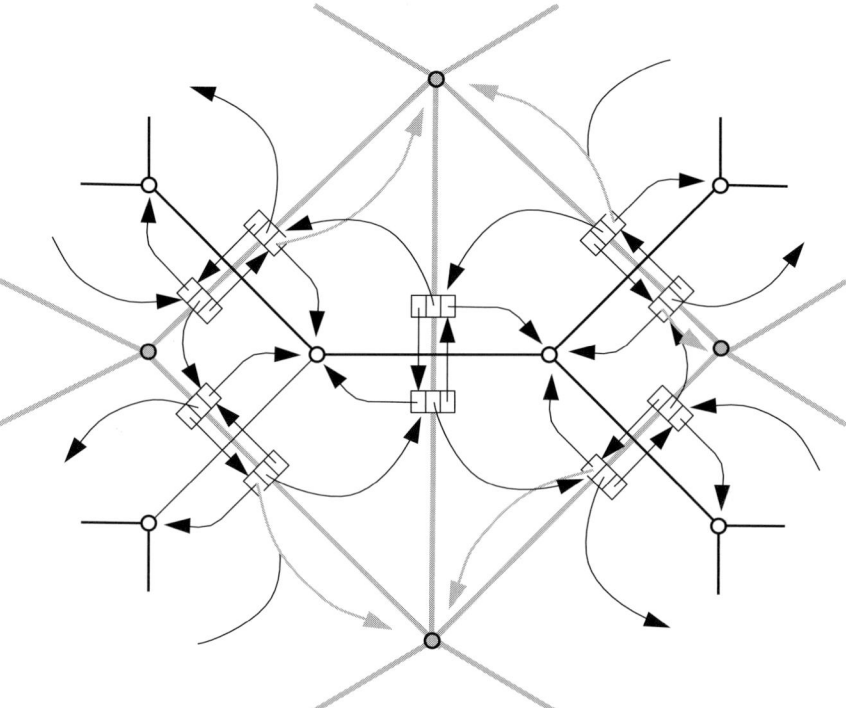

Fig. 16. Representing the dual of a mesh with a split-edge data structure. The original mesh and data structure are shown in black with the dual mesh shown in gray. The gray vertex links are the links to the face nodes in the dual structure. Note the counter-clockwise loops about the face and the clockwise loops about the vertex in the dual mesh. (See Color Plate 19 on page 360.)

Unlike the split-edge and half-edge operations, these operators can all be implemented in constant time except about the boundaries (see section 3.4 below). Along the boundary, cv is not available and in this case, ccf and ccv are linear in the number of edges in a face, or the number of edges radiating from a vertex, respectively.

Boundary Considerations Boundaries are represented in the corner structure by storing Null values in the *vertex_clockwise* links (see Figure 21). In this case, the cv operator may not be defined, and steps must be taken to use alternate operators to move about the mesh in the area of the boundary. However, the cf operator is always defined.

Data access primitives in the corner data structure The four basic queries are again $\mathcal{Q}_{\text{EF}}$, $\mathcal{Q}_{\text{EV}}$, $\mathcal{Q}_{\text{FE}}$, and $\mathcal{Q}_{\text{VE}}$, as they either return information stored explicitly in laths, or they return information stored in the two

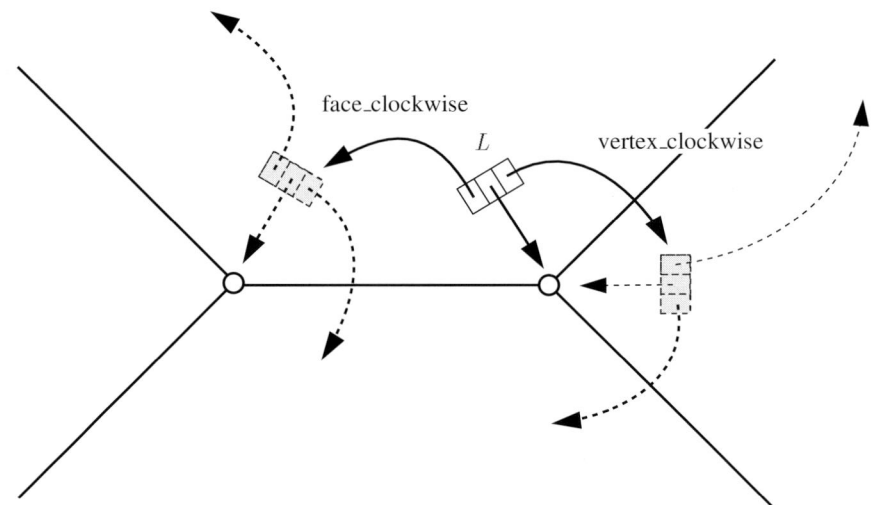

Fig. 17. The corner data element.

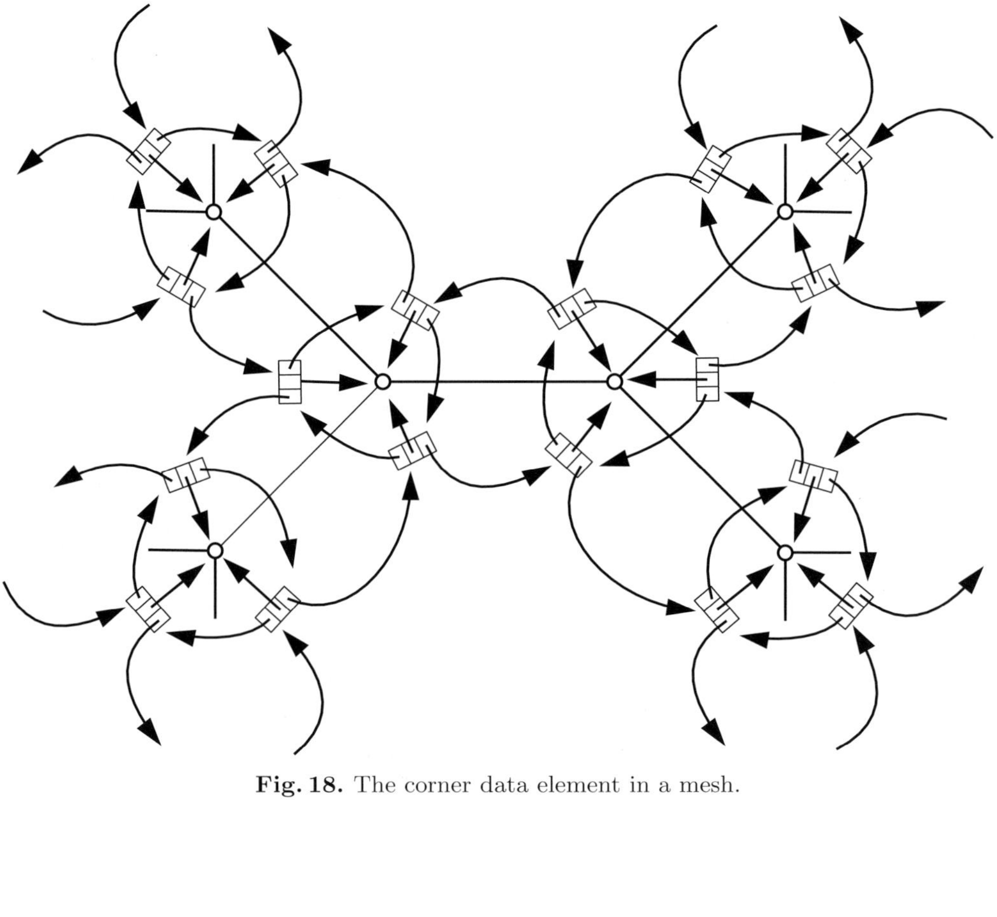

Fig. 18. The corner data element in a mesh.

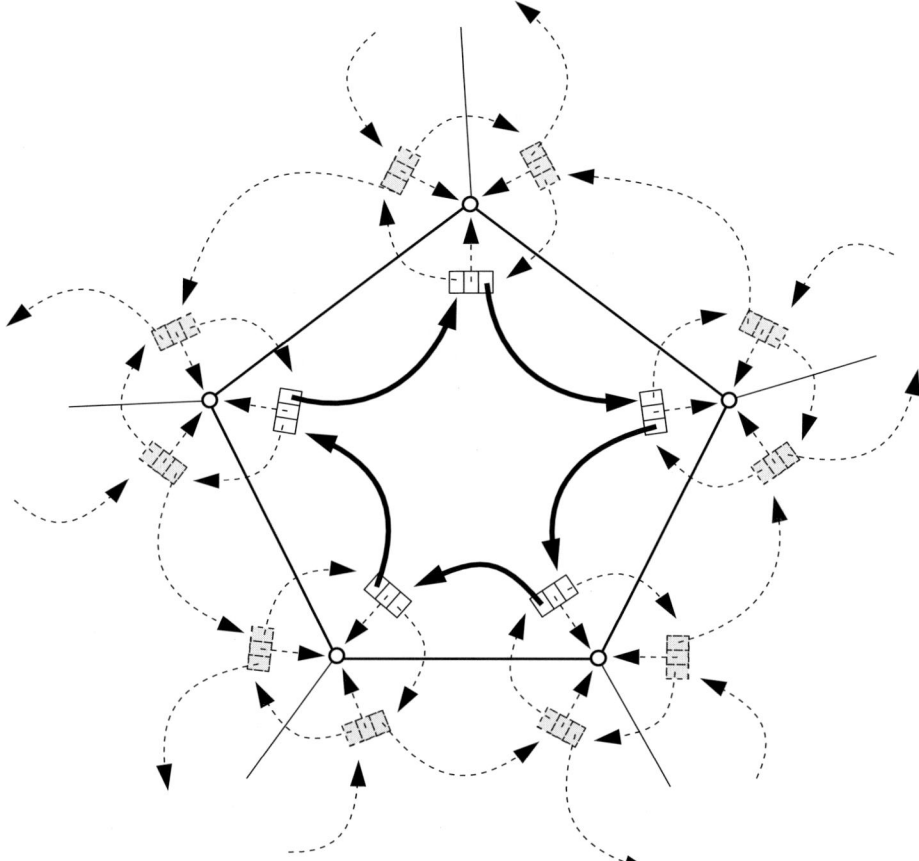

Fig. 19. A face loop in the corner representation. Starting with the lath L, and following *clockwise* links, a sequence of laths is generated that all reference the same face.

basic loops introduced by the data structure. The queries are basically the same here, except that we use cf to traverse the edges of the face, and cv to traverse the edges around the vertex. The boundaries are handled differently as only the cf operator is guaranteed to be defined, So, for a given lath L, we implement the queries as follows:

- $\mathcal{Q}_{\mathrm{EF}}$– Given a lath L representing an edge in the mesh, laths representing the two faces bounding the edge are given by L and $\mathsf{ec}(L)$. In the case the lath L represents a boundary edge, only L is returned, as $\mathsf{ec}(L)$ does not exist.
- $\mathcal{Q}_{\mathrm{EV}}$– Given a lath L representing an edge of the mesh, the two laths that represent the vertices that bound the edge are given by L and $\mathsf{cf}(L)$.

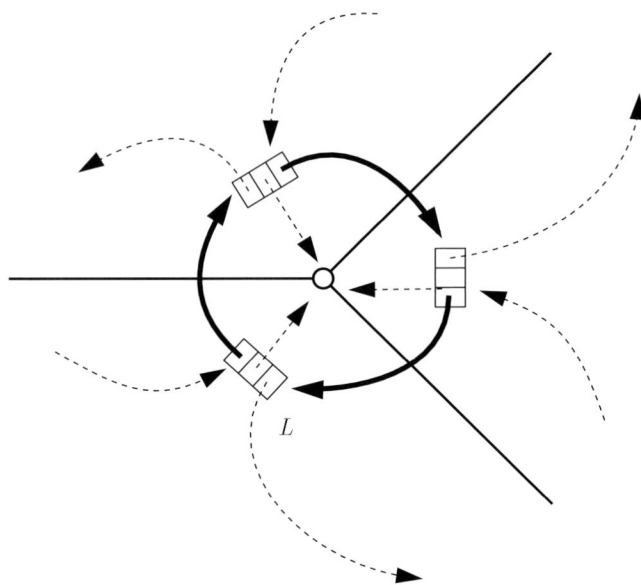

Fig. 20. A vertex loop in the corner representation. Starting with the lath L, and following *vertex_clockwise* links, a sequence of laths is generated that all reference the same vertex.

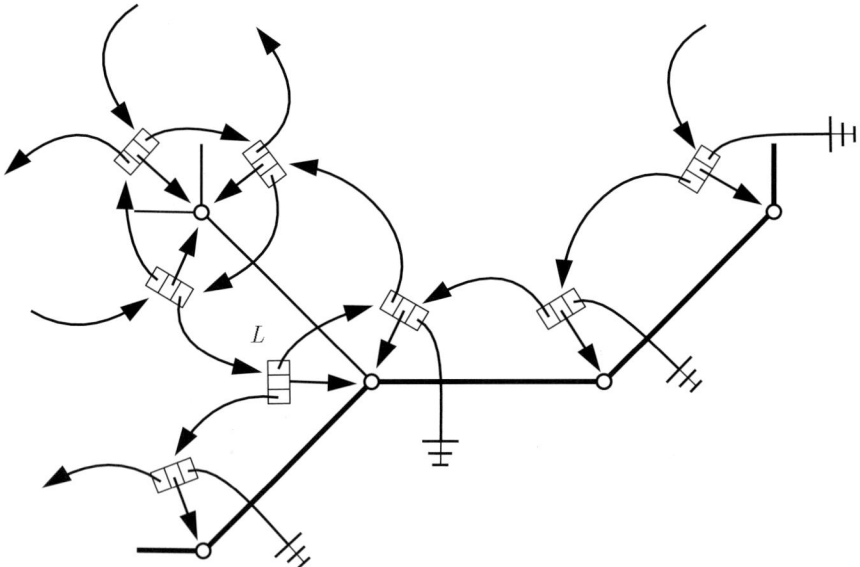

Fig. 21. The boundary of a mesh with an associated corner data structure. Here the *vertex_clockwise* links of the corner elements are `Null` on the boundary.

- $\mathcal{Q}_{\text{FE}}$– Given a lath L representing a face-edge pair in the mesh, laths representing the edges of the face are given by successively using the cf operator – following the clockwise loop about the face.
- $\mathcal{Q}_{\text{VE}}$– Given a lath L representing an edge-vertex pair, the laths representing edges that radiate from the vertex are obtained by successively using the cv operator – following the clockwise loop about the vertex.

The remaining queries are implemented in terms of these basic queries.

- $\mathcal{Q}_{\text{FV}}$– Given a lath L representing a face, the laths representing the vertices of the face are the same as those of $\mathcal{Q}_{\text{FE}}$.
- $\mathcal{Q}_{\text{VV}}$– Given an lath L, we can obtain the required laths by taking $\text{cf}(L)$ for each lath L in $\mathcal{Q}_{\text{EV}}$.
- $\mathcal{Q}_{\text{VF}}$ is identical with $\mathcal{Q}_{\text{VE}}$, as each lath identified by $\mathcal{Q}_{\text{VE}}$ belongs to a unique face.
- $\mathcal{Q}_{\text{EE}}$– Given a lath L representing an edge E, $\mathcal{Q}_{\text{EE}}$ produces a set of laths that correspond to the edges that radiate from the two vertices of E. This is implemented by taking the union of the results of $\mathcal{Q}_{\text{VE}}$ for both L and $\text{cf}(L)$. If $\text{ec}(L)$ exists, we remove it from the list so that the original edge is only referenced by only one lath.
- $\mathcal{Q}_{\text{FF}}$– Given a lath L representing a face-edge pair, all such faces are obtained by applying $\mathcal{Q}_{\text{EV}}$ to each lath in $\mathcal{Q}_{\text{FE}}$, however this produces multiple laths for each face in the output. To obtain a unique lath for each face in the output, we use a marking strategy that marks each lath output by $\mathcal{Q}_{\text{EV}}$. We only output those laths that are unmarked during the traversal.

It is possible that $\mathcal{Q}_{\text{VE}}$ (and $\mathcal{Q}_{\text{VF}}$, which is identical in this structure) fails on the boundary. For example, given a lath L as in Figure 21, $\mathcal{Q}_{\text{VE}}(L)$ returns two laths before encountering a `Null` link on the boundary. To retrieve all laths that represent edges that radiate from the vertex, the query must use cv to traverse around the vertex in the opposite direction.

Similar to the split-edge representation, if we have n edges radiating from the vertex represented by L, only $n-1$ can be obtained through cv and ccv. In this case we can obtain a lath that represents the nth edge by using $\text{cf}(L)$, but care must be taken with it's use as this lath does not point to the same vertex as L.

Duality If we form the dual of a mesh represented by a corner data structure (see Figure 22), we see that the dual mesh induced by the original mesh is also a corner data structure. The clockwise loop about a vertex transforms to a clockwise loop about a face in the dual mesh, and the clockwise loop about a face in the original mesh transforms to a clockwise loop about a vertex in the dual mesh. This makes the corner data structure quite useful for applications where finding and working with the dual mesh is necessary. Only one set of operators and queries need be implemented, and each works for both the original mesh and its dual.

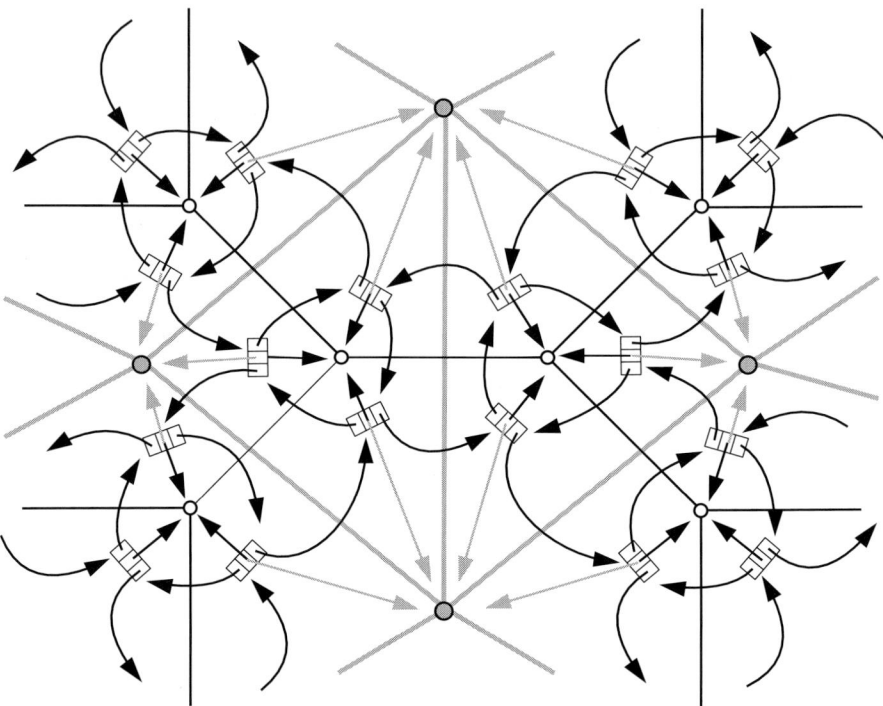

Fig. 22. Representing the dual of a mesh with a corner data structure. The original mesh and data structure are shown in black with the dual mesh shown in gray. The gray vertex links are the links to the face nodes in the dual structure. (See Color Plate 20 on page 361.)

4 Implementation

These data structures are straightforward to implement, as most of the operations are simple and only need to be modified on the boundaries. We have used them in a variety of applications and have found them to be most useful when "in-place" operations on large mesh structures are required. The corner structure is the most efficient for traversals as all basic operations can be implemented in constant time away from the boundary. It is also the most useful when dealing with the dual mesh, as it induces another corner data structure on the dual. The half-edge structure is most useful when frequent boundary operations are necessary. Here, the data structure is slightly larger, as not only are two laths kept for each interior edge of the structure, but two laths are kept for each edge on the boundary. This implies that only one of the basic operations is not available on the boundary and this this simplifies the operations around the boundary.

We should remark that there are other ways to connect laths. For example, we could specify a lath element that has a counter-clockwise face link

Table 1. Possible lath-based data structures.

Lath Links	Data Structure Type	Induced Dual Structure
edge_companion face_clockwise	Split-Edge	Half-Edge
edge_companion face_counter_clockwise	Split-Edge (reverse orientation)	Half-Edge (reverse orientation)
edge_companion vertex_clockwise	Half-Edge	Split-Edge
edge_companion vertex_counter_clockwise	Half-Edge (reverse orientation)	Split-Edge (reverse orientation)
face_clockwise face_counter_clockwise	Not a useful structure	
face_clockwise vertex_clockwise	Corner	Corner
face_clockwise vertex_counter_clockwise	Corner∗	Corner∗ (reverse orientation)
face_counter_clockwise vertex_clockwise	Corner∗ (reverse orientation)	Corner∗
face_counter_clockwise vertex_counter_clockwise	Corner (reverse orientation)	Corner (reverse orientation)
vertex_clockwise vertex_counter_clockwise	Not a useful structure	

and a counter-clockwise vertex link. Structurally this would be the same as the corner structure, just reversing the two loops. By using the five links (*edge_companion, face_clockwise, face_counter_clockwise, vertex_clockwise, vertex_counter_clockwise*), and taking two at a time, there are 10 possibilities of lath connection strategies available. As can be seen in Table 1, there is only one useful representation that we did not cover completely in this article (labeled corner∗ in the table). In this representation, where each lath has a *face_clockwise* link and a *vertex_counter_clockwise* link. The structure is very similar to the corner structure, however it does not have constant-time implementation of the basic operations, it has similar problems with the boundaries, and it induces a dual structure that has laths with *face_counter_clockwise* and *vertex_clockwise* links. We have not found an occasion where such a structure would be useful.

5 Conclusion

We have presented three data structures for unstructured meshes all based upon a single data type – the lath. We have analyzed three connection strategies for the lath elements, creating a split-edge, half-edge and corner data structure, respectively. Each of these structures has several common operations that allow movement about the data, and nine basic queries that return vertices, edges and faces in the neighborhood of a given lath. The corner data structure is new and appears to be the most promising for future use in a variety of algorithms that support multiresolution representations of data.

This paper represents a "minimalist" approach to these data structures. Certainly we can add additional links for faces, or from vertices to laths, or from faces to laths. This complicates the data structure and utilizes much more storage per element, but can make for efficient operations and queries.

Future work should clearly focus on lath-based structures for three-dimensional meshes, and eventually, multi-dimensional meshes. Each lath in these structures should represent an edge-face-cell, vertex-edge-cell, or vertex-face-cell triple, and each lath should be identified with exactly one vertex, edge, face and cell in the mesh.

6 Acknowledgments

This work was supported by the National Science Foundation through the Large Scientific and Software Data Set Visualization (LSSDSV) program under contract ACI 9982251, and through the National Partnership for Advanced Computational Infrastructure (NPACI); the Office of Naval Research under contract N00014-97-1-0222; the Army Research Office under contract ARO 36598-MA-RIP; the NASA Ames Research Center through an NRA award under contract NAG2-1216; the Lawrence Livermore National Laboratory under ASCI ASAP Level-2 Memorandum Agreement B347878 and under Memorandum Agreement B503159; the Lawrence Berkeley National Laboratory; the Los Alamos National Laboratory; and the North Atlantic Treaty Organization (NATO) under contract CRG.971628. We also acknowledge the support of ALSTOM Schilling Robotics and SGI. We thank the members of the Visualization and Graphics Research Group at the Center for Image Processing and Integrated Computing (CIPIC) at the University of California, Davis.

References

1. ALA, S. R. Design methodology of boundary data structures. *Internat. J. Comput. Geom. Appl. 1*, 3 (1991), 207–226.
2. BAUMGART, B. G. Geometric modeling for computer vision. AIM-249, STA-CS-74-463, CS Dept, Stanford U., Oct. 1974.

3. BERTRAM, M. *Multiresolution Modeling for Scientific Visualization.* PhD thesis, University of California, Davis, 2000.
4. BERTRAM, M., DUCHAINEAU, M. A., HAMANN, B., AND JOY, K. I. Bicubic subdivision-surface wavelets for large-scale isosurface representation and visualization. *Proceedings of IEEE Visualization 2000* (Oct. 2000), 389–396.
5. CATMULL, E., AND CLARK, J. Recursively generated B-spline surfaces on arbitrary topological meshes. *Computer-Aided Design 10* (Sept. 1978), 350–355.
6. DE FLORIANI, L., AND PUPPO, E. Hierarchical triangulation for multiresolution surface description geometric design. *ACM Transactions on Graphics 14*, 4 (Oct. 1995), 363–411.
7. DOO, D. A subdivision algorithm for smoothing down irregularly shaped polyhedrons. In *Proced. Int'l Conf. Ineractive Techniques in Computer Aided Design* (1978), pp. 157–165. Bologna, Italy, IEEE Computer Soc.
8. DOO, D., AND SABIN, M. Behaviour of recursive division surfaces near extraordinary points. *Computer-Aided Design 10* (Sept. 1978), 356–360.
9. EASTMAN, C. M. Introduction to Computer Aided Design, Course Notes, Carnegie-Mellon University, Pittsburg, PA, 1982.
10. GARLAND, M., AND HECKBERT, P. Surface simplification using quadric error metrics. *Proceedings of SIGGRAPH'97* (1997), 209–215.
11. GIENG, T., JOY, K. I., HAMANN, B., SCHUSSMAN, G., AND TROTTS, I. Smooth hierarchical surface triangulations. In *Proceedings of Visualization '97* (Oct. 1997), H. Hagen and R. Yagel, Eds., IEEE Computer Society, pp. 379–386.
12. GIENG, T. S., HAMANN, B., JOY, K. I., SCHUSSMAN, G. L., AND TROTTS, I. J. Constructing hierarchies for triangle meshes. *IEEE Transactions on Visualization and Computer Graphics 4*, 2 (Apr. 1998), 145–161.
13. HALL, M., AND WARREN, J. Adaptive polygonalization of implicitly defined surfaces. *IEEE Computer Graphics and Applications 10*, 6 (Nov. 1990), 33–42.
14. HOPPE, H. Progressive meshes. In *SIGGRAPH 96 Conference Proceedings* (Aug. 1996), H. Rushmeier, Ed., Annual Conference Series, ACM SIGGRAPH, Addison Wesley, pp. 99–108. held in New Orleans, Louisiana, 04-09 August 1996.
15. KALAY, Y. E. *Modeling Objects and Environments.* John Wiley and Sons, 1989.
16. MANTYLA, M. *An Introduction to Solid Modeling.* Computer Science Press, Rockville, Md, 1988.
17. RENZE, K. J., AND OLIVER, J. H. Generalized surface and volume decimation for unstructured tessellated domains. In *VRAIS '96 (IEEE Virtual Reality Annual Intl. Symp.)* (Mar. 1996), pp. 24–32.
18. SCHUSSMAN, S., BERTRAM, M., HAMANN, B., AND JOY, K. I. Hierarchical data representations based on planar voronoi diagrams. In *Proceedings of the Joint Eurographics and IEEE TVCG Conference on Visualization* (2000), R. van Liere, I. Hermann, and W. Ribarsky, Eds., pp. 63–72.
19. THOMPSON, J. F., AND HAMANN, B. A survey of grid generation techniques and systems with emphasis on recent developments. *Surveys on Mathematics for Industry 6* (1997), 289–310.
20. WEILER, K. J. *Topological structures for geometric modeling.* Ph.d. thesis, Rensselaer Polytechnic Institute, Aug. 1986.

21. Woo, T. C., AND WOLTER, J. D. A constant expected time, linear storage data structure for representing three-dimensional objects. *IEEE Transactions on Systems, Man and Cybernetics SMC-14* (May 1984), 510–515.
22. XIA, J. C., AND VARSHNEY, A. Dynamic view-dependent simplification for polygonal models. In *IEEE Visualization '96* (Oct. 1996), IEEE. ISBN 0-89791-864-9.

Scaling the Topology of Symmetric, Second-Order Planar Tensor Fields

Xavier Tricoche, Gerik Scheuermann, and Hans Hagen

University of Kaiserslautern, P.O. Box 3049, 67653 Kaiserslautern, Germany
E-mail: {tricoche|scheuer|hagen}@informatik.uni-kl.de

Abstract. Tensor fields can be found in most areas of physics and engineering sciences. Topology-based methods provide efficient means for visualizing symmetric, second-order, planar tensor fields. Yet, for tensor fields with complicated structure typically encountered in turbulent flows, topology-based techniques lead to cluttered pictures that confuse the interpretation and are of little help for analysts. In this paper, we present a hierarchical method that merges close degenerate points into one. This results in a simplified topology and a clarified depiction, though globally maintaining the qualitative properties of the original data. The whole process can be seen as a scaling of the topological information that neglects details of small scale and only retains their structural aspect in the large. We extend here previous work dealing with planar vector fields. Results are demonstrated on a computational fluid dynamics(CFD) dataset.

1 Introduction

Tensors are the language of mechanics. Therefore, tensor field visualization is a challenging issue for scientific visualization. Scientists and engineers need techniques that enable both qualitative and quantitative analysis of tensor data sets resulting from experiments or numerical simulations. A topology-based visualization method of symmetric, second-order tensor fields in two dimensions has been designed for that purpose [1]. It focuses on one of the two eigenvector fields corresponding to the minor or major eigenvalues. Like the vector case, the displayed graph consists of so-called degenerate points (where both eigenvalues are equal) connected by particular integral curves, the separatrices. This technique proved to be suitable for tensor fields with simple structure because the extracted topology contains few degenerate points and separatrices, leading to a comprehensible structure description. However, for tensor fields with complicated structure (like those encountered in turbulent flows, for instance), topology-based methods result in cluttered pictures that confuse the interpretation.

We present in this paper a method that merges close degenerate points into one, which simplifies the topology and clarifies its depiction, though globally maintaining the qualitative properties of the original data. It is the extension to the tensor case of previous work on vector fields [2]. We assume a planar, piecewise bilinear interpolated, symmetric, second-order tensor field over a curvilinear grid. The degenerate points are determined first. A clustering strategy is then applied to the resulting set of degenerate points. This leads to a grid partition into cell clusters such that the distance between degenerate points in each cluster is below a user-prescribed threshold. After this, we merge the degenerate points lying in each cluster to get the desired scaling effect. Finally, we determine and integrate the resulting separatrices.

The structure of the paper is as follows. Fundamental notions of tensor field topology are briefly presented in section 2. The formula governing eigenvalue and eigenvector computation are proposed in section 3. The clustering strategy is the same as for vector fields and has been presented earlier [2]. We bring back its basic principle in section 4. Once clusters have been defined, we achieve a local grid deformation which merges singular points, as explained in section 5. The a posteriori analysis of the resulting local topological structure is detailed in sections 6, 7. At last, results are shown in section 8 on a swirling jet simulation with evolving turbulence: The topology is clarified, the separatrices easier to track while the significant structural features have been preserved.

2 Topology of Tensor Fields

Based upon the concept of vector field topology, the topology of symmetric second-order 2D tensor fields has been introduced recently [1]. Basically, it is defined as the topology of one of the two (bidirectional) eigenvector fields. Since tensor field topology has been treated in very few publications so far, we give next the necessary definitions.

A real two-dimensional symmetric matrix M has always two (not necessarily distinct) real eigenvalues $\lambda_1 \leq \lambda_2$ with associated (orthogonal) eigenvectors $\boldsymbol{e_1}$ and $\boldsymbol{e_2}$:

$$\forall i \in \{1,2\}, M\boldsymbol{e_i} = \lambda_i \boldsymbol{e_i}, \text{with } \boldsymbol{e_i} \in \mathbb{R}^2 \text{ and } \boldsymbol{e_i} \neq \boldsymbol{0}.$$

(Note that an eigenvector as neither norm nor orientation for every multiplication of an eigenvector with a non-zero scalar results in another eigenvector.) Therefore, we write a symmetric, second order planar tensor field, i.e. a matrix-valued function that associates every position of a subdomain of $\mathbb{R}^2$ with a symmetric matrix:

$$T : (x,y) \in U \subset \mathbb{R}^2 \mapsto T(x,y) = \begin{pmatrix} \alpha(x,y) & \beta(x,y) \\ \beta(x,y) & \gamma(x,y) \end{pmatrix}.$$

One can define a major (resp. minor) *eigenvector field*, at each position of the domain, as the eigenvector related to the major (resp. minor) eigenvalue of the tensor field. One defines major (resp. minor) *tensor lines* as the curves everywhere tangent to the major (resp. minor) eigenvector field. Consequently, these curves have no orientation. Yet, this definition only holds outside locations where both eigenvalues are equal and thus the eigenvectors cannot be

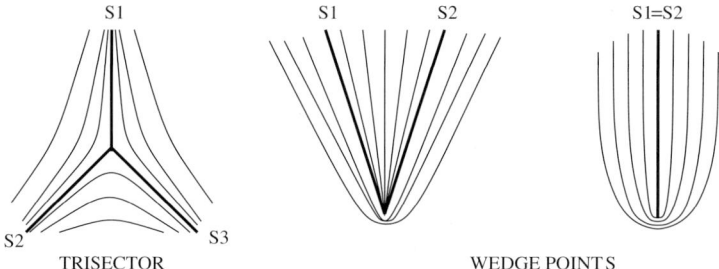

Fig. 1. First Order Degenerate Points

uniquely determined. These positions are called *degenerate points* (or *singular points*). Thus, at a degenerate point, the tensor value is always a diagonal matrix of the form αI_2, which is used in practice to compute the position of the singularities. In the linear case, they appear in two possible types: *Trisector* or *wedge point* (see Fig. 1). The classification of a degenerate point can be done by the computation of its *index* (see [4], p. 105): The index of a degenerate point of a symmetric tensor field is the number of counterclockwise rotations made by the eigenvectors along a closed, non self-intersecting curve around the degenerate point that encloses no other singularity, when traveling once in a counterclockwise direction along this curve. A trisector point has index $-\frac{1}{2}$ while a wedge point has index $+\frac{1}{2}$.

In the general (non-linear) case, a degenerate point can have two generic types: *center* type (no tensor line reaches the singularity, so every tensor line in its vicinity is closed and rotates around the singularity location) or *non-center* type (at least, one tensor line reaches the singularity, forming one or more curvilinear sectors): See Fig. 2. In the non-center case, the characterization is based upon position and type of the curvilinear sectors. These sectors have three possible natures: *Hyperbolic*, *parabolic* or *elliptic* (see Fig. 3). Separatrices are defined as streamlines bounding hyperbolic sectors. Now, with the

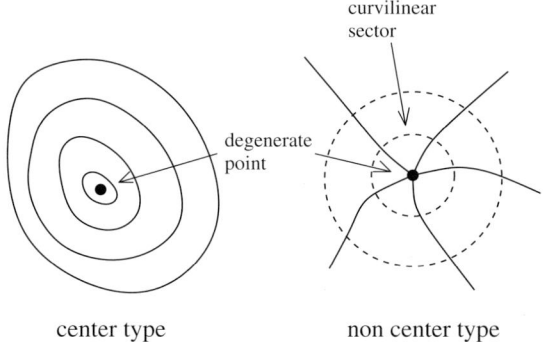

Fig. 2. Center and Non-Center Type for a Degenerate Point

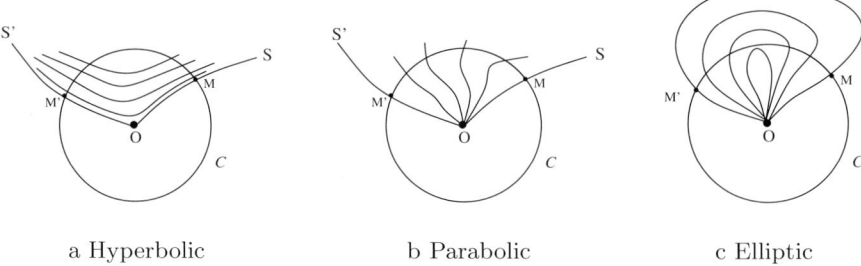

Fig. 3. Sector Natures

definitions above, the topology of an eigenvector field is defined as the graph formed by all degenerate points and by the separatrices connecting them. (Closed orbits, known from vector field topology, are very rare in practice).

3 Eigenvectors and Eigenvalues Computation

As said previously, the depiction of the topology of a symmetric, second-order tensor field is based upon the computation of its eigenvalues and eigenvectors. As a preliminary step, we first compute the so-called deviator [3] of the considered tensor field, defined as follows. Let T be a symmetric, second-order planar tensor field

$$T := \begin{pmatrix} \alpha & \beta \\ \beta & \gamma \end{pmatrix},$$

where α, β and γ are scalar functions in the coordinates x and y of the Euclidean space $\mathbb{R}^2$. Denoting $tr(T) = \alpha + \gamma$ the trace of T, the isotropic part of T is given by

$$J := \frac{1}{2} tr(T) \ I_2$$

and the deviator of T is then defined by

$$D := T - J = \begin{pmatrix} \delta & \beta \\ \beta & -\delta \end{pmatrix}, \text{ where } \delta = \frac{\alpha - \gamma}{2}$$

Now, the topology of a symmetric, second-order tensor field is the same as the topology of its deviator field because both fields have the same eigenvector fields, as we show next. By definition of an eigenvalue λ and its corresponding eigenvector $\boldsymbol{e} \neq \boldsymbol{0}$, one has

$$T \boldsymbol{e} = \lambda \boldsymbol{e}. \tag{1}$$

Using the notations above, this is equivalent to

$$\left(D + \frac{1}{2}tr(T) I_2\right) \mathbf{e} = \lambda \mathbf{e},$$

that is

$$D \mathbf{e} = \left(\lambda - \frac{1}{2}tr(T)\right) \mathbf{e}.$$

Thus, setting $\mu = \lambda - \frac{1}{2}tr(T)$, μ is eigenvalue of the deviator D and $\mathbf{e}$ is the corresponding eigenvector. Furthermore, if one denotes by μ_i the eigenvalue related to λ_i, one gets $\mu_1 < \mu_2$ if $\lambda_1 < \lambda_2$. This proves the equivalence of both topologies.

For this reason, we will only consider matrices of deviator type in the following, which simplifies the calculus.

Now, the eigensystem (1) is singular (for there exist non-zero solutions) which is equivalent to

$$det\,(T - \lambda I_2) = 0.$$

For each position on the plane, this is a quadratic polynomial equation that we use in practice to compute the eigenvalues. With the deviator, one gets

$$\lambda_{1,2} = \pm\sqrt{\delta^2 + \beta^2}$$

Except in the special case where the deviator is diagonal (i.e. $\beta = 0$), solving the singular eigensystem (1) leads to the following expressions (with the same notations). The eigenvectors are $\mathbf{e}_{1,2} := (u, v_{1,2})^T$ with

$$u = \beta \quad \text{and} \quad v_{1,2} = -\delta \pm \sqrt{\delta^2 + \beta^2}$$

If the deviator is diagonal, the eigensystem becomes trivial: The eigenvalues are δ and $-\delta$ and the corresponding eigenvectors are the basis vectors of the Euclidean coordinate system: $(1,0)^T$ and $(0,1)^T$.

4   Clustering

We recall now the scheme we used in previous work[2] for grouping grid cells together in order to get a domain decomposition into cell clusters that contain only close singularities. The clustering process is monitored by the proximity of the singularities contained in a cluster. For simplicity, we deal in this section with a curvilinear grid mapped in computational space to a rectilinear grid. We denote by $P_1, ..., P_m$ the positions of the m singularities lying inside a particular cluster. We want to minimize the approximation error of these m singularities by a single point, where this point (or cluster center) Q is chosen to be the best approximation (for a particular norm) of the singularities (see Fig. 4).

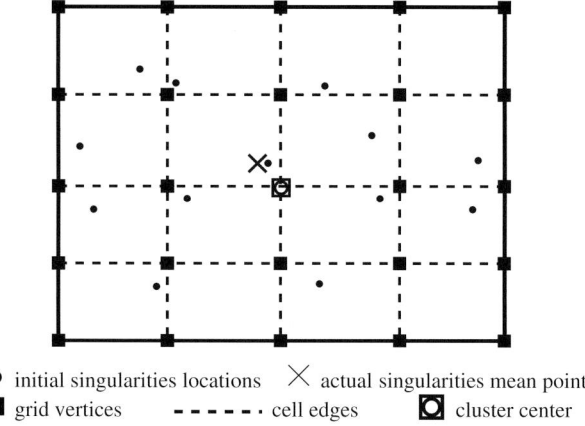

Fig. 4. Cluster singularities and cluster center

The corresponding error is

$$S = \frac{\sum_{j=1}^{m} \omega_j \|P_j - Q\|}{\sum_{j=1}^{m} \omega_j}$$

where ω_i is the weight associated with the ith singularity: We have set these weights to 1 in the following for simplicity but these values can be used to introduce more flexibility in the method, in particular by privileging degenerate points with interesting numerical or qualitative properties, based upon user-defined criteria. The aim of our clustering scheme is now to get a domain decomposition into clusters that all fulfill an approximation criterion, that is that all have an error value smaller than a specified threshold (which is the only parameter of our method).

If a cluster does not satisfy the given error criterion, we split it into two subclusters. To do this, we need to introduce the projected variances associated with a given cluster:

$$V_i = \sum_{j=1}^{m} \omega_j (P_j^i - Q^i)$$

where $i \in 0, 1$ is the considered coordinate axis ($P_j = (P_j^0, P_j^1)$).

Now, considering the whole grid as initial cluster, the method is as follows.

Step 1. Take as cluster center the best
    vertex approximation of all cluster
    singular points.

Step 2. Compute the approximation error S.
    If (S > THRESHOLD) go to step 3.
    Otherwise stop.

Step 3. Compute the coordinate axis with
largest projected variance
(i.e. max(V0, V1)).

Step 4. Create 2 subclusters by splitting the
cluster at an edge polyline through Q
perpendicular[1] to the selected
coordinate axis.
For each cluster, go to step 1.

Step 4 justifies the need for a cluster center to be a grid vertex. As a matter of fact, when splitting a cluster, we keep processing cell groups in the form of Fig. 5. To ensure that the algorithm terminates, we require the existence of a best singularity mean point approximation by a grid vertex that does not lie on the cluster boundary, as additional criterion.

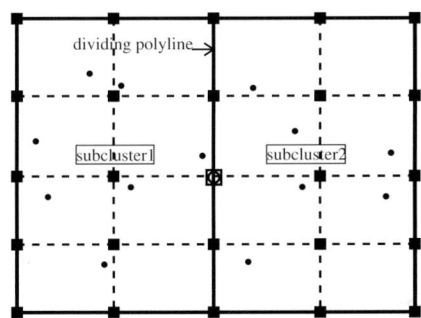

Fig. 5. Cluster splitting

5   Local Topology Deformation

In each cluster, we aim at locally simulating the fusion of all preexisting singularities while preserving consistency with the original global topology. Practically, we must remove the degenerate points, replace them by a single one and let the cluster boundary unchanged. Like the vector case, we achieve it as follows: We first remove all quadrilateral cells in the cluster. Then we add a new vertex at the cluster center, associated with a degenerate tensor value. At last, we cover the resulting empty domain by linear interpolated triangles joining the new vertex to those on the cluster boundary. Yet, contrary to the vector case, a degenerate tensor value is not unique and may be any *isotropic* tensor (c.f. section 3). Now, as we chose to deal with a deviator tensor field, the only possible choice in our case is a zero matrix that we associate with the new vertex.

---
[1] in computational space

6 "Parallel" Positions Localization

Once such an artificial degenerate point has been created, we need to determine its local structure to find the positions of its separatrices. As a matter of fact, separatrices are integral curves that bound the so-called hyperbolic sectors of a singularity (see section 2). For that reason, boundary curves and sector types must be found. As the singularity lies inside a piecewise linear domain, one can show that its structure may be fully identified on the piecewise linear edges on the boundary of its containing cluster. Consider Fig. 6. For each such edge, one has the following configuration: One is given a seg-

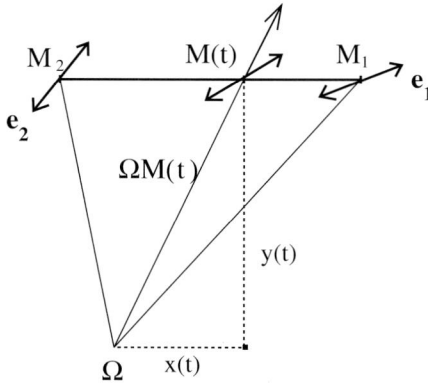

Fig. 6. Saddle point

ment $\overline{|M_1 M_2|}$ with linear parametrization $t \in [0,1]$ on which a linear varying, symmetric, second-order (deviator) tensor field is defined. Furthermore, one is given the cluster center position Ω that does not lie on $\overline{|M_1 M_2|}$. Now, we seek on $\overline{|M_1 M_2|}$ the positions $M(t)$ where the vector $\boldsymbol{\Omega M}$ is parallel to the eigenvectors, that is

$$\boldsymbol{\Omega M} \wedge \boldsymbol{e} = \boldsymbol{0}$$

with a linear parametrization of the considered edge that can be written

$$\boldsymbol{\Omega M}(t) = (x(t), y(t))^T = (1-t)\,\boldsymbol{\Omega M_1} + t\,\boldsymbol{\Omega M_2}, \quad t \in [0,1].$$

Supposing $\beta(t) \neq 0$, the "parallel" condition becomes then (with the notations of section 3)

$$\begin{vmatrix} x(t) & \beta(t) \\ y(t) & -\delta(t) \pm \sqrt{\delta(t)^2 + \beta(t)^2} \end{vmatrix} = 0$$

(β and δ being linear functions in t). Finally, the expression above is equivalent to the following system.

$$\begin{cases} \beta(t) = 0 \\ \text{or} \\ x(t)^2 \beta - y(t)^2 \beta(t) - 2x(t)y(t)\delta(t) = 0 \end{cases}$$

As $x(t)$, $y(t)$, $\beta(t)$ and $\delta(t)$ are all linear functions in t, the second equation is a cubic polynomial that we solve with a standard analytic method to obtain the required "parallel" positions.

7   Sector Type Identification

Because of the sign indeterminacy, the sector type identification cannot be obtained by the method described in the vector case. As a matter of fact, the distinction between hyperbolic and elliptic type is impossible without additional information. For that reason, we use the tensor index (see section 2): Between two consecutive "parallel" positions, we compute the angle variation of the considered eigenvector field (see Fig. 7). Depending on the sector type, one gets

- $\Delta \alpha = \theta$ in the *parabolic* case
- $\Delta \alpha = \theta - \pi$ in the *hyperbolic* case
- $\Delta \alpha = \theta + \pi$ in the *elliptic* case

which enables a sector type identification. When a hyperbolic sector is found, its boundary curves are then integrated to form the topological graph.

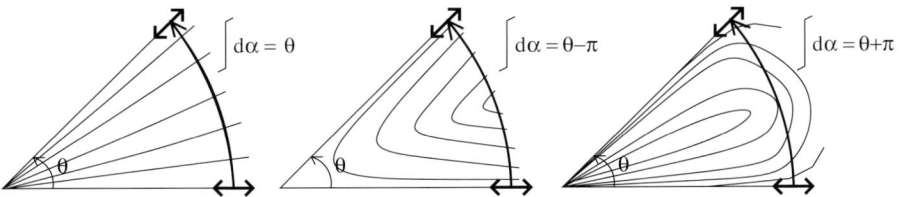

Fig. 7. Angle variation in the parabolic, hyperbolic and elliptic case

8   Results

We show the results of our method applied on the rate of strain tensor field of a swirling jet simulation. The dataset is almost axisymmetric, a property that is well preserved by our algorithm. The grid is structured and has 124 x 101 cells ranging from 0 to 9.87 in x, resp. -3.85 to 3.85 in y. The original

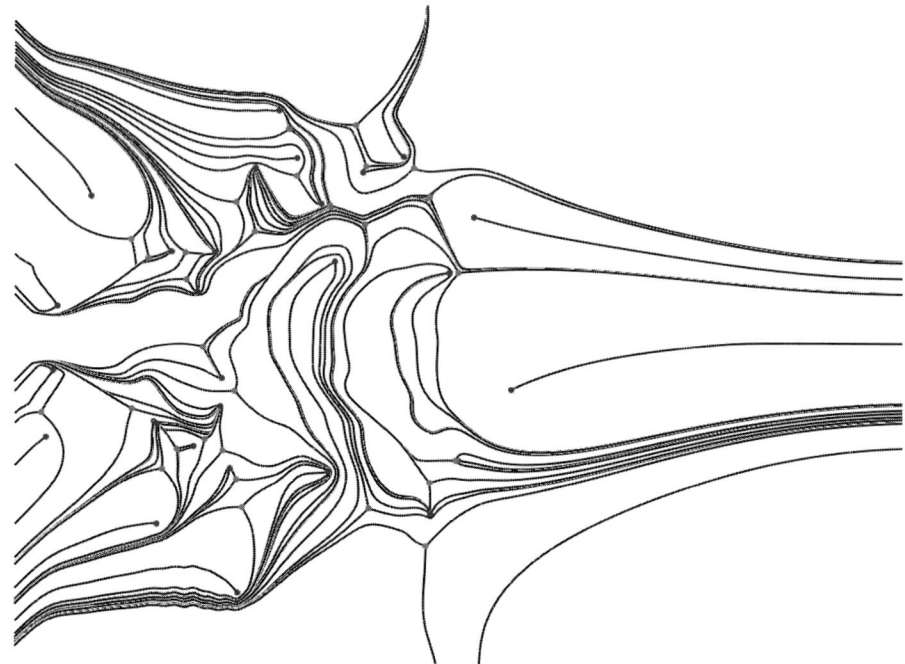

Fig. 8. Original topology

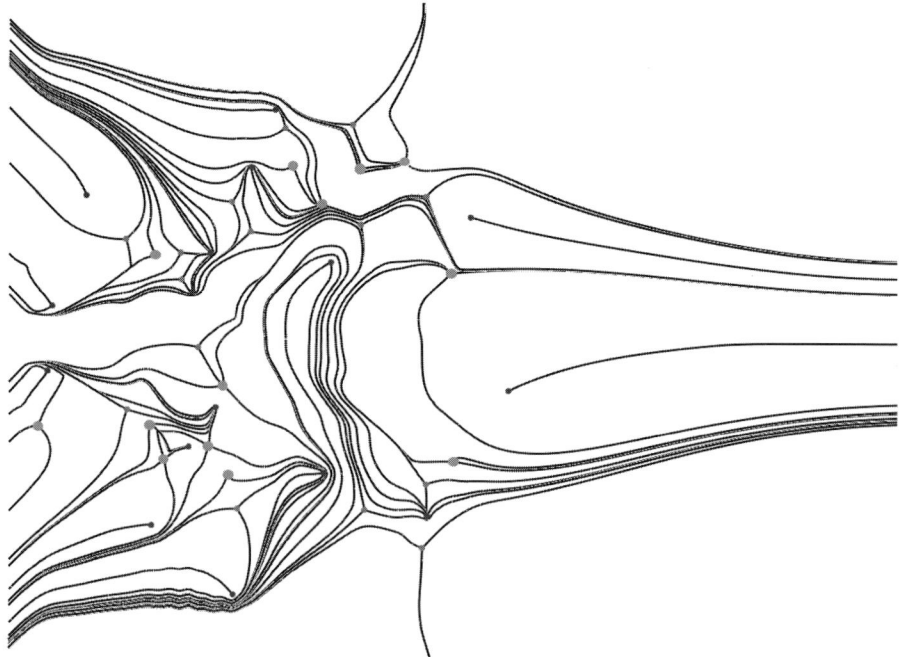

Fig. 9. Scaled topology (*threshold* = 0.2)

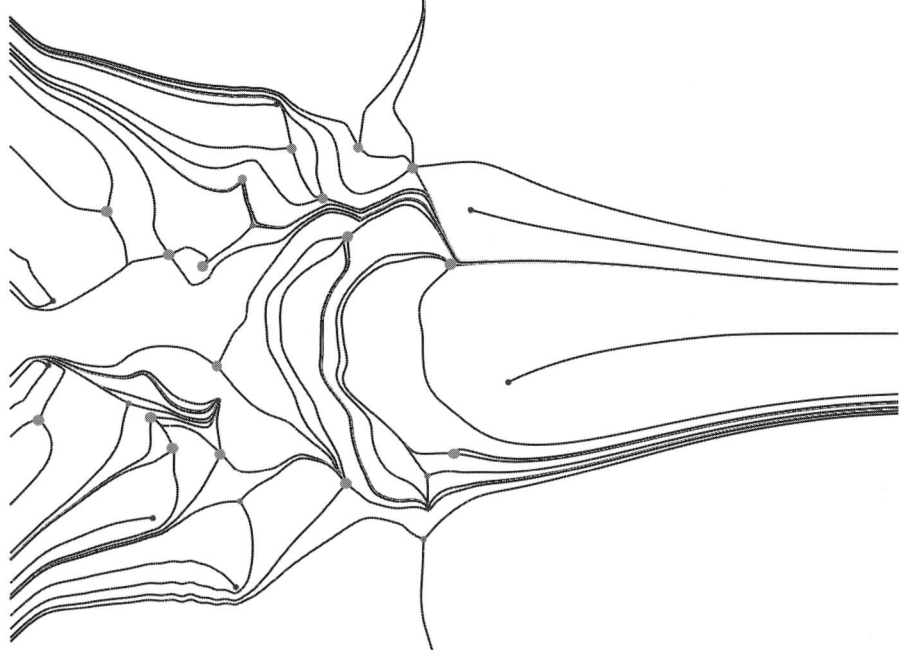

Fig. 10. Scaled topology (*threshold* = 0.4)

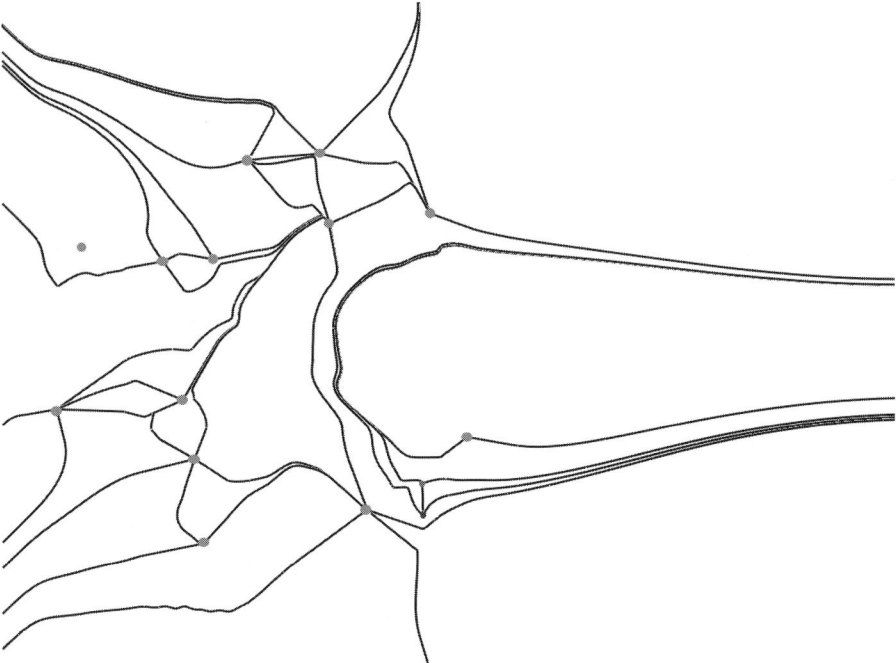

Fig. 11. Scaled topology (*threshold* = 0.8)

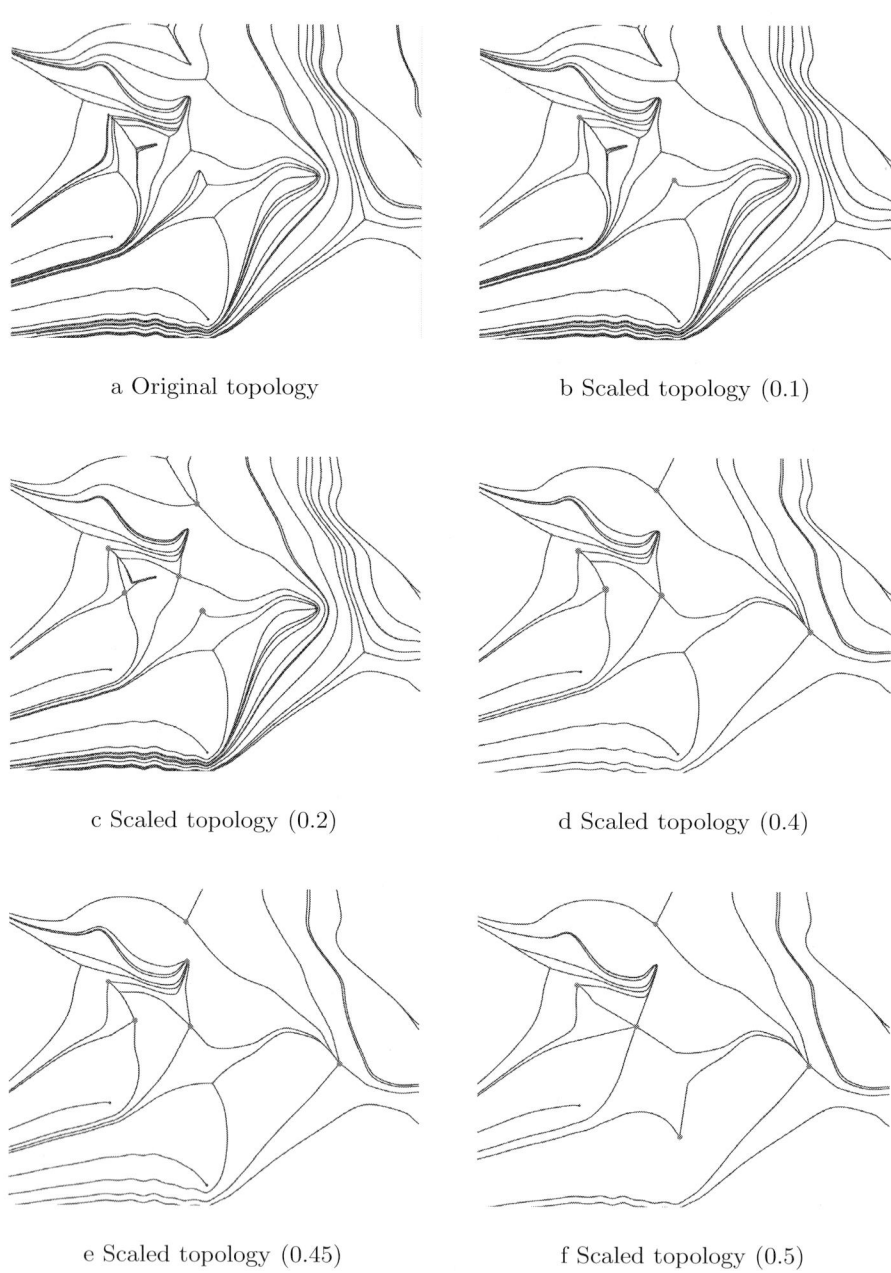

Fig. 12. Samples of local topology scaling

topology presents 61 degenerate points and 131 separatrices. It is shown in Fig. 8.

We start scaling with *threshold* = 0.2. The resulting topology contains 44 singularities (13 of which being non-linear) and 101 associated separatrices: See Fig. 9. (Non-linear singularities are depicted by big dots.) Increasing the threshold (*threshold* = 0.4), we simplify the topology further. There are now 31 degenerate points (17 being non-linear) and 76 separatrices in the graph: See Fig. 10. The last stage of our scaling process is obtained with *threshold* = 0.8. We stop here because the resulting graph already appears very coarse. This last topology has 15 degenerate points (only 2 of them are original ones!) and 42 separatrices. The result is shown in Fig. 11.

To focus on the local effect of the method on close singularities, we propose samples of a close-up corresponding to increasing thresholds in Fig. 12.

9 Conclusion

We have presented a method that scales the topology of symmetric, second-order planar tensor fields defined over curvilinear grids. Such a scaling is necessary when dealing with turbulent tensor fields because their structural complexity results in cluttered depictions with standard techniques which inconveniences interpretation. Our method basically replaces several close features of small scale by a single one, with non-linear structure, using their topological equivalence in the large. Close features are first supplied by a cell-based clustering strategy that is monitored by the proximity of the degenerate points contained in a same cluster. In a second step, the internal cell structure of each retained cluster is changed to produce the desired local topological merging. At last, the resulting structure is identified mathematically. This eventually permits the clarification of the topological graph by the reduction of the number of singular points and associated separatrices. This was illustrated by our example, provided by a swirling jet simulation from the CFD: Global and local effects of the method were shown with increasing scaling factors.

Acknowledgment

The authors wish to thank Wolfgang Kollmann, MAE Department of the University of California at Davis, for providing the swirling jet dataset. Furthermore, we would like to thank Tom Bobach, Holger Burbach, Stefan Clauss, Jan Frey, Aragorn Rockstroh, René Schätzl and Thomas Wischgoll for their programming efforts.

References

1. Delmarcelle T., Hesselink L., *The Topology of Symmetric, Second-Order Tensor Fields*. Proceedings IEEE Visualization'94, 1994.

2. X. Tricoche, G. Scheuermann, H. Hagen, *A Topology Simplification Method for 2D Vector Fields*. Proceedings IEEE Visualization'00, 2000.
3. Lavin Y., Levy Y., Hesselink L., *Singularities in Nonuniform Tensor Fields*. Proceedings IEEE Visualization'97, 1997.
4. Delmarcelle T., *The Visualization of Second-Order Tensor Fields*. PhD Thesis, Stanford University, 1994.

Simplification of Nonconvex Tetrahedral Meshes

Martin Kraus and Thomas Ertl

Visualization and Interactive Systems Group, Institut für Informatik,
Universität Stuttgart, Breitwiesenstr. 20–22, 70565 Stuttgart, Germany,
E-mail: {Martin.Kraus | Thomas.Ertl}@informatik.uni-stuttgart.de

Abstract. The simplification of nonconvex tetrahedral meshes with the help of edge collapses (edge contractions) is considerably more complex than the corresponding simplification of convex meshes. In particular, edge collapses in nonconvex meshes may cause intersections of cells that are hard to detect, and therefore hard to avoid. More precisely spoken, the problem is quadratic in the number of cells.

However, we show how to reduce the complexity of this problem by employing a pre-processing step, which was originally proposed in order to sort nonconvex meshes. Moreover, our method is able to handle meshes with topologically non-trivial boundaries and to control the modification of the topology of the mesh's boundary.

1 Introduction

In order to visualize today's huge data sets, hierarchical representations of data meshes are often employed. The production of such representations requires a simplification of an original, fine mesh to a coarser mesh, e.g. by resampling a mesh to a lower resolution with less vertices. For unstructured tetrahedral meshes, one way of performing this simplification is to apply a sequence of edge collapses, which are discussed in some detail in Sect. 2. In particular, we will describe problems of edge collapses that are specific to nonconvex tetrahedral meshes.

Our solution to these problems in nonconvex meshes consists of two parts: A preprocessing step – originally proposed by Williams in the context of sorting nonconvex meshes [7] – is briefly described in Sect. 3. The second part – edge collapses in the resultant meshes – is discussed in Sect. 4 with special attention being paid to modifications of the topology of the mesh's boundary. The algorithm is demonstrated with an example in three dimensions in Sect. 5. Section 6 presents our conclusions and plans for future work on this subject.

2 Background and Related Work

2.1 Edge Collapses

In the following we will discuss edge collapses in fair tetrahedral meshes, i.e. each face of a tetrahedral cell is either part of the boundary of the mesh or shared by (at most) two cells. (Note that the term *edge collapse* in computer graphics corresponds to the term *edge contraction* in computational geometry. We will use the former in order to be consistent with the majority of our references.) Intersections of cells are not allowed; in fact avoiding them is our primary concern. As edge collapses in triangular meshes in two dimensions are very similar to the three-dimensional case of tetrahedral meshes, we will illustrate our method with triangular meshes in Figs. 1–8 before presenting the method in three dimensions in Figs. 9–13.

It is useful for the description of our method to define some particular terms. We call a tetrahedron a *vertex neighbor* of a vertex if the vertex is shared by the tetrahedron. A tetrahedron is an *edge neighbor* of an edge if the edge is shared, and a *vertex neighbor* of an edge if one of the vertices of the edge is shared. Finally, a tetrahedron is a *face neighbor* of another tetrahedron if the tetrahedra share one face. The definitions of *vertex* and *edge neighbors* can also be applied to triangles in triangular meshes.

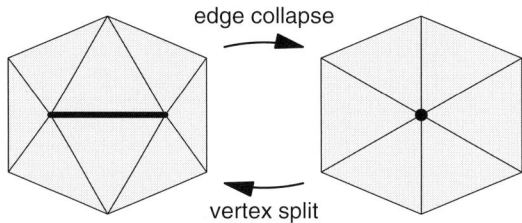

Fig. 1. The collapse (contraction) of an edge (*thick line*) to a vertex (*dot*) and the inverse vertex split

The effect of an edge collapse (see Fig. 1 for a two-dimensional example) is to remove all edge neighbors of the collapsing edge and to join the two vertices of the collapsing edge in a new vertex. Figure 1 also indicates the inverse operation, which is called a *vertex split*.

Edge collapses are one of the most powerful tools to simplify triangular or tetrahedral meshes. They can be employed to remove vertices or edges [5] and also to remove triangles or tetrahedra by successive edge collapses [2,6].

Moreover, a sequence of edge collapses can be applied in order to produce hierarchical representations of triangular and tetrahedral meshes as demonstrated in many publications, for example [2,4–6].

2.2 Avoiding Intersections of Cells

As demonstrated in Fig. 2 an edge collapse can cause an intersection of cells in a triangular or tetrahedral mesh. In order to avoid such self-intersections of a mesh, edge collapses are tested before they are performed [2,5,6].

In convex meshes, i.e. meshes the boundary of which are convex polytopes, the test for intersections is particularly simple because any intersection of cells is accompanied by an *inversion* of at least one cell, i.e. a sign flip of the signed

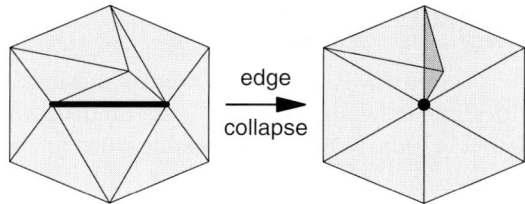

Fig. 2. An edge collapse, which causes several intersections of cells and one inversion of a cell (*dark gray*)

volume of a cell. (The inverted cell is marked gray in Fig. 2.) Therefore, it is sufficient to test all vertex neighbors of a collapsing edge for inversions in order to avoid self-intersections. This test is *local* as only vertex neighbors are involved.

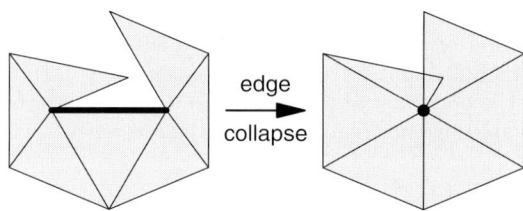

Fig. 3. An edge collapse, which causes an intersection of two cells without causing an inversion of any cell

However, if a collapsing edge in a nonconvex mesh has vertex neighbors that are cells at the boundary (i.e. one of the faces of the cell is part of the boundary of the mesh) then the edge collapse can cause self-intersections of the mesh without causing an inversion of a cell as shown in Figs. 3 and 10.

A naive procedure to avoid such intersections is to test the vertex neighbors of the collapsing edge for intersections with all boundary cells of the mesh. As the wrost-case time complexity of this *global* test depends linearly on the number of cells in the whole mesh, it is usually too expensive to be performed without additional auxiliary data structures. A more elaborated implementation of this test is discussed in [2] and [5] while the system described in [6] tries to preserve the boundary of the tetrahedral mesh.

However, performance issues are not the only problem of edge collapses in nonconvex meshes. Additional problems occur in disconnected meshes as edge collapses are obviously not able to join clusters of disconnected meshes in order to simplify them. More generally spoken, it is desirable to modify the topology of the mesh's boundary in a controlled way when performing edge collapses.

Before presenting our approach to solve these problems in Sect. 4, we describe the necessary preprocessing step in the following section.

3 Convexification of Nonconvex Meshes

In [7] Williams proposed to convert nonconvex meshes to convex meshes by triangulating all voids and cavities and marking the cells generated by this triangulation as *imaginary*. (*Virtual* is today's more fashionable word for the same idea.)

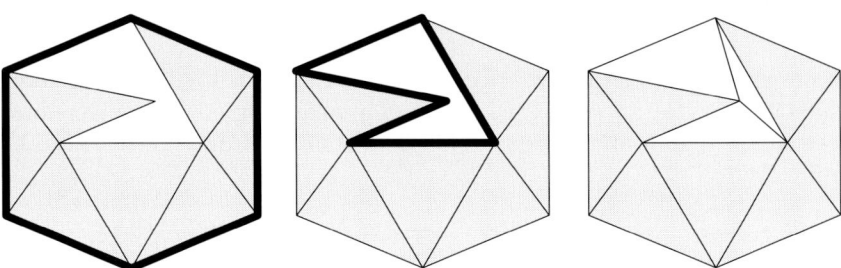

Fig. 4. A step-by-step convexification of the mesh shown in the left-hand side of Fig. 3. From left to right: the convex hull (*thick line*), the nonconvex polygon (*thick line*) between the convex hull and the boundary of the mesh, and the mesh together with imaginary cells (*white*) generated by the triangulation of the nonconvex polygon

Figures 4 and 11 summarize the basic steps of this process. Firstly, the convex hull of the mesh is computed; then all voids and cavities are identified and triangulated; finally, the new imaginary cells are attached to the existing mesh. As will be shown in Sect. 4 the number of imaginary cells generated by the triangulation is not relevant in the context of edge collapses. An optimal algorithm (with respect to the number of generated tetrahedra) for the tetrahedralization of nonconvex polyhedra was published in [1].

We call this preprocessing step a *convexification* of a nonconvex mesh as it allows us to apply (slightly modified) algorithms for convex meshes to a nonconvex mesh. (In this sense the convexification might be called a *meta-algorithm*.) The following section shows how to overcome problems of edge collapses introduced by nonconvexities (including non-trivial topologies of the boundary) with the help of this preprocessing step.

4 Edge Collapses in Convexified Meshes

4.1 Geometric Tests

As mentioned in Sect. 2 and shown in Fig. 3, edge collapses in nonconvex meshes can cause intersections of cells without causing an inversion of any cell. In convexified meshes, however, such edge collapses will always cause an

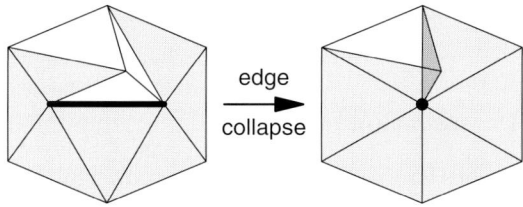

Fig. 5. The edge collapse of Fig. 3 in the convexified mesh of Fig. 4 causes an inversion of an imaginary cell (*dark gray*). (Compare also with Fig. 2)

inversion of at least one imaginary cell as shown in Fig. 5 for the same edge collapse as in Fig. 3 in a convexified version of the same triangular mesh.

Therefore, convexified meshes allow us to test for self-intersections of cells by simply testing all vertex neighbors (including imaginary cells) of the collapsing edge for sign flips of the signed cell volume, which is a local, geometric test as in the case of convex meshes. Thus, the total number of new imaginary cells generated by the convexification is not relevant for the efficiency of this test.

4.2 Preservation of the Convex Hull

Not only are edge collapses in nonconvex meshes more complicated than in convex meshes, they can also transform a convex mesh into a nonconvex mesh.

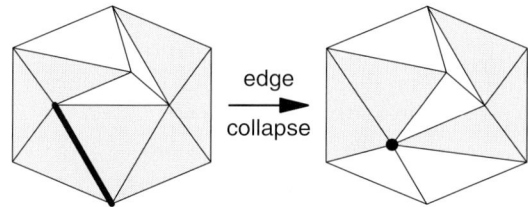

Fig. 6. An edge collapse, which could generate a nonconvexity. Two new imaginary cells are inserted in order to preserve the original convex hull

An example is depicted in Fig. 6, which also shows our solution: Instead of recomputing the convex hull (a global operation if implemented naively), we insert imaginary cells between the new vertex and the convex hull in order to preserve the original convex hull. (The effect can also be seen at the bottom of Fig. 12.)

This is an efficient, local operation. However, it will also insert some new edges; therefore, a simplification process might run into an endless loop by collapsing edges which are instantly reconstructed by the insertion of imaginary cells. In order to avoid this problem, a simple test for edge collapses has to be added. Edge collapses are avoided if the following three conditions are met: All edge neighbors of the collapsing edge are imaginary, one vertex is part of the boundary of the mesh, and all vertex neighbors of this vertex are imaginary. (For an example see the edge between the new imaginary cells in the right-hand side of Fig. 6.)

4.3 Incomplete Topology Preservation

Edge collapses in convexified meshes are considerably more powerful than edge collapses in the original meshes. For example, disconnected meshes can be joined in order to be simplified, tunnels in the boundary of a mesh can be closed, bridges between meshes can be broken, etc. (Figure 12 shows a new connection between originally disconnected parts of the mesh in the top-right corner and a disconnection of cells at the bottom.)

However, not all of these features are always welcome; instead, it is more appropriate to have full control over the modifications of the topology of the mesh's boundary. Here we present two very simple tests for edge collapses in order to avoid topological changes of the mesh. The tests guarantee the preservation of a certain *edge connectivity*, but do not avoid all possible topological changes. Both tests involve only vertex neighbors of the collapsing edge. Before presenting these topological tests, we have to define some basic terms.

Def.: The *type* of a cell is either imaginary or non-imaginary.
Def.: A cell T_1 is *connected* to a cell T_2 of the same type if the cells share a vertex (direct connection), or if T_1 is connected to a third cell T_3 of the same type that is connected to T_2 (indirect connection).
Def.: Two cells are *disconnected* if they are not connected.

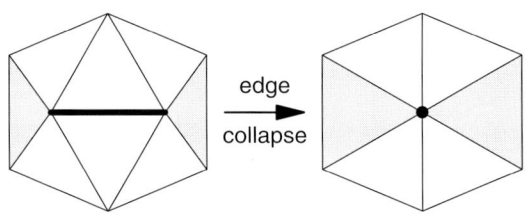

Fig. 7. An edge collapse, which connects two non-imaginary cells (*gray*)

Figure 7 shows an example of an edge collapse that establishes a new connection between two non-imaginary cells. In general, the collapse of an edge e between two vertices v_1 and v_2 *can* connect two previously disconnected cells if all of the following three conditions are met: All of the edge neighbors of e are of the same type t; at least one of the vertex neighbors of v_1 is not of type t; and at least one of the vertex neighbors of v_2 is not of type t. This is a necessary condition; therefore, it is sufficient to avoid edge collapses that fulfill it in order to avoid new connections between cells. The test is the same for triangular and tetrahedral meshes.

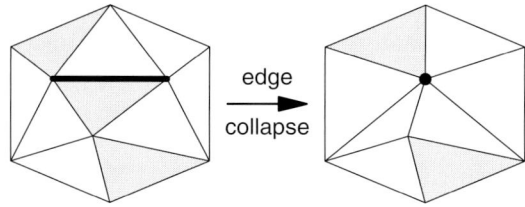

Fig. 8. An edge collapse, which disconnects two non-imaginary cells (*gray*)

Edge collapses are also able to disconnect cells; an example is presented in Fig. 8. In order to formulate a test for disconnecting edge collapses, we need one more definition:

Def.: An edge neighbor T of type t of an edge e is *isolated* if none of the faces of T that do not share e are shared by a face neighbor of T of type t.

Using this definition we can state that the collapse of an edge e *can* disconnect two previously connected cells if at least one of the edge neighbors of e is isolated. This is again a necessary condition, which works for triangular and tetrahedral meshes. (Note that the faces of a triangular cell are its edges.)

These two topological tests allow us to avoid new connections and/or disconnections simply by avoiding edge collapses that fulfill the conditions stated above. An example in three dimensions will be presented in the Sect. 5.

4.4 Complete Topology Preservation

In the preceding subsection, we were only concerned with a certain edge connectivity. It is possible to extend this concept to faces of cells by defining a corresponding *face connectivity*:

Def.: A cell T_1 is *face connected* to a cell T_2 of the same type if the cells share a face (direct face connection), or if T_1 is face connected to a third cell T_3 of the same type that is face connected to T_2 (indirect face connection).
Def.: Two cells are *face disconnected* if they are not face connected.
Def.: An edge neighbor T of type t of an edge e is *actively isolated* if all of the faces of T that do not share e are shared by face neighbors of T that are not of type t.

The collapse of an edge e can face connect previously face disconnected cells if at least one of the edge neighbors of e is actively isolated.

It is necessary to test all edge neighbors of the collapsing edge for face disconnections. This is a sufficient test because face disconnections of cells that are not edge neighbors of the collapsing edge imply the existence of a face disconnection between edge neighbors of the collapsing edge.

This is again a simple, local test, which is easily implemented. However, even the combined tests for edge and face connectivity do not preserve the topology completely. Fortunately, a rigorous solution to this problem has been published by Dey et al. in [3]. In fact, our tests turn out to be special cases of the *Link Conditions* for 3-complexes. (See Theorem C in [3].)

5 Example in Three Dimensions

After presenting our algorithm in Sects. 3 and 4, we will now briefly demonstrate our algorithm in three dimensions. Figure 9 shows the topological non-trival mesh that is the starting point for the calculations depicted in Figs. 10–13.

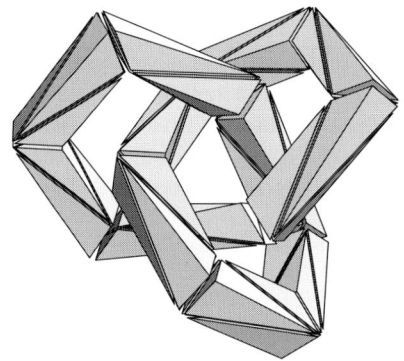

Fig. 9. A tetrahedral mesh with the shape of a trefoil knot

The problem of cell intersections caused by edge collapses is illustrated in Fig. 10: An edge collapse in the original mesh of Fig. 9 has caused two self-intersections of the nonconvex mesh. (One of these self-intersections is magnified.)

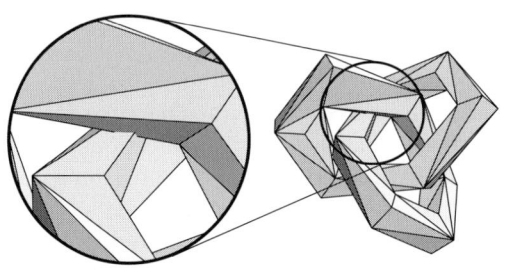

Fig. 10. The tetrahedral mesh of Fig. 9 after one edge collapse

In analogy to Fig. 4, the convexification of the trefoil mesh is depicted in Fig. 11. In order to simplify the original mesh, edge collapses have to be performed in the resultant, convexified mesh.

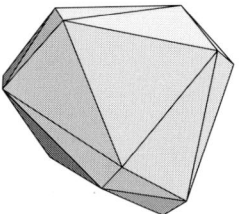 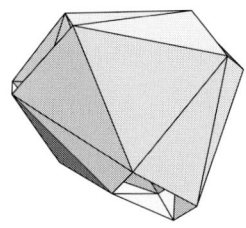 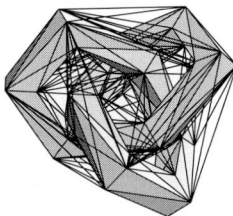

Fig. 11. (Intermediate) results of the convexification of the mesh shown in Fig. 9. From left to right: the convex hull of the mesh; the nonconvex polyhedron between the convex hull and the boundary of the mesh, which has to be tetrahedralized; and the mesh together with imaginary tetrahedra generated by the tetrahedralization (only edges of imaginary cells are shown)

As mentioned, the convexification requires the computation of the convex hull of the mesh, the computation of the space between the mesh's boundary and its convex hull, and the tetrahedralization of this space. The most difficult problem of these three steps is the tetrahedralization. Although a general algorithm for this problem exists [1], we did not implement it yet, but decided to use a strongly simplified variant of this algorithm in order to process our specific example.

Figure 12 shows a simplified mesh after several edge collapses. The employed geometric tests for cell inversions guarantee that there are no self-intersections and hamper further edge collapses. Note that more edge collapses were possible, if the new vertex is not placed in the center of the collapsed edge. However, we did not exploit this freedom in our example.

The topology of the mesh in Fig. 12 is obviously different from the original mesh as the knot was disconnected. Topological tests in a simplification of the convexified mesh of Fig. 11 guarantee the preservation of edge connectivity as shown in Fig. 13; therefore, the simplification process is halted earlier than without these tests. Again, further simplification steps are possible with more general edge collapses.

6 Conclusions

As we have demonstrated, the idea of Williams to convert nonconvex into convex meshes can be successfully applied to solve the problems of edge collapses in nonconvex tetrahedral meshes with topologically non-trival boundaries, which may be non-manifolds. Intersections of cells, new nonconvexities

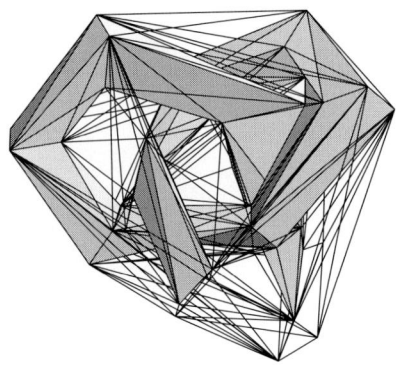

Fig. 12. The result of a simplification of the convexified mesh depicted in Fig. 11 without topological tests

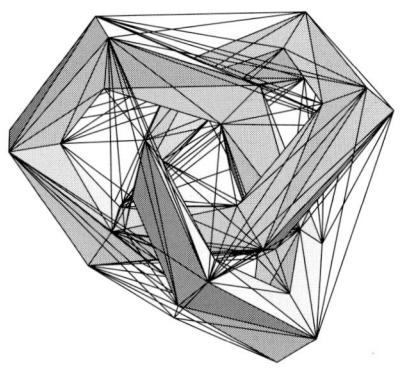

Fig. 13. Same as Fig. 12 with additional topological tests

of the boundary, and modifications of the topology are identified and avoided by efficient, local tests. Therefore, the particular problems of simplifying nonconvex meshes by edge collapses are in principle solved.

However, many problems are still open: an efficient and robust computation of the convex hull of an arbitrary tetrahedral mesh; an efficient and robust computation of the tetrahedralization of an arbitrary polyhedron; the integration of complete topology preservation into our system; and, of course, applications to real-world data sets.

References

1. Chazelle B., Palios L. (1990) Triangulating a Nonconvex Polytope. Discr Comp Geom 5:505–526

2. Cignoni P., Costanza D., Montani C., Rocchini C., Scopigno R. (2000) Simplification of Tetrahedral Meshes with Accurate Error Evaluation. In: Ertl T., Hamann B., Varshney A. (Eds.) Visualization 2000, Salt Lake City, Utah, USA, October 8–13, 2000. IEEE Computer Society Press, Piscataway, and ACM Press, New York, 85–92
3. Dey T.K., Edelsbrunner H., Guha S., Nekhayev D.V. (1999) Topology Preserving Edge Contraction. Publ Inst Math (Beograd) 66:23–45
4. Popović J., Hoppe H. (1997) Progressive Simplicial Complexes. In: Whitted T., Mones-Hattal B. (Eds.) SIGGRAPH 97, Los Angeles, California, USA, August 3–8, 1997. ACM Press, New York, 217–224
5. Staadt O.G., Gross M.H. (1998) Progressive Tetrahedralizations. In: Evert D., Hagen H., Rushmeier H. (Eds.) Visualization '98, Research Triangle Park, North Carolina, USA, October 18–23, 1998. IEEE Computer Society Press, Piscataway, and ACM Press, New York, 397–402
6. Trotts I.J., Hamann B., Joy K.I., Wiley D.F. (1998) Simplification of Tetrahedral Meshes. In: Evert D., Hagen H., Rushmeier H. (Eds.) Visualization '98, Research Triangle Park, North Carolina, USA, October 18–23, 1998. IEEE Computer Society Press, Piscataway, and ACM Press, New York, 287–295
7. Williams P.L. (1992) Visibility Ordering Meshed Polyhedra. ACM Trans Graph 11(2):103-126

A Framework for Visualizing Hierarchical Computations

Terry J. Ligocki[1], Brian Van Straalen[2], John M. Shalf[1], Gunther H. Weber[3,4], and Bernd Hamann[1,3]

[1] Lawrence Berkeley National Laboratory, National Energy Research Scientific Computing Center, 1 Cyclotron Road, M/S 50F, Berkeley, CA 94720, USA
[2] Lawrence Berkeley National Laboratory, National Energy Research Scientific Computing Center, 1 Cyclotron Road, M/S 50A-1148, Berkeley, CA 94720, USA
[3] Center for Image Processing and Integrated Computing, Department of Computer Science, University of California, 1 Shields Avenue, Davis, CA 95616-8562, USA
[4] AG Graphische Datenverarbeitung und Computergeometrie, FB Informatik, Universitaet Kaiserslautern, Postfach 3049, D-67653 Kaiserslautern, Germany

Abstract. Researchers doing scientific computations are attempting to accurately model physical phenomena. When these physical phenomena take place at a variety of different spatial and temporal scales it can be more efficient and accurate to model them at different levels of detail in an adaptive, hierarchical manner. We present a framework for visualizing adaptive, hierarchical computations – a conceptual framework and an implementation framework. Given that researchers have already defined a hierarchical structure for their data and are performing their computations using this structure, it has become important to provide a visualization tool which accurately represents this data and visualizes it directly. The tool we have designed for this purpose was built using the Visualization Toolkit, VTK, and one of its interpretive interfaces, Tcl/Tk. In addition to creating a visualization tool, we are developing extensions of visualization techniques and algorithms to hierarchical data (e.g., seamless isosurface generation).

1 Introduction

Researchers doing scientific computations are attempting to accurately model physical phenomena. When those physical phenomena take place at a variety of different scales it can computationally be more efficient and accurate to model them at different levels of detail in an adaptive manner. Two groups in the National Energy Research Scientific Computing center, NERSC [14], at the Lawrence Berkeley National Laboratory, LBNL [12], are doing just that. One group is headed by John Bell (Center for Computational Sciences and Engineering, CCSE [4]), and the other is headed by Phil Colella (Applied Numerical Algorithms Group, ANAG [2]). Both groups are doing computations using similar adaptive mesh refinement, AMR, techniques. Since the term "AMR" can mean a variety of things to researchers it should be clarified that we use it to refer exclusively to block-structured AMR data as defined in Berger and Colella [3].

Given that researchers have already defined a hierarchical structure for their data and are performing their computations using this structure, it has become important to provide a visualization tool that accurately represents and visualizes this data. This work has been a joint effort between ANAG and the LBNL/NERSC Visualization Group [17]. It includes extending and modifying visualization algorithms (e.g., isosurface computation, streamline generation) to correctly generate results while taking advantage of inherent computational efficiencies supported by the original AMR data structure. We have chosen ANAG's AMR computational library, Chombo [5], and built an extensible visualization tool, ChomboVis [6], for the data sets Chombo produces. This tool was built using the Visualization Toolkit, VTK [18], and one of its interpretive interfaces, Tcl/Tk [16]. By using VTK we have been able to use a broad foundation of existing algorithms and infrastructure while benefiting from ongoing extensions and improvements to VTK, e.g. [15]. Since VTK is an extensible, object-oriented library, we can add functionality to existing algorithms and add new algorithms relatively easily. The interpretive interface has allowed us to rapidly prototype ideas, add new functionality, benefit from the work other groups are doing with Tcl/Tk, and provide researchers using our tool with the option of directly extending the tool in an interactive fashion.

The development of this tool has proceeded in several directions. First, we have implemented and released ChomboVis, which provides researchers with many of the tools they need to view and investigate their data. This version provides 2D and 3D visualization capabilities including the ability to look at selected data directly using spreadsheets. Second, we have been developing extensions of visualization techniques and algorithms to AMR data (e.g., seamless isosurface generation). These extensions can be integrated into VTK and ChomboVis. Finally, we have been using recent extensions to VTK to handle AMR data sets in a more natural, efficient and direct manner. We believe that the new framework we have been developing provides researchers with a tool that meets their immediate needs, can and will be extended to meet future needs, and will benefit from other work being done by the visualization research community.

2 Past and Current Work

The following is a brief description of the AMR data generated by Chombo. It is intended to give the reader an idea of the structure of the data but not to be a precise or complete definition. The AMR data produced by computations using Chombo consists of a set of regular (structured) grids that are grouped by level. All grids on a given level have the same cell size or resolution. All grids on a given level are completely covered by grids on the next coarser level. The ratio of the cell size on one level to the cell size on the next finer level is always an integer. Finally, data values are cell-centered values and the computations done on AMR grids are finite-difference approximations to partial differential equations, PDEs.

There are many approaches to extending visualization algorithms to AMR data. One of the primary difficulties is that there may be multiple data values at a given point – each value being associated with a different level in the AMR hierarchy. There are several approaches that deal with this difficulty:

1. Treat all grids (and their values) independently.
2. Use the data value(s) from the finest grid available and ignore data value(s) from the coarser grids.
3. Combine the data in some way that is physically meaningful and use the result for visualization.

Each of these approaches can be useful depending on what the user is looking for in the data sets. If users are debugging computational algorithms, the first approach allows them to look at all the data. If they are trying to understand computations and/or present results, the second or third approaches may be better.

2.1 Visualization Tool

We started developing a visualization tool by treating all grids and data independently since it was straightforward to implement and was of immediate value to ANAG researchers as they worked on the Chombo library. Also, it is of considerable value to the users of the Chombo library to have a visualization tool that can be extended. The result was ChomboVis. The first version was released in early 2000, and the second version was released in late 2000.

ChomboVis was built using VTK and its Tcl/Tk interface. VTK was chosen because it is a freely available library that includes source code and thus can be modified and extended directly. ChomboVis contains several new VTK objects – for example, an object that reads the HDF5 output of Chombo [11] and converts it into VTK data objects. Tcl/Tk was used to develop the user interface and to combine VTK objects into a working system. In addition, we envision users directly interacting with ChomboVis via its Tcl/Tk interface and creating custom extensions.

The second release of ChomboVis can be used to visualize 2D and 3D AMR data sets and provides the following capabilities (see Fig. 1 for an illustration of the tool in operation):

- Data selection by scalar variable and level
- Orthogonal data slicing and display
- Multiple isosurface/contour generation
- Spreadsheet viewing of individual grids
- Interactive selection of grids from the visualization (i.e., picking)
- Display of grid bounds, cell size, etc.

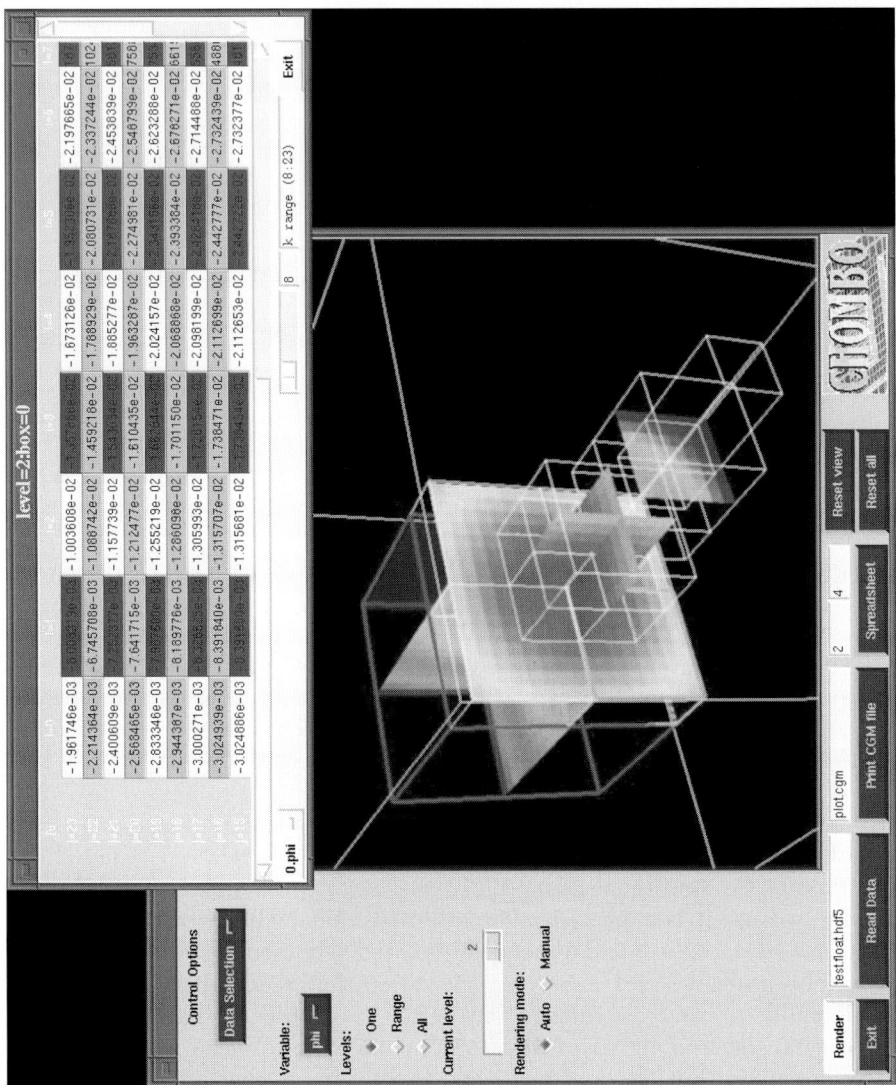

Fig. 1. Visualization of an AMR data set using ChomboVis. (See Color Plate 21 on page 362.)

- User specified colormaps
- Output in Computer Graphics Metafile, CGM, format

These are fairly modest capabilities, and yet they required a substantial amount of development. This was due to the fact that VTK (like many widely available visualization packages) does not directly support multiple grids. Much of the work we did was to attain the goal of providing a mechanism for sending multiple grids through a given VTK pipeline and collecting the results for rendering.

Initially we planned to use some of the recent VTK extensions to handle groups/arrays of grids [1]. This was not possible due to differences between the task the extensions were addressing (domain decomposition) and our task (handling overlapping sets of grids). We then turned to a novel, streaming, out-of-core technique base on work done by Matthew Hall [10] with VTK in order to minimize memory requirements and pipeline overhead.

2.2 Visualization Research

While the work on ChomboVis continued we began to collaborate with researchers from the Center for Image Processing and Integrated Computing, CIPIC [7], in the Department of Computer Science at the University of California, Davis. This collaboration combined nicely the work on AMR data sets done by NERSC researchers with the work done by CIPIC in hierarchical representations for and visualizations of large data sets. The goal of this collaboration was to extend visualization algorithms and techniques to AMR grids. Specifically, the following areas were and are being investigated:

1. Generation and rendering of isosurfaces from AMR data sets with no artifacts due to overlap, gaps, or cracks between grids at different levels.
2. Visualization of vector fields defined in AMR data sets.
3. Visualization of embedded boundaries, EBs, that was being developed by ANAG in conjunction with AMR data structures.
4. Interactive, immersive visualization of large AMR data sets using a variety of techniques (e.g., "seeded isosurface generation").
5. Interactive previewing (fast, with artifacts) and high-quality (slow, without artifacts) volume rendering techniques for AMR data sets.

One goal was to take advantage of the regular AMR grid structure wherever possible and only do additional work where it was necessary. This was one of the general advantages of this type of AMR representation in scientific computations. Substantial progress was made on each project. The accomplishments in each area were:

1. A technique was implemented for handling multiple, overlapping grids [8] [9]. It removes portions of grids that overlap finer grids, handles the rest using marching cubes [13], and generates stitching cells and geometry at the boundaries to create a seamless result.
2. Initial system infrastructure was developed to allow AMR vector data to be manipulated, and a technique was implemented to compute integral curves of AMR vector field data.
3. The infrastructure to represent EB data for boundary reconstruction and its visualization was implemented.

4. Interactive, seeded isosurface algorithms were developed for regular grids, curvilinear grids, and general tetrahedral meshes. Time budgets and work queues are used to achieve interactivity. This work was extended to AMR data sets.
5. Several promising techniques for interactive volume rendering were explored – including a technique that ignores much of the detailed grid structure of AMR data by viewing 3D data as point sets.

3 Future Work

There are ample opportunities for continued work both in the form of direct extensions to current work and more subtle new directions. Some of the successful visualization research will be integrated into ChomboVis, and use of ChomboVis will suggest new areas of visualization research as well.

3.1 Visualization Tool

ChomboVis is a maturing visualization tool for AMR data sets produced by Chombo. It will be used by researchers and improved to meet their needs. There are several aspects to this process. The user interface(s), documentation, and general usability will be improved. This can be done by working with users to get a better understanding of how they use ChomboVis. The performance of the tool will need to be improved for large data sets. This will require substantial modifications to VTK to handle sets of grids in a more direct fashion. There is ongoing work in this area by other users of VTK, and we plan to use this work to help us address performance issues.

We plan to integrate some of the successful AMR visualization research being done into VTK and ChomboVis to provide users with the most advanced visualization tools. Specifically, seamless isosurface generation, AMR vector visualization, and interactive volume rendering are candidates for integration.

3.2 Visualization Research

AMR visualization problems lead to many opportunities for additional research. All visualization techniques being developed could be studied in the interactive, immersive context being developed at CIPIC and NERSC. Researchers will be combining AMR and EB technologies and thus research in visualizing AMR and EB data sets will need to be combined and extended. The computing and visualizing of scalars and vectors derived from quantities originally computed is of considerable interest to researchers. Finally, applying and extending all the visualization tools and research to time-varying AMR data sets provides difficult challenges.

4 Conclusions

By working closely with researchers doing AMR computations and developing visualization tools that AMR simulation scientists use, we have developed a framework for visualizing AMR computations. This has given rise to a number of visualization research questions and problems. Work on these problems has lead to extensions of our original framework and tools. This, in turn, will lead to more fundamental visualization research.

5 Acknowledgements

This work was supported by the Directory, Office of Science, Office of Basic Energy Sciences, of the U.S. Department of Energy under Contract No. DE-AC03-76SF00098; by the National Science Foundation, through the Large Scientific and Software Data Set Visualization (LSSDSV) program and through the National Partnership for Advanced Computational Infrastructure (NPACI); the Office of Naval Research; the Army Research Office; the NASA Ames Research Center through an NRA award; the Lawrence Livermore National Laboratory under an ASCI ASAP Level-2 contract; the Los Alamos National Laboratory; and the North Atlantic Treaty Organization (NATO). We also acknowledge the support of ALSTOM Schilling Robotics and Silicon Graphics, Inc.

In addition, we would like to thank ANAG, the LBNL/NERSC Visualization Group, and the members of the Visualization Thrust at the Center for Image Processing and Integrated Computing (CIPIC) at the University of California, Davis, for their help.

References

1. Ahrens J., Law C., Schroeder W., Martin K., and Papka M. A Parallel Approach for Efficiently Visualizing Extremely Large, Time-Varying Datasets, Los Alamos National Laboratory – Tech. Report #LAUR-00-1620
2. http://seesar.lbl.gov/anag/
3. Berger M. and Colella P. (1989) Local adaptive mesh refinement for shock hydrodynamics. Journal of Computational Physics, 82:64-84, May 1989. Lawrence Livermore Laboratory Report No. UCRL-97196
4. http://seesar.lbl.gov/ccse/
5. http://seesar.lbl.gov/anag/chombo/
6. http://seesar.lbl.gov/anag/chombo/chombovis.html
7. http://graphics.cs.ucdavis.edu/
8. Weber G.H., Kreylos O., Ligocki T.J., Shalf J.M., Hagen H., Hamann B., and Joy K.I. (2001) Extraction of crack-free isosurfaces from adaptive mesh refinement data. Proceedings of the Joint EUROGRAPHICS and IEEE TCVG Symposium on Visualization, Ascona, Switzerlang, May 28-31, 2001, Springer Verlag, Wien, Austria, May 2001

9. Weber G.H., Hagen H., Hamann B., Joy K.I., Ligocki T.J., Ma K.L., and Shalf J.M. (2001) Visualization of adaptive mesh refinement data. Proceedings of the SPIE (Visual Data Exploration and Analysis VIII, San Jose, CA, USA, Jan. 22-23), Volume 4302, SPIE – The International Society for Optical Engineering, Bellingham WA, January 2001
10. http://zeus.ncsa.uiuc.edu/~mahall/
11. http://hdf.ncsa.uiuc.edu/HDF5/
12. http://www.lbl.gov/
13. Lorensen W.E. and Cline H.E. (1987) Marching Cubes: A high resolution 3D surface construction algorithm. Computer Graphics (SIGGRAPH '87 Proceedings), 21(4):163-169, July 1987
14. http://www.nersc.gov/
15. Norman M.L., Shalf J., Levy S., and Daues G. (1999) Diving deep: Data management and visualization strategies for adaptive mesh refinement simulations. Computing in Science and Engineering, 1(4):36-47, July/August 1999
16. Ousterhout J.K. (1994) Tcl and the Tk Toolkit. Addison-Wesley
17. http://www-vis.lbl.gov/
18. Schroeder W., Martin K., and Lorensen B. (1998) The Visualization Toolkit, 2nd Edition. Prentice-Hall

Virtual-Reality Based Interactive Exploration of Multiresolution Data

Oliver Kreylos[1], E. Wes Bethel[2], Terry J. Ligocki[3], and Bernd Hamann[1]

[1] Center for Image Processing and Integrated Computing (CIPIC), Department of Computer Science, University of California, Davis, One Shields Avenue, Davis, CA 95616–8562, USA
[2] Applied Numerical Algorithms Group, National Energy Research and Scientific Computing Center (NERSC), Ernest Orlando Lawrence Berkeley National Laboratory, One Cyclotron Road, Berkeley, CA 94720, USA
[3] Visualization Group, National Energy Research and Scientific Computing Center (NERSC), Ernest Orlando Lawrence Berkeley National Laboratory, One Cyclotron Road, Berkeley, CA 94720, USA

Abstract. We describe a system supporting the interactive exploration of three-dimensional scientific data sets in a virtual reality (VR) environment. This system aids a scientist in understanding a data set by interactively placing and manipulating visualization primitives, e. g., isosurfaces or streamlines, and thereby finding features in the data and understanding its overall structure.

We discuss how the requirement of interactivity influences the architecture of the visualization system, and how to adapt standard visualization techniques to work under real-time interaction constraints.

Though we have implemented our visualization system to work with multiple types of data sets structures – cartesian, tetrahedral, curvilinear-hexahedral and adaptive mesh refinement (AMR) – we will focus on AMR grids and show how their inherent multiresolution structure is useful for interactive visualization.

1 Introduction

Contemporary scientific research is performed by gathering and analyzing vast amounts of data. Regardless of whether this data is the result of real-world measurements or numerical simulations, extracting information from data becomes more and more difficult as the amount of data grows. Scientific visualization is becoming a major tool for research: The human visual system is unparalleled in its capacity to see patterns or detect features in data.

We concentrate on scalar- or vector-valued trivariate functions, i. e., functions of the type $f_s\colon \Omega \to \mathbf{R}$ or $f_v\colon \Omega \to \mathbf{R}^3$, where $\Omega \subset \mathbf{R}^3$ is some three-dimensional domain. We will also consider time-varying data sets, where the functions' domains are $\Omega \times [a,b]$, where $[a,b] \subset \mathbf{R}$ is some interval of time.

1.1 Visual Exploration

Typically, visualization is used for two different purposes: First, it can be used to gain understanding of phenomena underlying data; second, it can be used to communicate this understanding to an audience. We concentrate on the first purpose, *visual exploration*. A scientist is typically confronted with a data set from an experiment or a simulation, and the task is to gain insight into the data. Often, especially when data is generated by a simulation, it is also necessary to show that the simulation system generated meaningful data at all; in this context, visual exploration becomes a debugging tool.

Visual exploration is most useful when applied early during data generation. Simulations generating large data sets typically run for long periods of time, even when using a supercomputer. If a visual exploration system can be used to visualize preliminary or evolving simulation results, it might be possible to detect errors in the simulation early, and to fix them by changing simulation parameters "on the fly." This interplay between visualization and simulation, also referred to as *computational steering*, can be very helpful in increasing the performance of a simulation system and its value for the overall scientific investigation process. To allow coupling visualization and simulation, the visualization system should be able to read the data generated by the simulation with as little pre-processing as possible. Since simulation systems create data in a variety of different formats (tetrahedral, cartesian, (curvilinear) hexahedral, AMR, etc.), the visualization system must support all these data structures directly, without having to resort to converting them to a standard format by re-sampling.

When working directly on the native data structure used by a simulation system, the debugging power of the visual exploration system is greatly increased. For example, an AMR simulation program creates the grid hierarchy without intervention [1], by considering certain properties of the data itself. If the visualization system can display the AMR data directly, it can also display the grid structure, providing the scientist with clues about the simulation process. We explain how to isolate the data set structure from the rest of the visualization program in Sec. 3.

1.2 Benefits of VR Methods

A standard visualization system that addresses both uses of visualization (exploration and communication) usually sacrifices interactivity for image quality. This fact can reduce the usefulness of a general-purpose system for visual exploration. For example, a data set might be the result of a numerical simulation of air flow around an aircraft wing, and the task is to determine whether the design fulfills the required aerodynamic objectives and how the wing design could be improved. In a standard visualization system, this task might be handled in the following way:

1. Place some visualization primitives, e. g., streamlines, at points considered "interesting."
2. Generate an image or multiple images.
3. If the imagery does not reveal anything exciting, repeat from step 1.

There are several problems with such an approach. First, a user has to specify at which points to place primitives. Specifying positions in 3D space using a 2D input device, e. g., a mouse, is awkward. Second, the process of image generation is time-consuming, and with placement of primitives essentially being a trial-and-error process, the overall time spent on analysis can grow prohibitively large. Even worse, merely seconds of delay between placing a primitive and seeing a resulting image can make intuitive exploration of a data set almost impossible. Third, even when an image reveals some features, it is only a 2D projection of a 3D data set and can be misleading concerning spatial relationships of features.

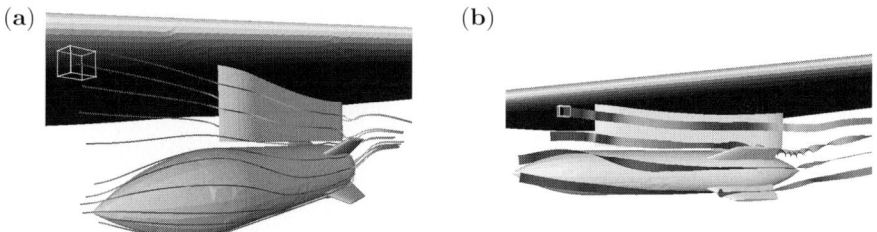

Fig. 1. Exploration of vector field on tetrahedral mesh with embedded geometry. (**a**) Visualization with streamlines (**b**) Visualization with streamribbons. (See Color Plate 22 on page 363.)

VR can attack all three of these problems, making the exploration process much more productive. There are probably more contradicting definitions of VR in existence than there are VR researchers; but for the purposes of this paper, a system is considered a VR system if it offers at least the following functionalities:

- True 3D (stereoscopic) display with head tracking, i. e., support of a display that is dependent on a user's true eye position in space, updated at least at 30 Hz to create the illusion that the visualization is "real,"
- support of six-degree-of-freedom (6-DOF) input devices to allow a user to directly interact with and manipulate the visualization, and
- interactivity, meaning that the system's response time to user actions is small enough to seem immediate (within 0.05 s to 0.1 s).

With a VR system like this, the basic exploration loop remains the same; but with immediate response, a user is able to move a visualization primitive directly through a 3D data set until something interesting is visualized. The immediate update of displayed imagery, leading to an animation of a visualization primitive's behaviour, provides valuable clues as to where to move the primitive next in order to "close in" on a feature.

1.3 Real-time Constraints

The goal of implementing an interactive, real-time visualization system imposes several constraints on the visualization methods that can be supported. A severly limiting constraint is the *display constraint*, which requires that all imagery can be rendered at a frame rate of at least 30 Hz per eye. If the frame rate drops below 30 Hz, the display will appear "choppy," which can lead to eye strain and even motion sickness. Since users cannot hold their heads perfectly still, displayed stereo images have to be updated continuously, even if the visualization itself does not change. This forbids using very costly visualization methods, e. g., high-quality volume rendering [7]. Visualization methods of choice are the ones based on graphics primitives supported by available graphics hardware, e. g., isosurfaces, streamlines or streamribbons.

Even when considering only visualization methods supported by available graphics hardware, generating the visualizations can still be time-consuming. Though it is not necessary to update the visualization for every frame, should updating require more than 0.1 s, the system is "lagging," defeating the goal of interactivity. We refer to this constraint as the *update constraint*. This implies, for example, that a standard implementation of the Marching-Cubes (MC) isosurfacing algorithm [8–10] is not appropriate for visual exploration: Though the resulting isosurface triangulation might consist of a relatively small number of triangles, small enough to satisfy the display constraint, the time to generate the triangles might require seconds or minutes for large data sets, violating the update constraint. In Sec. 3.2, we explain how to adapt standard visualization methods to an interactive environment.

1.4 Related Work

Data visualization based on VR is not new. Several early important contributions were made by NASA scientists: The *NASA Virtual Wind Tunnel*, described, for example, by Bryson and Levit [5], is a visualization system for time-varying vector-valued data. Meyer and Globus [6] describe an interactive isosurface generator for scalar-valued data.

2 User Interface

User interfaces for virtual-reality environments are still an area of active research, and an interface paradigm agreed upon by most researchers has still to be developed. For our system, we implemented the simple gesture- and menu-driven user interface described in the following sections.

2.1 Available VR Devices

The design of the user interface is dependent on the underlying VR hardware, especially on the number of input devices and their degrees of freedom, and the number and arrangement of buttons available. We implemented our system for two classes of VR hardware:

Virtual Workbench A virtual workbench is a semi-immersive VR system consisting of a large display screen, a pair of head-tracked shutter glasses, and three 6-DOF input devices: One *stylus* with a single button, and two *pinch gloves* with four buttons each, that can be activated by pinching the thumb and one of the other fingers.

CAVE A CAVE is a semi-immersive VR system consisting of one or more large display screens, a pair of head-tracked shutter glasses, and a single 6-DOF input device: A *wand* with three buttons and a pressure-sensitive joystick on it. CAVE systems come in different configurations, from a single-screen desk-like environment (ImmersaDesk) to a ten-by-ten-by-ten foot cube with five display screens as walls ("classic" CAVE). For the purposes of our system, we can treat all these identically, because they provide the same functionality in input devices, and the differences of the display hardware are hidden by a common application program interface, the *CAVE library*.

2.2 VR Toolkit

To simplify the task of writing VR applications, we implemented a VR toolkit to isolate the application program from the VR hardware and from basic interaction techniques. The toolkit's programming paradigm is similar to that of the Motif window environment toolkit: It supplies callback mechanisms, actions and simple widgets. The toolkit's architecture is shown in Fig. 2.

Using this VR toolkit, we are able to port a VR application between several different classes of VR hardware without changing the application's source code. Currently, the VR toolkit is available for the two VR systems listed above, and a version for desktop workstations is used for application testing purposes.

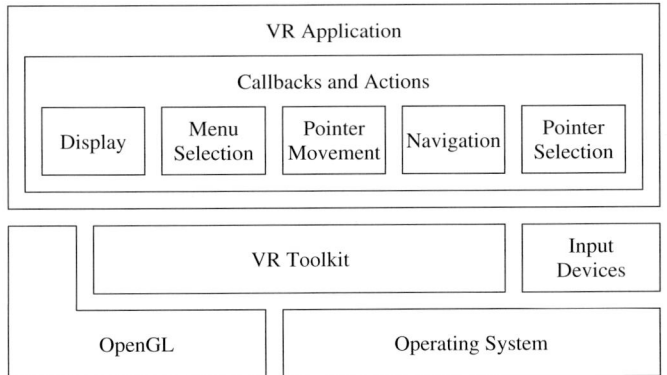

Fig. 2. Overview of the VR toolkit's architecture

2.3 3D Pointer

The main selection and interaction is done with the *3D pointer*. This equivalent to a single-button mouse in a 2D window system is used to inquire data values at any point inside a data set, to place and drag new primitives, and to select existing ones. In the Virtual Workbench environment, we use the stylus as 3D mouse; in the CAVE environment, we use the wand and its leftmost button, the *mouse button*.

2.4 Navigation

In order to explore large data sets, a user must be able to freely navigate through the data sets. It must be possible to get an overview of the whole data set, and it must also be possible to zoom in to a small region of space. Furthermore, navigation must be as intuitive as possible to not interfere with the exploration process. We chose to implement navigation by *direct manipulation*: Instead of using indirect navigation tools, e. g., scrollbars in 2D window systems, a user can directly "grab" a point in space using a 6-DOF input device. Grabbing will fix the data set's coordinate system with respect to the input device's coordinate system, and any movement (translation or rotation) of the input device while a point is grabbed will move the whole data set accordingly, allowing panning and rotating.

In the Virtual Workbench environment, we employ pinch gloves for navigation. A point can be grabbed with either hand, by pinching thumb and index finger. This two-handed navigation suits both left- and right-handed users, and it also allows to navigate with one hand even while the other hand is interacting with the data set using the stylus. In the CAVE environment, the wand is switched into navigation mode using its middle button, the *navigation button*.

Though it is possible to grab any point in space, inexperienced users will intuitively only grab parts of displayed geometry, either existing visualization primitives or models embedded in the data set. Experiments have shown that this is not a serious limitiation; users get adjusted to the navigation paradigm quickly (in a matter of minutes) and are able to pan/rotate data sets according to their wishes.

While navigation involving panning and rotating is very intuitive, zooming is more difficult to implement. This is probably due to the fact that zooming is not possible in the real world: While every child learns how to pick up an object and move it around to inspect it from all sides from a very early age on, there is no natural way to enlarge an object. We implemented zooming in different ways, depending on the available input devices. In an environment with two input devices, a user can grab one point in space with each device. Then, pulling the two devices apart will zoom in, and pushing them together will zoom out. We decided to make the zoom factor proportional to the two device's distance[1]; following this principle, the apparent motion of the data set will relate to the input devices' motion in a natural way. In a single input device environment, we reserved a special button combination on the input device for switching into zoom mode. A user can zoom in by pulling the input device towards her, and zoom out by pushing the input device away.

In the Virtual Workbench environment, both pinch gloves are used for zooming. First, the user will grab one point with one glove, enabling panning and rotating. Then the user will grab another point with the second glove, enabling zooming. The data set's coordinate system is still fixed to the first glove while zooming; this enables panning, rotating and zooming simultaneously. In the CAVE environment, zooming is activated while the wand is already in navigation mode. Pressing the mouse button while the navigation button is held stops panning and rotating and enables zooming. Luckily, the two buttons are close enough to allow even inexperienced users to press both at the same time using one thumb. Still, the restriction to only one input device limits the usability of the system. Experiments have shown that the ambidexterous Virtual Workbench is more intuitive and makes it easier to navigate to a desired viewpoint.

Even in an environment with two devices, experiments show that inexperienced users will hardly use the zoom feature. Instead, they tend to pull a part of the data set they want to examine closer to their eyes, making it appear larger. This is a natural reaction, but it clashes with the limitations of the graphics system. Due to the stereo rendering, objects close to the user's eyes will be out of focus, and due to inaccuracies of head tracking, objects will not appear to be stable but move erratically. Correctly using the zoom feature to inspect an area more closely requires some training.

2.5 Higher-level Functions

To access higher-level program functions, e.g., selecting tools or changing visualization parameters, we provide a simple system of cascading popup menus. The main menu can be popped up at any position and orientation in space. While it is displayed, the 3D pointer is used to select menu entries or pop up submenus.

In the Virtual Workbench environment, the main menu is popped up by pinching thumb and middle finger of one hand. While the menu is displayed, it will move with the hand, allowing the user to move it to a position where it is convenient to select entries using the stylus and the other hand. In the CAVE environment, the rightmost wand button, the *menu button*, will pop up the main menu at the current wand position. As soon as the menu is displayed, the wand is used to activate entries or popup submenus. Releasing the menu button selects the currently active menu item.

---

[1] $z = \|p_1 - p_2\|/d_0$, where p_i are the device's positions, and d_0 is their initial distance at the time grabbing occured.

3 System Architecture

In order to fulfill the requirements of simulation debugging and computational steering, the visual exploration system has to be able to work as closely as possible with the applications generating the data it has to visualize. This means that the system must support a wide variety of data set structures. On the other hand, the system has to support a wide variety of different visualization primitives to address different needs. To make the system robust under changes or additions to both parts, it is necessary to isolate the generation of visualization primitives from the underlying data set structure, and to allow the two modules only to communicate through two well-defined interfaces. An overview of the complete system's architecture is shown in Fig. 3; the following sections describe the modules in detail.

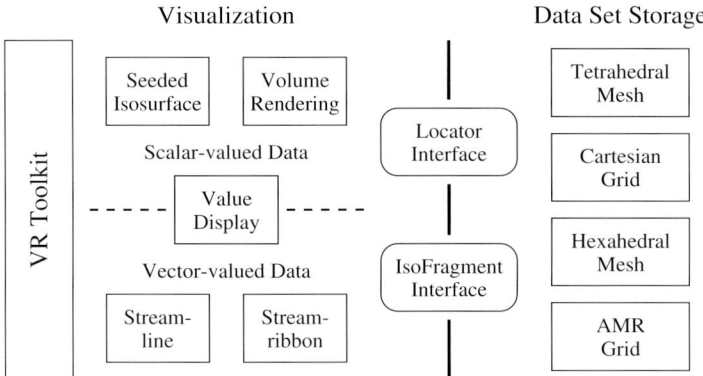

Fig. 3. Overview of the visual exploration system's architecture

3.1 Interfaces

The following sections describe the two interfaces connecting the generation of visualization primitives and the underlying data set structure.

The Locator Interface

The most basic function of any visualization system is to evaluate the function represented by a given data set at any point inside its domain. For most types of data sets, this involves first locating a cell that contains the query point, and then applying some interpolation scheme to calculate the value at the query point's position. To allow maximal efficiency, we defined the Locator interface to contain both functions separately:

- `bool locatePoint(p, bool traceHint)` prepares a locator to interpolate the data set's value at position p. The function's return value indicates whether the given point p is inside the data set's domain. If the return value is `false`, it is not possible to determine the data set's value.

  For some data set structures, locating a point "from scratch" might be very expensive; it might become much less difficult when a locator can use the fact that the new point it has to locate is close to the last located point. The parameter `traceHint` is a hint, provided by the visualization system, that the given point is close to the last point, and that using this information might be beneficial. Several typical visualization primitives, e. g., streamlines, evaluate a data set at a long sequence of points being very close to each other. In some cases, using the `traceHint` parameter resulted in speed-ups of an order of magnitude.

- `ValueType interpolateValue(void)` calculates the data set's value at the last located position.

With these two functions, iterators serve a dual purpose: They are not only used to evaluate a data set, but also to identify positions in a data set that can be passed between data structures and algorithms. In this way, they resemble the iterator mechanism used in the C++ Standard Template Library. The two steps involved in evaluating a data set at an arbitrary point are illustrated in Fig. 4.

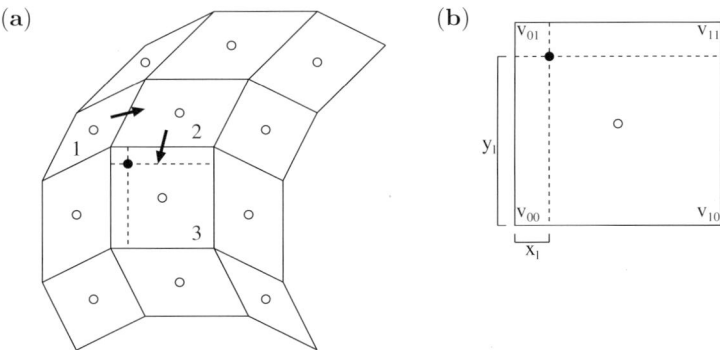

Fig. 4. Steps involved in evaluating a data set. (**a**) Locating cell containing query point (solid dot) and determining its local coordinates. Cell 1 is tried first because its centroid (hollow dots) is closest to the query point. The algorithm then crosses into cells 2 and finally 3 (**b**) Bilinear interpolation used to evaluate the cell at local coordinates (x_l, y_l)

The IsoFragment Interface

The IsoFragment interface has a more specific purpose than the Locator interface: It is used to generate seeded isosurfaces, as will be explained in Sec. 3.2. An IsoFragment identifies a cell in a data set, and it provides the following two functions:

- setCell(Locator loc) associates an IsoFragment with the data set cell containing the point most recently located by the Locator. This function is used to seed an isosurface.
- expandSurface(ValueType isovalue, Queue expandNext) creates the fragment of the surface connecting all points with value isovalue inside the associated cell, and puts all the cell's neighbours sharing faces that are intersected by the isosurface in the expansion queue.

3.2 Supported Visualization Primitives

The choices of visualization primitives for visual exploration systems are limited by the two constraints discussed in Sec. 1.3: The graphics system must be able to display all active primitives at a framerate of at least 30 Hz per eye (display constraint), and the visualization system must be able to update all primitives the user is interacting with at a rate of at least 10 Hz (update constraint). Some common primitives, e. g., streamlines, are almost directly applicable to visual exploration, whereas others, e. g., isosurfaces, have to be adapted to be usable.

Data Value Display

The most basic tool, hardly a visualization primitive, is to display the data value at the current 3D pointer location. It can be applied to both scalar- and vector-valued data sets. The only operation required by this tool is calculation of a function value for an arbitrary point, given in the data set's coordinate system. This operation is directly supported by the Locator interface described in Sec. 3.1. Since the 3D pointer is traced by the system at a rate of 30 Hz, it usually only moves a short distance between two point location requests. Communicating this fact to the Locator interface using the traceHint parameter mentioned in Sec. 3.1 can increase system performance, depending on the underlying data set structure.

Primitives for Scalar-valued Data Sets

The following primitives apply only to scalar-valued data sets, i. e., those defining functions $f\colon \Omega \to \mathbf{R}$.

Seeded Isosurfaces The most often used visualization primitive for scalar-valued data is the isosurface, connecting all points in space having identical function value [8]: For a scalar-valued function $f\colon \Omega \subset \mathbf{R}^3 \to \mathbf{R}$, the isosurface $I(v)$ for isovalue v is defined by $I(v) = \{\, x \in \Omega \mid f(x) = v \,\}$. In standard systems, isosurfaces are typically generated using some variation of the Marching-Cubes algorithm [9]. In this algorithm, isosurface fragments are generated independently for each cell of the data set. This algorithm has a runtime proportional to the number of cells, and generally runs several seconds to minutes for large data sets.

To use isosurfaces in an interactive visualization system, we have to change their generation to satisfy both display and update constraint, see Sec. 1.3. The basic idea is to generate isosurfaces incrementally starting from a given point in space, instead of globally starting from a given function value. In our system, a user can move the 3D pointer to an arbitrary position in space and *seed* an isosurface there. A seeded isosurface is generated using the following main steps:

1. Determine the cell containing the 3D pointer and calculate the function value at the pointer's position. Store the found value as the isovalue to use, and put the found cell in the *expansion queue*.
2. Take the next cell from the expansion queue. Create an *isosurface fragment* for the stored isovalue inside the cell, and determine the cell faces intersected by the isosurface fragment. Put all cells sharing those faces that have not yet been visited back in the expansion queue. An illustration of this process is shown in Fig. 5.
3. While there are cells in the expansion queue, repeat from step 2.

This algorithm will generate an isosurface for the function value at the 3D pointer's position. It can be used in an interactive system, because the expansion loop (step 2 in the algorithm) can be stopped at any time. To satisfy the update constraint, the system will start a timer before starting expansion, and will stop expanding as soon as the timer expires. To satisfy the display constraint, the loop will be stopped as soon as a certain amount of geometry has been generated. This amount depends on the underlying graphics system's performance; typical values are between 100,000 and 200,000 triangles. The described algorithm needs both interfaces described in Sec. 3.1. The Locator interface is needed to calculate the isovalue at the 3D pointer's position, and the two main operations in the algorithm, surface fragment generation and queueing of neighbouring cells, are provided by the IsoFragment interface.

The benefit of seeded isosurfaces is that they can always be generated at interactive rates, independently of data set size and performance of the underlying hardware. They also scale directly with hardware performance; on a faster CPU, the program will automatically generate larger isosurface parts. Furthermore, we found out that animation of isosurfaces, by moving

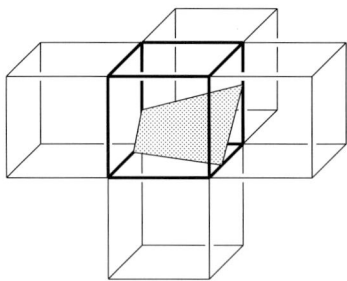

Fig. 5. Expanding a seeded isosurface. The IsoFragment interface generates a fragment of an isosurface inside a cell, and puts all neighbouring cells the isosurface continues into in an expansion queue

the 3D pointer while continuosly seeding, enables a user to quickly gain understanding of the behaviour of the data set in a region of space.

Their main drawback is, that in most cases only a part of the complete isosurface for a given isovalue is generated. This prohibits the user from getting a "big picture" overview of a complete data set. To offset this drawback, we decided to continue growing seeded isosurfaces up to the limits set by the display constraint once the user stops moving the 3D pointer. In this way, larger portions of an isosurface can be created. Figure 6 shows a seeded isosurface during animation and after the user stopped animation.

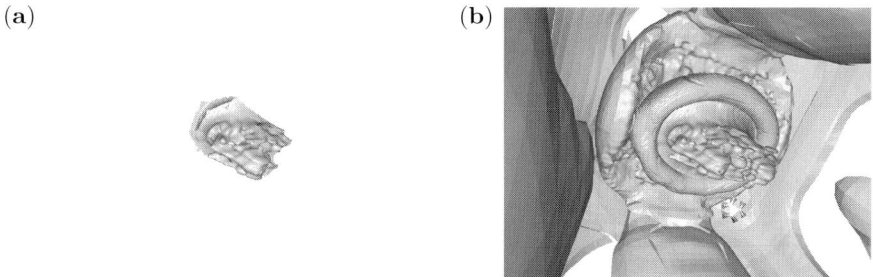

Fig. 6. A seeded isosurface. (**a**) Isosurface while being animated (**b**) Isosurface after it has been released

Primitives for Vector-valued Data Sets

The following primitives apply only to vector-valued data sets, i.e., those defining functions $f: \Omega \to \mathbf{R}^3$.

Streamlines Streamlines are probably the most commonly used visualization primitive for vector-valued functions. By definition, a streamline for a vector-valued function $f: \Omega \subset \mathbf{R}^3 \to \mathbf{R}^3$ is an integral curve $p: [t_0, t_1] \to \Omega$ defined by an initial value problem $p(t_0) = x_0$, $\dot{p}(t) = f(p(t))$. A streamline directly visualizes the local vector directions of a vector-valued function. Fig. 1 (a) shows a visualization of a tetrahedral vector field using streamlines.

In our system, a user can create a streamline by selecting a starting point $x_0 \in \Omega$ inside the data set's domain. The points defining the streamline are then generated from the initial value problem using an adaptive step-size fourth-order Runge–Kutta method. This generation algorithm satisfies both real-time constraints; streamlines are generated one point at a time, and the generation can be interrupted when a computation timer runs out. As all interactive visualization primitives, the user can animate a streamline by moving the 3D pointer while the streamline is continuously regenerated. We found out that this animation is very helpful in understanding the behaviour of the function represented by a data set. By observing a streamline's "reaction" to moving the start point, a user can intuitively "home in" to critical points, and can get an understanding of the function's vector field topology [3].

When using a Runge–Kutta method to generate streamlines, the only functionality needed is to evaluate the given data set at a sequence of positions inside its domain. This functionality is provided by the Locator interface. We observe that subsequent evaluation positions in a Runge–Kutta computation are very close to each other. Thus, to maximize performance, our system uses the `traceHint` parameter to the `locatePoint` function to communicate this fact to the data set storage.

Streamribbons Streamlines as described above only visualize the local vector directions of a vector-valued functions. Often it is desirable to also directly visualize derived properties of a vector-valued function. Streamribbons are a generalization of streamlines that directly visualize the local vector direction and the local vorticity of a vector-valued function [4]. Instead of being a single curve, a streamribbon is rendered as a thin strip, whose rotation around its longitudinal axis is proportional to the dot product of the visualized function's local vorticity and the streamribbon's direction. Fig. 1 (b) shows a visualization of a tetrahedral vector field using streamribbons.

Streamribbons are generated similarly to streamlines: A fourth-order Runge–Kutta method is used to iteratively solve the defining initial value problem, and the data set's vorticity is evaluated directly through the Locator interface. As for streamlines, the computation can be interrupted at any time to satisfy the real-time constraints.

3.3 Data Set Structures

In this section, we describe the different data set structures implemented in our system, and how the two interfaces are implemented for each.

Cartesian Grids

For our purposes, cartesian grids are the simplest data set structure. A cartesian grid consists of a three-dimensional rectangular arrangement of $i \times j \times k$ rectangular hexahedral cells, where each cell is of identical size $s_x \times s_y \times s_z$.

Locating a point inside a cartesian grid merely involves multiplication and modulo division operations, and interpolating the function value at a point inside a cell is done using trilinear interpolation.

To generate an isosurface fragment inside a cell, we use a standard Marching-Cubes case table; to enumerate all intersected neighbours of a cell, we use the implicit neighbourhood relation imposed by the grid structure.

Tetrahedral Meshes

In tetrahedral meshes, locating a point involves finding the tetrahedron containing it, and calculating the point's barycentric coordinates inside that tetrahedron. To locate a point, we use the following two-stage approach:

1. Find the tetrahedron whose centroid is closest to the point in question. We implemented this by computing all cell centroids upon loading a data set, and storing them in an octree to later retrieve them using a closest-point query. If the `traceHint` parameter is set, skip this step and use the tetrahedron containing the last located point.
2. Determine the barycentric coordinates of the query point with respect to the current tetrahedron. If the query point is not inside the tetrahedron, i.e., at least one barycentric coordinate is negative, move to the neighbour sharing the face whose barycentric coordinate is smallest, and repeat this step. Neighbourhood relations between tetrahedra are stored explicitly in the data set, see Fig. 7; thus, finding the tetrahedron sharing a given face is an $O(1)$ operation.

This algorithm, illustrated in Fig. 8, works very well in practice. From-scratch point locations take on the order of $O(\log n)$ steps for n tetrahedra to locate an initial tetrahedron, but they occur rather infrequently. Tracing a point from the initial tetrahedron to the final one containing it usually requires no more than two or three steps, which is sufficiently fast to generate streamlines or streamribbons interactively. Once a point is located, interpolation is done as a convex combination of vertex values using the query point's barycentric coordinates.

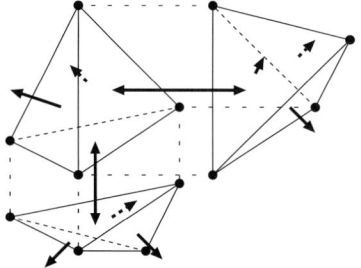

Fig. 7. "Exploded" view of a tetrahedral mesh. Each tetrahedron stores pointers to its four vertices and pointers to the up to four tetrahedra sharing its faces. This allows crossing of tetrahedron faces in $O(1)$ time during point tracing.

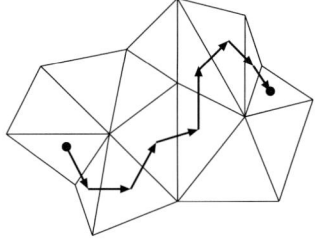

Fig. 8. Tracing a point in a tetrahedral mesh. If the point is not contained in the current tetrahedron, the face associated with the smallest (negative) barycentric weight is crossed

To generate isosurface fragments, we use a standard marching tetrahedron case table. Finding neighbours just involves traversing the explicit neighbour pointers stored with each tetrahedron.

Curvilinear Hexahedral Meshes

Point location in curvilinear hexahedral meshes works very similarly to point location in tetrahedral meshes. We store cell centroids in an octree to quickly locate cells "close" to a point; then we perform the tracing loop to find the final cell containing the query point. Due to the curvilinear cells' almost-cubic structure, tracing usually takes even less steps than for tetrahedral meshes. Similarly to tetrahedral meshes, we store the neighbourhood information explicitly for each cell. This allows $O(1)$ traversal between neighbouring cells, and also allows "stitching" one or more curvilinear meshes together to form a larger one without having to treat traversal between different grids as a special case. An example where stitching is necessary is a fusion plasma data set, shown in Fig. 9, which closes around its x and z axes to form a hollow torus.

The difficult part in point location is calculating a point's local coordinates with respect to a cell. The transformation from cell coordinates to world coordinates is defined by trilinear interpolation of a cell's vertex positions; to transform the other way, the inversion of trilinear interpolation has to be calculated.

Let a hexahedral cell be defined by its eight vertices $\boldsymbol{v}_0, \ldots, \boldsymbol{v}_7 \in \Omega \subset \mathbf{R}^3$. Then trilinear interpolation converts a point $\boldsymbol{p} = (x, y, z)^\mathrm{T} \in [0, 1]^3$ in local cell coordinates to a point

$$\begin{aligned}
\boldsymbol{p}_w = T(\boldsymbol{p}) = \quad & \boldsymbol{v}_0 \cdot (1-x)(1-y)(1-z) + \boldsymbol{v}_1 \cdot x\,(1-y)(1-z) \\
+ & \boldsymbol{v}_2 \cdot (1-x)\,y\,(1-z) + \boldsymbol{v}_3 \cdot x\,y\,(1-z) \\
+ & \boldsymbol{v}_4 \cdot (1-x)(1-y)\,z + \boldsymbol{v}_5 \cdot x\,(1-y)\,z \\
+ & \boldsymbol{v}_6 \cdot (1-x)\,y\,z + \boldsymbol{v}_7 \cdot x\,y\,z
\end{aligned}$$

To invert this transformation, we rewrite the vector equation $\boldsymbol{p}_w = T(\boldsymbol{p})$ as $T(\boldsymbol{p}) - \boldsymbol{p}_w = \boldsymbol{0}$ and solve for the unknown $\boldsymbol{p}$ using Newton–Raphson iteration [11]. The derivative of $T(\boldsymbol{p})$ needed for the Newton–Raphson method

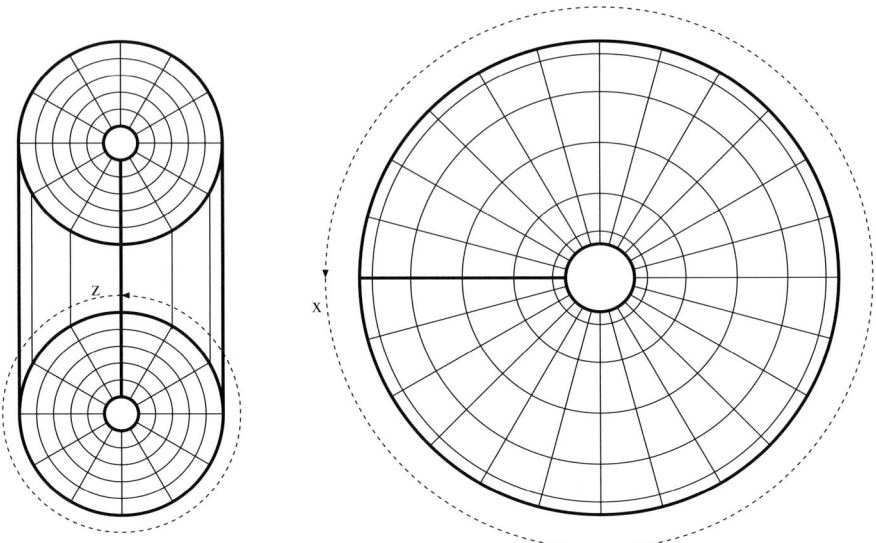

Fig. 9. A curvilinear hexahedral grid forming a hollow torus. Left: A crossection along the main torus axis; right: A view along the main torus axis.

is given by the 3×3-matrix $D(\boldsymbol{p}) := (\frac{\partial}{\partial x}, \frac{\partial}{\partial y}, \frac{\partial}{\partial z}) \cdot T(\boldsymbol{p})$, given by

$$\begin{aligned}
\frac{\partial}{\partial x} T(\boldsymbol{p}) = \ & (\boldsymbol{v}_1 - \boldsymbol{v}_0) \cdot (1-y)(1-z) + (\boldsymbol{v}_3 - \boldsymbol{v}_2) \cdot y\,(1-z) \\
& + (\boldsymbol{v}_5 - \boldsymbol{v}_4) \cdot (1-y) \quad z \quad + (\boldsymbol{v}_7 - \boldsymbol{v}_6) \cdot y \quad z \\
\frac{\partial}{\partial y} T(\boldsymbol{p}) = \ & (\boldsymbol{v}_2 - \boldsymbol{v}_0) \cdot (1-x)(1-z) + (\boldsymbol{v}_3 - \boldsymbol{v}_1) \cdot x\,(1-z) \\
& + (\boldsymbol{v}_6 - \boldsymbol{v}_4) \cdot (1-x) \quad z \quad + (\boldsymbol{v}_7 - \boldsymbol{v}_5) \cdot x \quad z \\
\frac{\partial}{\partial z} T(\boldsymbol{p}) = \ & (\boldsymbol{v}_4 - \boldsymbol{v}_0) \cdot (1-x)(1-y) + (\boldsymbol{v}_5 - \boldsymbol{v}_1) \cdot x\,(1-y) \\
& + (\boldsymbol{v}_6 - \boldsymbol{v}_2) \cdot (1-x) \quad y \quad + (\boldsymbol{v}_7 - \boldsymbol{v}_3) \cdot x \quad y
\end{aligned}$$

Using $T(\boldsymbol{p})$ and $D(\boldsymbol{p})$, the iteration step can now be written as $\boldsymbol{p}_{i+1} := \boldsymbol{p}_i - D(\boldsymbol{p})^{-1} T(\boldsymbol{p})$. As long as the cells are not too oddly shaped, the iteration will converge to a solution after only a few iterations.

Generating isosurface fragments works exactly as for cartesian grids. We use a Marching-Cubes case table to generate fragments, and the explicitly stored neighbourhood information to find neighbouring cells.

AMR Grids

AMR grids [1] are difficult to handle for our visualization system due to their complicated structure, but they offer benefits for interactive visualization due to their inherent multiresolution structure. The AMR grids supported by our system consist of several *levels*, where each level consists of several cartesian grids of identical cell sizes. The grids inside each level are non-overlapping, and the union of their domains forms the domain of the level. The domain of the whole data set is the domain of the first grid level. Let L_i and L_{i+1} be two adjacent levels; then the cell sizes of all grids in L_i must be an integer multiple of the cell sizes of all grids in L_{i+1}. Furthermore, each cell of each grid in L_{i+1} must be completely contained in a cell of some grid of L_i. This implies that the domain of level L_{i+1} is a subset of the domain of level L_i. Figure 10 shows an illustration of an AMR data set's grid structure.

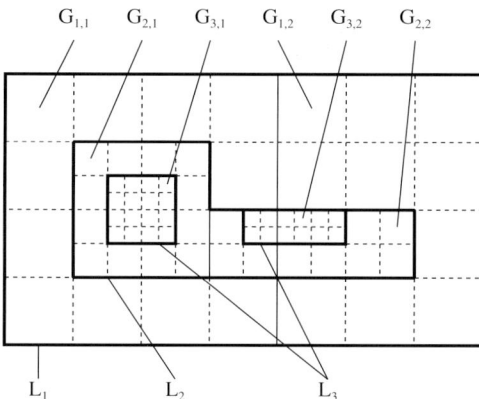

Fig. 10. Structure of an AMR grid with three levels L_1, L_2 and L_3. Each level L_i consists of two grids $G_{i,1}$ and $G_{i,2}$

These restrictions stem from the AMR simulation algorithm: After each iteration step, those cells that have to be subdivided in order to guarantee certain error limits are subdivided independently, and all the subdivided cells are clustered together to form a set of cartesian grids.

The problem in locating points in AMR grids is to find the smallest cell in the hierarchy containing the query point. We implemented this by overlaying the AMR grid's domain with an octree structure. Octree leaves are restricted to only contain cells from a single grid of the most refined grid level whose domain intersects the octree leaves. Octree nodes not satisfying this property are subdivided until their children do, see Fig. 11. From this construction, we know that every point inside an octree leaf must be located in the level and grid associated with the leaf, and that it cannot be overlayed by a grid on a more refined level. Thus, locating a point inside an octree leaf is reduced to location in cartesian grids. When the `tracingHint` parameter to the `locatePoint` function is `true`, the program first checks if the new point position is inside the same octree leaf as the old one. In that case, the point is relocated using the cartesian grid algorithm. Otherwise, the program will traverse the octree structure upwards from the leaf to find an interior node containing the new position, and then downward again to find a leaf

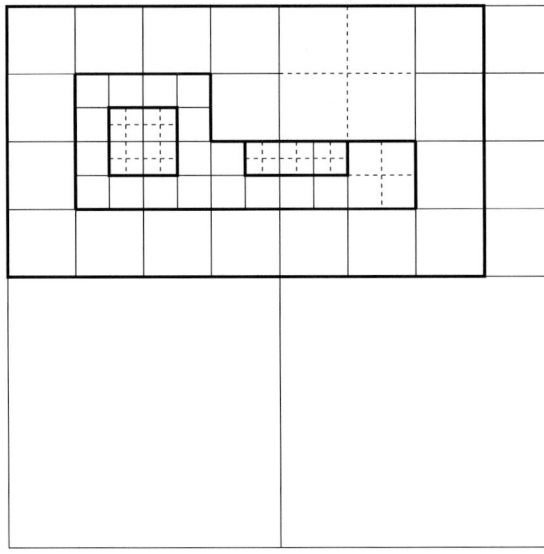

Fig. 11. AMR grid with overlayed octree structure used to locate points. Octree leaves only contain complete cells from a single grid of the most refined level whose domain intersects the octree leaf

containing it. Inside this leaf, the point is finally located using the cartesian grid algorithm again.

Creating isosurfaces inside AMR grids is even more complicated. Generating the isosurface fragment itself is done by using a Marching-Cubes case table, since all finest-level cells are cartesian grid cells. The problem arises when cell neighbours have to be enumerated. As long as a cell is not touching the border of the octree leaf containing it, its neighbours are determined as for cartesian grids. Otherwise, the octree is traversed upwards and then downwards again to find all cells sharing the originating cell's traversed face. Depending on the local refinement, the possible configurations can be one neighbour of identical or larger size, or several smaller neighbours. Both cases are handled by the recursive octree traversal mechanism.

The major problem with this approach is that isosurfaces can exhibit "cracks" when they cross boundaries between differently refined grids. This problem might be fixed in the future by using the grid structures and algorithms described in [2]

Multiresolution Visualization AMR grids in the form described here are inherently a multiresolution representation of their underlying functions. Since grid cells that are overlayed by grids in more refined levels still have meaningful values, usually generated by some subsampling technique, one can visualize a lower-resolution version of an AMR grid by just ignoring several of the more refined levels. This can be done on-the-fly by tagging each octree node with the level of the grid they are contained in. This way, interior nodes can be treated like leaves when their grid level is at least as refined as the maximum level that should be considered at the selected level of resolution. The other parts of the AMR grid code do not have to be changed to accomodate multiresolution visualization in this way.

4 Conclusions and Future Work

We presented an interactive VR visual exploration system that can be used by scientists to explore large data sets of different structures. It is specifically designed to work closely with the simulation systems generating the visualized data, to allow exploration of preliminary or evolving data and to enable "computational steering." We discussed how the requirement of interactivity has influenced the architecture of our system and the choice of supported visualization primitives and their implementation. We described how to isolate the two major system components, visualization and data set storage, from each other by introducing small, well-defined interfaces, and how these interfaces are implemented for the supported data set structures. We concentrated on visualizing AMR data, and on how to exploit the inherent multiresolution structure of AMR grids for interactive visualization.

The main areas for future work are increasing the range of visualization tools available, especially by adding localized volume rendering, implementing a crack-free isosurface algorithm for AMR grids, and implementing a feedback mechanism that allows our system to be used for computational steering, by visualizing "live" data from a running simulation and allowing to change simulation parameters from within the visualization program.

5 Acknowledgments

This work was supported by the Lawrence Berkeley National Laboratory; the National Science Foundation under contract ACI 9624034 (CAREER Award), through the Large Scientific and Software Data Set Visualization (LSSDSV) program under contract ACI 9982251, and through the National Partnership for Advanced Computational Infrastructure (NPACI); the Office of Naval Research under contract N00014–97–1–0222; the Army Research Office under contract ARO 36598–MA–RIP; the NASA Ames Research Center through an NRA award under contract NAG2–1216; the Lawrence Livermore National Laboratory under ASCI ASAP Level–2 Memorandum Agreement B347878 and under Memorandum Agreement B503159; the Los Alamos National Laboratory; and the North Atlantic Treaty Organization (NATO) under contract CRG.971628. We also acknowledge the support of ALSTOM Schilling Robotics and SGI. The data sets depicted here were provided by Paresh Parikh at Vigyan, Inc., by Zhihong Lin at the Princeton Plasma Physics Laboratory, and by the Center for Computational Sciences and Engineering at the Lawrence Berkeley National Laboratory. We thank the members of the Visualization Group at the Lawrence Berkeley National Laboratory and the members of the Visualization and Graphics Research Group at the Center for Image Processing and Integrated Computing (CIPIC) at the University of California, Davis.

References

1. Berger, M., and Colella, P., *Local Adaptive Mesh Refinement for Shock Hydrodynamics*, in: Journal of Computational Physics, 82:64–84, May 1989. Lawrence Livermore Laboratory Report No. UCRL–97196
2. Weber, G. H., Kreylos, O., Ligocki, T. J., Shalf J. M., Hagen H., Hamann, B., and Joy, K. I., *Extraction of Crack-Free Isosurfaces from Adaptive Mesh Refinement Data*, Proceedings of the Joint EUROGRAPHICS and IEEE TCVG Symposium on Visualization, Ascona, Switzerland, May 28–31, 2001, Springer Verlag, Wien, Austria, May 2001
3. Helman, J. L., and Hesselink, L., *Representation and Display of Vector Field Topology in Fluid Flow Data Sets*, in: Computer 22(8) (1989), pp. 27–36
4. Ueng, S.-K., Sikorski, C., and Ma, K.-L., *Efficient Streamline, Streamribbon, and Streamtube Constructions on Unstructured Grids*, in: IEEE Transactions on Visualization and Computer Graphics 2(2) (1996), pp. 100–110
5. Bryson, S. and Levit, C., *The Virtual Windtunnel: An Environment for the Exploration of Three-Dimensional Unsteady Flows*, in: Proc. of Visualization '91 (1991), IEEE Computer Society Press, Los Alamitos, CA, pp. 17–24
6. Meyer, T. and Globus, A., *Direct Manipulation of Isosurfaces and Cutting Planes in Virtual Environments*, technical report CS–93–54 (1993), Brown University, Providence, RI
7. Drebin, R. A., Carpenter, L. and Hanrahan, P., *Volume Rendering*, in: Proc. SIGGRAPH '88 (1988), pp. 65–74
8. Bloomenthal, J., *Polygonization of Implicit Surfaces*, in: Computer Aided Geometric Design 5(4) (1988), pp. 341–356
9. Lorensen, W. E. and Cline, H. E., *Marching Cubes: A High Resolution 3D Surface Construction Algorithm*, in: Proc. of SIGGRAPH '87 (1987), pp. 163–169
10. Nielson, G. M., and Hamann, B., *The Asymptotic Decider: Resolving the Ambiguity in Marching Cubes*, in: Proc. of Visualization '91, (1991), IEEE Computer Society Press, Los Alamitos, CA, pp. 83–91
11. Press, W. H., Teukolsky, S. A., Vetterling, W. T., and Flannery, B. P. *Numerical Recipes in C*, 2nd ed. (1992), Cambridge University Press, Cambridge, MA

Hierarchical Indexing for Out-of-Core Access to Multi-Resolution Data

Valerio Pascucci and Randall J. Frank

Center for Applied Scientific Computing, Lawrence Livermore National Laboratory

Abstract. Increases in the number and size of volumetric meshes have popularized the use of hierarchical multi-resolution representations for visualization. A key component of these schemes has become the adaptive traversal of hierarchical data-structures to build, in real time, approximate representations of the input geometry for rendering. For very large datasets this process must be performed out-of-core. This paper introduces a new global indexing scheme that accelerates adaptive traversal of geometric data represented as binary trees by improving the locality of hierarchical/spatial data access. Such improvements play a critical role in the enabling of effective out-of-core processing.

Three features make the scheme particularly attractive: (i) the data layout is independent of the external memory device blocking factor, (ii) the computation of the index for rectilinear grids is implemented with simple bit address manipulations and (iii) the data is not replicated, avoiding performance penalties for dynamically modified data.

The effectiveness of the approach was tested with the fundamental visualization technique of rendering arbitrary planar slices. Performance comparisons with alternative indexing approaches confirm the advantages predicted by the analysis of the scheme.

1 Introduction

The real time processing of very large volumetric meshes introduces unique algorithmic challenges due to the impossibility of fitting the data in the main memory of a computer. The basic assumption (RAM computational model) of uniform-constant-time access to each memory location is not valid because part of the data is stored out-of-core or in external memory. The performance of many algorithms does not scale well in the transition from the in-core to the out-of-core processing conditions. This performance degradation is due to the high frequency of I/O operations that start dominating the overall running time (thrashing).

Out-of-core computing [22] addresses the issues of algorithm redesign and data layout restructuring, necessary to enable data access patterns with minimal out-of-core processing performance degradation. Research in this area is also valuable in parallel and distributed computing, where one has to deal with the similar issue of balancing processing time with data migration time.

The solution of the out-of-core processing problem is typically divided into two parts:

(i) algorithm analysis to understand its data access patterns and, when possible, redesign to maximize data access locality;

(ii) storage of the data in secondary memory with a layout consistent with the access patterns of the algorithm, making it possible to amortize the cost individual I/O operations over several memory access operations.

In the case of hierarchical visualization algorithms for volumetric data, the 3D input hierarchy is traversed to build derived geometric models with adaptive levels of detail. The shape of the output models are then modified dynamically with incremental updates of their level of detail. The parameters that govern this continuous modification of the output geometry are dependent on runtime user interaction, making it impossible to determine, *a priori*, what levels of detail will be constructed. For example, parameters can be external, such as the viewpoint of the current display window or internal, such as the isovalue of an isocontour or the position of an orthogonal slice. The general structure of the access pattern can be summarized into two main points: (i) the input hierarchy is traversed coarse to fine, level by level so that the data in the same level of resolution is accessed at the same time and (ii) within each level of resolution the data is primarily traversed coherently in regions that are in close geometric proximity.

In this paper we introduce a new static indexing scheme that induces a data layout satisfying both requirements (i) and (ii) for the hierarchical traversal of n-dimensional regular grids. The scheme has three key features that make it particularly attractive. First, the order of the data is independent of the out-of-core blocking factor so that its use in different settings (e.g. local disk access or transmission through a network) does not require any large data reorganization. Second, conversion from the standard Z-order indexing to the new indexing scheme can be implemented with a simple sequence of bit-string manipulations making it appealing for possible hardware implementations. Third, there is no data replication, avoiding any performance penalty for dynamic updates or any inflated storage costs typical of most hierarchical and out-of-core schemes.

Beyond the theoretical interest in developing hierarchical indexing schemes for n-dimensional space filling curves, our approach targets practical applications in out-of-core visualization algorithms. In this paper, we report algorithmic analysis and experimental results for the case of slicing large volumetric datasets.

The remainder of this paper is organized as follows. Section 2 discusses briefly previous work in related areas. Section 3 introduces the general framework for the computation of the new indexing scheme. Section 4 discusses the implementation of the approach for binary tree hierarchies. Section 5 analyzes the application of the scheme for progressive computation of orthogonal slices reporting experimental timings for memory mapped files. Section 6

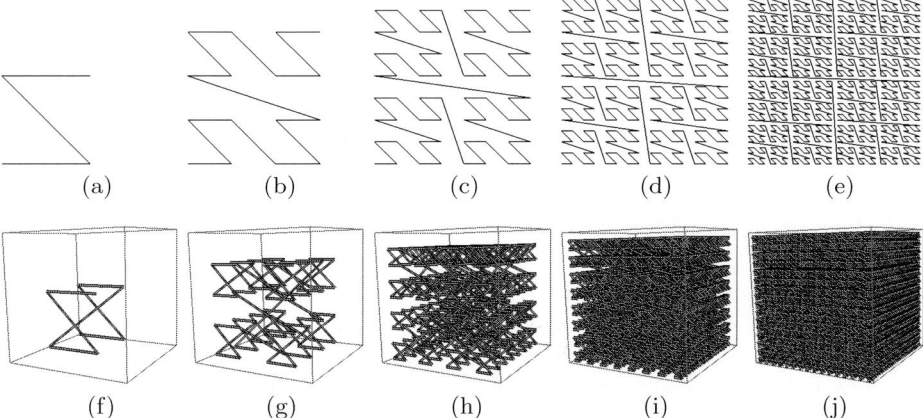

Fig. 1. (a–e) The first five levels of resolution of the 2D Lebesgue's space filling curve. (f–j) The first five levels of resolution of the 3D Lebesgue's space filling curve.

presents the structure of the I/O system and practical results obtained with compressed data. Concluding remarks and future directions are discussed in section 7.

2 Related Previous Work

External memory algorithms [22], also known as out-of-core algorithms, have been rising to the attention of the computer science community in recent years as they address, systematically, the problem of the non-uniform memory structure of modern computers (e.g. L1/L2 cache, main memory, hard disk, etc). This issue is particularly important when dealing with large datastructures that do not fit in the main memory of a single computer since the access time to each memory unit is dependent on its location. New algorithmic techniques and analysis tools have been developed to address this problem in the case of geometric algorithms [1,2,9,15] and scientific visualization [4,8]. Closely related issues emerge in the area of parallel and distributed computing where remote data transfer can become a primary bottleneck in the computation. In this context space filling curves are often used as a tool to determine, very quickly, data distribution layouts that guarantee good geometric locality [10,18,20]. Space filling curves [21] have been also used in a wide variety of applications [3] because of their hierarchical fractal structure as well as for their well known spatial locality properties. The most popular is the Hilbert curve [11] which guarantees the best geometric locality properties [19]. The pseudo-Hilbert scanning order [7,6,12] generalizes the scheme to rectilinear grids that have different number of samples along each coordinate axis.

Recently Lawder [13,14] explored the use of different kinds of space filling curves to develop indexing schemes for data storage layout and fast retrieval in multi-dimensional databases.

Balmelli at al. [5] use the Z-order (Lebesgue) space filling curve to navigate efficiently a quad-tree data-structure without using pointers.

In the approach proposed here a new data layout is used to allow efficient progressive access to volumetric information stored in external memory. This is achieved by combining interleaved storage of the levels in the data hierarchy while maintaining geometric proximity within each level of resolution (multidimensional breadth first traversal). One main advantage is that the resulting data layout is independent of the particular adaptive traversal of the data. Moreover the same data layout can be used with different blocking factors making it beneficial throughout the entire memory hierarchy.

3 Hierarchical Subsampling Framework

This section discusses the general framework for the efficient definition of a hierarchy over the samples of a dataset.

Consider a set S of n elements decomposed into a hierarchy $\mathcal{H}$ of k levels of resolution $\mathcal{H} = \{S_0, S_1, \ldots, S_{k-1}\}$ such that:

$$S_0 \subset S_1 \subset \cdots \subset S_{k-1} = S$$

where S_i is said to be coarser than S_j if $i < j$. The order of the elements in S is defined by a cardinality function $I : S \to \{0 \ldots n-1\}$. This means that the following identity always holds:

$$S[I(s)] \equiv s$$

where square brackets are used to index an element in a set.

One can define a derived sequence $\mathcal{H}'$ of sets S'_i as follow:

$$S'_i = S_i \backslash S_{i-1} \qquad i = 0, \ldots, k-1$$

where formally $S_{-1} = \emptyset$. The sequence $\mathcal{H}' = \{S'_0, S'_1, \ldots, S'_{k-1}\}$ is a partitioning of S. A derived cardinality function $I' : S \to \{0 \ldots n-1\}$ can be defined on the basis of the following two properties:

- $\forall s, t \in S'_i : I'(s) < I'(t) \Leftrightarrow I(s) < I(t)$;
- $\forall s \in S'_i, \forall t \in S'_j : i < j \Rightarrow I'(s) < I'(t)$.

If the original function I has strong locality properties when restricted to any level of resolution S_i then the cardinality function I' generates the desired global index for hierarchical and out-of-core traversal. The scheme has strong locality if elements with close index values are also close in geometric position. These locality properties are well studied in [17].

The construction of the function can be achieved in the following way: (i) determine the number of elements in each derived set S'_i and (ii) determine a cardinality function $I''_i = I'|_{S'_i}$ restriction of I' to each set S'_i. In particular if c_i is the number of elements of S'_i one can predetermine the starting index of the elements in a given level of resolution by building the sequence of constants $C_0, \ldots, C_{k-1}$ with

$$C_i = \sum_{j=0}^{i-1} c_j. \tag{1}$$

Next, one must determine a set of local cardinality functions $I''_i : S'_i \to \{0 \ldots c_i - 1\}$ so that:

$$\forall s \in S'_i : I'(s) = C_i + I''_i(s). \tag{2}$$

The computation of the constants C_i can be performed in a preprocessing stage so that the computation of I' is reduced to the following two steps:

- given s determine its level of resolution i (that is the i such that $s \in S'_i$);
- compute $I''_i(s)$ and add it to C_i.

These two steps must be performed very efficiently as they will be executed repeatedly at run time.

4 Binary Trees And the Lebesgue Space Filling Curve

This section reports the details of how to derive from the Z-order space filling curve the local cardinality functions I''_i for a binary tree hierarchy in any dimension and its remapping to the new index I'.

4.1 Indexing the Lebesgue Space Filling Curve

The Lebesgue space filling curve, also called Z-order space filling curve for its shape in the 2D case, is depicted in figure 1(a-e). The Z-order space filling curve can be defined inductively by a base Z shape of size 1 (figure 1a) whose vertices are replaced each by a Z shape of size $\frac{1}{2}$ (figure 1b). The vertices obtained are then replaced by Z shapes of size $\frac{1}{4}$ (figure 1c) and so on. In general, the ith level of resolution is defined as the curve obtained by replacing the vertices of the $(i-1)^{th}$ level of resolution with Z shapes of size $\frac{1}{2^i}$. The 3D version of this space filling curve has the same hierarchical structure with the only difference that the basic Z shape is replaced by a connected pair of Z shapes lying on the opposite faces of a cube as shown in Figure 1(f). Figure 1(f-j) show five successive refinements of the 3D Lebesgue space filling curve. The d-dimensional version of the space filling curve has also the same hierarchical structure, where the basic shape (the Z of the 2D case) is defined as a connected pair of $(d-1)$-dimensional basic shapes lying on the opposite faces of a d-dimensional cube.

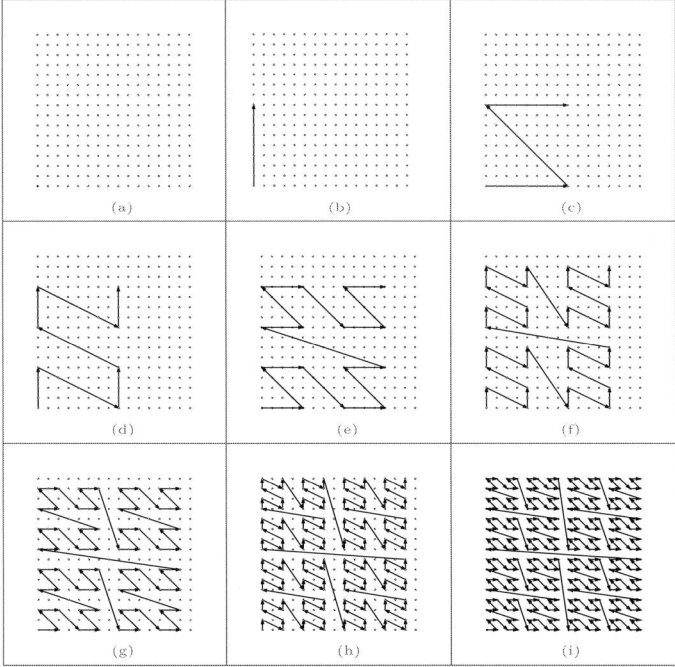

Fig. 2. The nine levels of resolution of the binary tree hierarchy defined by the 2D space filling curve applied on 16×16 rectilinear grid. The coarsest level of resolution (a) is a single point. The number of points that belong to the curve at any level of resolution (b-i) is double the number of points of the previous level.

The property that makes the Lebesgue's space filling curve particularly attractive is the easy conversion from the d indices of a d-dimensional matrix to the 1D index along the curve. If one element e has d-dimensional reference $(i_1, \ldots, i_d)$ its 1D reference is built by interleaving the bits of the binary representations of the indices $i_1, \ldots, i_d$. In particular if i_j is represented by the string of h bits "$b_j^1 b_j^2 \cdots b_j^h$" (with $j = 1, \ldots, d$) then the 1D reference I of e is represented the string of hd bits $I = $ "$b_1^1 b_2^1 \cdots b_d^1 b_1^2 b_2^2 \cdots b_d^2 \cdots b_1^h b_2^h \cdots b_d^h$".

Table 1. Structure of the hierarchical indexing scheme for binary tree combined with the order defined by the Lebesgue space filling curve.

level	0	1	2	3	4
Z-order index (2 levels)	0	1			
Z-order index (3 levels)	0	2	1 3		
Z-order index (4 levels)	0	4	2 6	1 3 5 7	
Z-order index (5 levels)	0	8	4 12	2 6 10 14	1 3 5 7 9 11 13 15
hierarchical index	0	1	2 3	4 5 6 7	8 9 10 11 12 13 14 15

The 1D order can be structured in a binary tree by considering elements of level i, those that have the last i bits all equal to 0. This yields a hierarchy where each level of resolution has twice as many points as the previous level. From a geometric point of view this means that the density of the points in the d-dimensional grid is doubled alternatively along each coordinate axis. Figure 2 shows the binary hierarchy in the 2D case where the resolution of the space filing curve is doubled alternatively along the x and y axis. The coarsest level (a) is a single point, the second level (b) has two points, the third level (c) has four points (forming the Z shape) and so on.

4.2 Index Remapping

The cardinality function discussed in section 3 for a binary tree case has the structure shown in table 1. Note that this is a general structure suitable for out-of-core storage of static binary trees. It is independent of the dimension d of the grid of points or of the Z-order space filling curve.

The structure of the binary tree defined on the Z-order space filling curve allows one to easily determine the three elements necessary for the computation of the cardinality. They are: (i) the level i of an element, (ii) the constants C_i of equation (1) and (iii) the local indices I_i''.

i - if the binary tree hierarchy has k levels then the element of Z-order index j in the Z-order belongs to the level $k - h$, where h is the number of trailing zeros in the binary representation of j;

C_i - the total number of elements in the levels coarser than i, with $i > 0$, is $C_i = 2^{i-1}$ with $C_0 = 0$;

I_i'' - if an element has index j and belongs to the set S_i' then $\frac{j}{2^{k-i}}$ must be an odd number, by definition of i. Its local index is then:

$$I_i''(j) = \left\lfloor \frac{j}{2^{k-i+1}} \right\rfloor.$$

The computation of the local index I_i'' can be explained easily by looking at the bottom right part of table 1 where the sequence on indices (1, 3, 5, 7, 9, 11, 13, 15) needs to be remapped to the local index (0, 1, 2, 3, 4, 5, 6, 7). The original sequence is made up of a consecutive series of odd numbers. A right shift of one bit (or rounded division by two) turns them into the desired sequence.

These three elements can be put together to build an efficient algorithm that computes the hierarchical index $I'(s) = C_i + I_i''(s)$ in the two steps shown in the diagram of Figure 3:

1. set to 1 the bit in position $k + 1$;
2. shift to the right until a 1 comes out of the bit-string.

Clearly this diagram could have a very simple and efficient hardware implementation. A software C++ version can be implemented as follows:

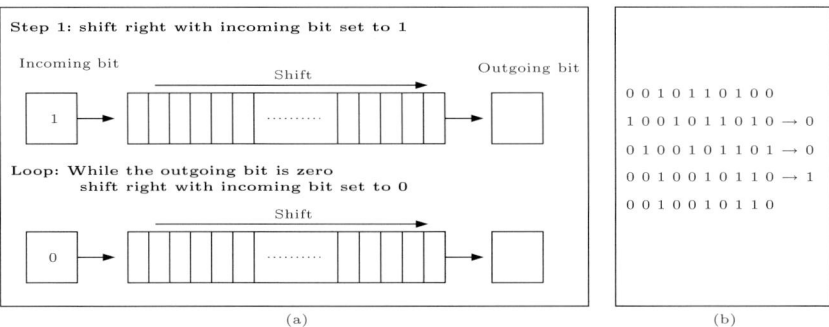

Fig. 3. (a) Diagram of the algorithm for index remapping from Z-order to the hierarchical out-of-core binary tree order. (b) Example of the sequence of shift operations necessary to remap an index. The top element is the original index the bottom is the output remapped index.

```
inline adhocidex remap(register adhocindex i){
   i |= last_bit_mask;  // set leftmost one
   i /= i&-i;           // remove trailing  zeros
   return (i>>1);       // remove rightmost one
}
```

This code will work only on machines with two's complement representation of numbers. In a more portable version one needs to replace `i /= i&-i` with `i /= i&((~i)+1)`.

Figure 4 shows the data layout obtained for a 2D matrix when its elements of are reordered following the index I'. The data is stored in this order and divided into blocks of constant size. The inverse image of such decomposition

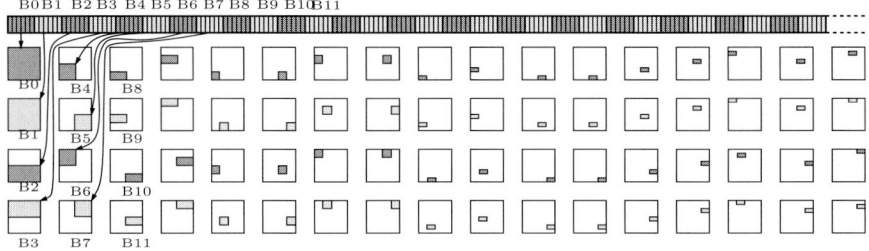

Fig. 4. Data layout obtained for a 2D matrix reorganized using the index I' (1D array at the top). The inverse image of the block decomposition of the 1D array is shown below. Each gray region shows where the block of data is distributed in the 2D array. In particular the first block is the set of coarsest levels of the data distributed uniformly on the 2D array. The next block is the next level of resolution still covering the entire matrix. The next two levels are finer data covering each half of the array. The subsequent blocks represent finer resolution data distributed with increasing locality in the 2D array.

has the first block corresponding to the coarsest level of resolution of the data. The following blocks correspond to finer and finer resolution data that is distributed more and more locally.

5 Computation of Planar Slices

This section presents some experimental results based on the simple, fundamental visualization technique of computing orthogonal slices of a 3D rectilinear grid. The slices can be computed at different levels of resolution to allow real time user interactivity independent of the size of the dataset. The data layout proposed here is compared with the two most common array layouts: the standard row major structure and the $h \times h \times h$ brick decomposition of the data. Both practical performance tests and complexity analysis lead to the conclusion that the data layout proposed allows one to achieve substantial speedup both when used at coarse resolution and traversed in a progressive fashion. Moreover no significant performance penalty is observed if used directly at the highest level of resolution.

5.1 External Memory Analysis for Axis-Orthogonal Slices

The out-of-core analysis reports the number of data blocks transferred from disk under the assumption that each block of data of size b is transferred in one operation independently of how much data in the block is actually used. At fine resolution the simple row major array storage achieves the best and worst performances depending on the slicing direction. If the overall grid size is g and the size of the output is t, then the best slicing direction requires one to load $O(t/b)$ data blocks (which is optimal) but the worst possible direction requires one to load $O(t)$ blocks (for $b = \Omega(\sqrt[3]{g})$). In the case of simple $h \times h \times h$ data blocking (which has best performance for $h = \sqrt[3]{b}$) the number of blocks of data loaded at fine resolution are $O(\frac{t}{\sqrt[3]{b^2}})$. Note that this is much better than the previous case because the performance is close to (even if not) optimal, and independent of the particular slicing direction. For a subsampling rate of k the performance degrades to $O(\frac{tk^2}{\sqrt[3]{b^2}})$ for $k < \sqrt[3]{b}$. This means that at coarse subsampling, the performance goes down to $O(t)$. The advantage of the scheme proposed here is that independent of the level of subsampling, each block of data is used for a portion of $\sqrt[3]{b^2}$ so that, independently of the slicing direction and subsampling rate, the worst case performance is $O(\frac{t}{\sqrt[3]{b^2}})$. This implies that the fine resolution performance of the scheme is equivalent to the standard blocking scheme while at coarse resolutions it can get orders of magnitude better. More importantly, this allows one to produce coarse resolution outputs at interactive rates independent of the total size of the dataset.

A more accurate analysis can be performed to take into account the constant factors that are hidden in the big O notation and determine exactly

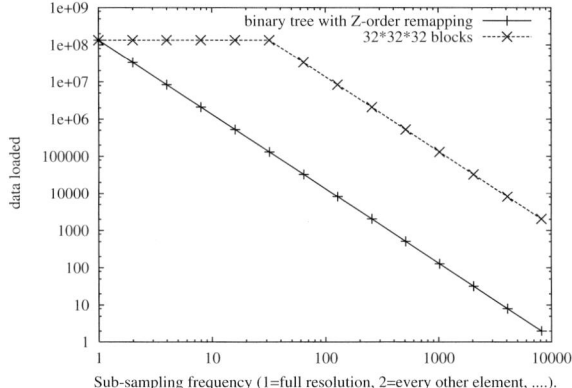

Fig. 5. Maximum data loaded from disk (vertical axis) per slice computed depending on the level of subsampling (horizontal axis) for an 8G dataset. (a) Comparison of the brick decomposition with the binary tree with Z-order remapping scheme proposed here. The values on the vertical axis are reported in logarithmic scale to highlight the difference in orders of magnitude at any level of resolution.

which approach requires one to load into memory more data from disk. We can focus our attention to the case of an 8GB dataset with disk pages on the order of 32KB each as shown in diagram of Figure 5. One slice of data is 4MB in size. In the brick decomposition case, one would use $32 \times 32 \times 32$ blocks of 32KB. The data loaded from disk for a slice is 32 times larger than the output, that is 128MB bytes. As the subsampling increases up to a value of 32 (one sample out of 32) the amount of data loaded does not decrease because each $32 \times 32 \times 32$ brick needs to be loaded completely. At lower subsampling rates, the data overhead remains the same: the data loaded is 32768 times larger than the data needed. In the binary tree with Z-order remapping the data layout is equivalent to a KD-tee constructing the same subdivision of an octree. This maps on the slice to a KD-tree with the same decomposition as a quadtree. The data loaded is grouped in blocks along the hierarchy that gives an overhead factor in number of blocks of $1 + \frac{1}{2} + \frac{1}{4} + \frac{1}{16} + \cdots < 2$ while each block is 16KB. This means that the total amount of data loaded at fine resolution is the same. If the block size must be equal to 32KB the data located would twice as much as the previous scheme. The advantage is that each time the subsampling rate is doubled the amount of data loaded from external memory is reduced by a factor of four.

5.2 Tests with Memory Mapped Files

A series of basic tests were performed to verify the performance of the approach using a general purpose paging system. The out-of-core component of the scheme was implemented simply by mapping a 1D array of data to a file on disk using the `mmap` function. In this way the I/O layer is implemented by

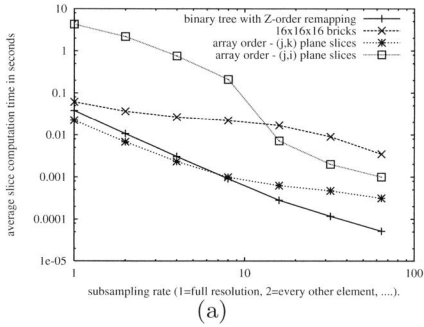
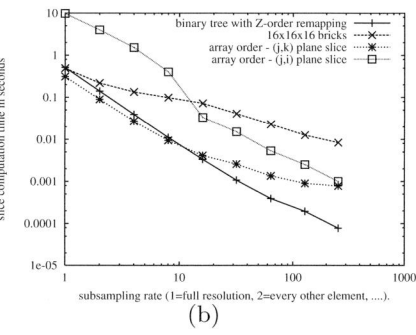

Fig. 6. Two comparisons of slice computation times of four different data layout schemes. The horizontal axis is the level of subsampling of the slicing scheme (test at the finest resolution are on the left).

the operating system virtual memory subsystem, paging in and out a portion of the data array as needed. No multi-threaded component is used to avoid blocking the application while retrieving the data. The blocks of data defined by the system are typically 4KB. Figure 6(a) shows performance tests executed on a Pentium III Laptop. The proposed scheme shows the best scalability in performance. The brick decomposition scheme with 16^3 chunks of regular grids shows the next best compromise in performance. The (i,j,k) row major storage scheme has the worst performance compromise because of its dependency on the slicing direction: best for (j,k) plane slices and worst for (j,i) plane slices. Figure 6(b) shows the performance results for a test on a larger, 8GB dataset, run on an SGI Octane. The results are similar.

6 Budgeted I/O and Compressed Storage

A portable implementation of the indexing scheme based on standard operating system I/O primitives was developed for Unix and Windows. This implementation avoids several application level usability issues associated with the use of `mmap`. The implemented memory hierarchy consists of a fixed size block cache in memory and a compressed disk format with associated meta-data. This allows for a fixed size runtime memory footprint, required by many applications.

The input to this system is a set of sample points, arbitrarily located in space (in our tests, these were laid out as a planar grid) and their associated level in the index hierarchy. Points are converted into a virtual block number and a local index using the hierarchical Z order space filling curve. The block number is queried in the cache. If the block is in the cache, the sample for the point is returned, otherwise, an asynchronous I/O operation for that block

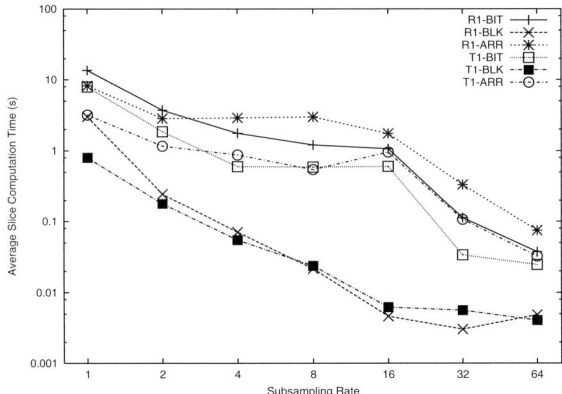

Fig. 7. Average slice computation time for 512^2 sample slices on a Linux laptop with a Intel Pentium III at 500Mhz using a 20MB fixed data cache. Results are given for two different plane access patterns R1 and T2 as well as three different data layouts BIT, BLK and ARR. The input data grid was $2048 \times 2048 \times 1920$.

is added to an I/O queue and the point marked as pending. Point processing continues until all points have been resolved (including pending points) or the system exceeds a user defined processing time limit. The cache is filled asynchronously by I/O threads which read compressed blocks from disk, decompresses them into the cache, and resolve any sample points pending on that cache operation.

The implementation was testing using the same set of indexing schemes noted in the previous section: (BIT) our hierarchical space filling curve, (BLK) decomposition of the data in cubes of size equal to the disk pages and (ARR) storage of the data as a standard row major array. The dataset was one 8GB ($2048 \times 2048 \times 1920$) time-step of the PPM dataset [16] shown in Figure 11 (same results were achieved with the visible female dataset shown in Figure 12). Since the dataset was not a power of two it was conceptually embedded in a 2048^3 grid and reordered. The resulting 1D array was blocked into 64KB segments and compressed segment-wise using zlib. Entirely empty blocks resulting from the conceptual embedding were not stored as they would never be accessed. The re-ordered, compressed data was around 6 percent of the original dataset size, including the extra padding.

Two different slicing patterns were considered. Test R1 is a set of one degree rotations over each primary axis. Test T1 is a set of translations of the slice plane parallel to each primary axis, stepping throught every slice sequentially. Slices were sampled at various levels of resolution.

In the baseline test on a basic PC platform, shown in Figure 7, with a very limited cache allocation, the proposed indexing scheme was clearly superior (by orders of magnitude), particularly as the sampling factor was

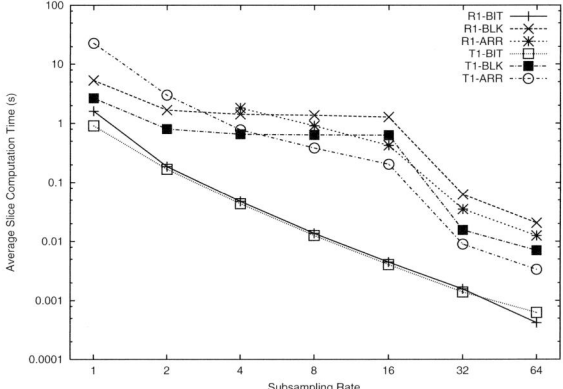

Fig. 8. Average slice computation time for 512^2 sample slices on an SGI Onyx2 with 300Mhz MIPS R12000 CPUs using a 20MB fixed data cache. Results are given for two different plane access patterns R1 and T2 as well as three different data layouts BIT, BLK and ARR. The input data grid was $2048 \times 2048 \times 1920$.

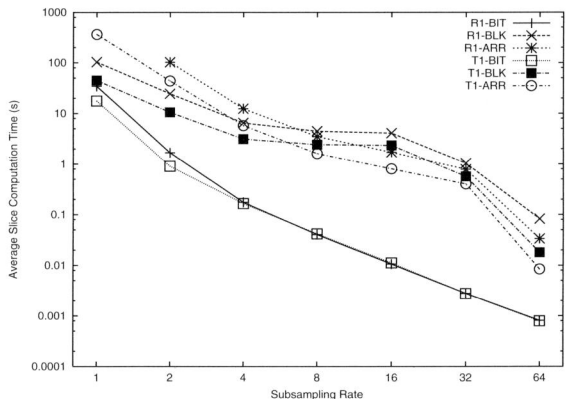

Fig. 9. Average slice computation time for 1024^2 sample slices on an SGI Onyx2 with 300Mhz MIPS R12000 CPUs using a 20MB fixed data cache. Results are given for two different plane access patterns R1 and T2 as well as three different data layouts BIT, BLK and ARR. The input data grid was $2048 \times 2048 \times 1920$.

increased. Our scheme allows one to maintain real-time interaction rates for large datasets using very modest resources (20MB).

We repeated the same test on an SGI Onyx2 system with higher performance disk arrays, the results are shown in Figure 8. The results are essentially equivalent, with slightly better performance being achived at extreme sampling levels on the SGI. Thus, the even the hardware requirements for the algorithm are very conservative.

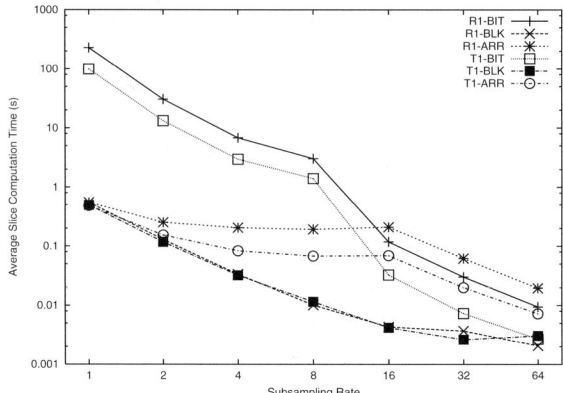

Fig. 10. Average slice computation time for 512^2 sample slices on an SGI Onyx2 with 300Mhz MIPS R12000 CPUs using a 60MB fixed data cache. Results are given for two different plane access patterns R1 and T2 as well as three different data layouts BIT, BLK and ARR. The input data grid was $8192 \times 8192 \times 7680$.

To test the scalability of the algorithm, we ran tests with increased output slice size and input volume sizes. When the number of slice samples was increased by a factor of four (Figure 9) we note that our BIT scheme is the only one that scales running times linearly with the size of the output for subsampling rates of two or higher.

Finally, a 0.5TB dataset ($8192 \times 8192 \times 7680$ grid) formed by replicating the 8GB timestep 64 times was run on the same SGI Onyx2 system using a larger 60MB memory cache. Results are shown in Figure 10. Interactive rates are certainly achievable using our indexing scheme on datasets of this extreme size and are very comparable to those obtained for a the $2k^3$ grid case (Figure 8) with this scheme.

Overall, results generally parallel those illustrated in the mmap experiments. Major differences stem from the increases in access time caused by the use of computationally expensive compression schemes and the potential for cache thrashing caused by the selection of (relatively) small local cache sizes, particularly with schemes lacking the degree of data locality provided by our scheme.

7 Conclusions and Future Directions

This paper introduces a new indexing and data layout scheme that is useful for out-of-core hierarchical traversal of large datasets. Practical tests and theoretical analysis for the case of multi-resolution slicing of rectilinear grids illustrate the performance improvements that can be achieved with this approach, particularly within the context of a progressive computational frame-

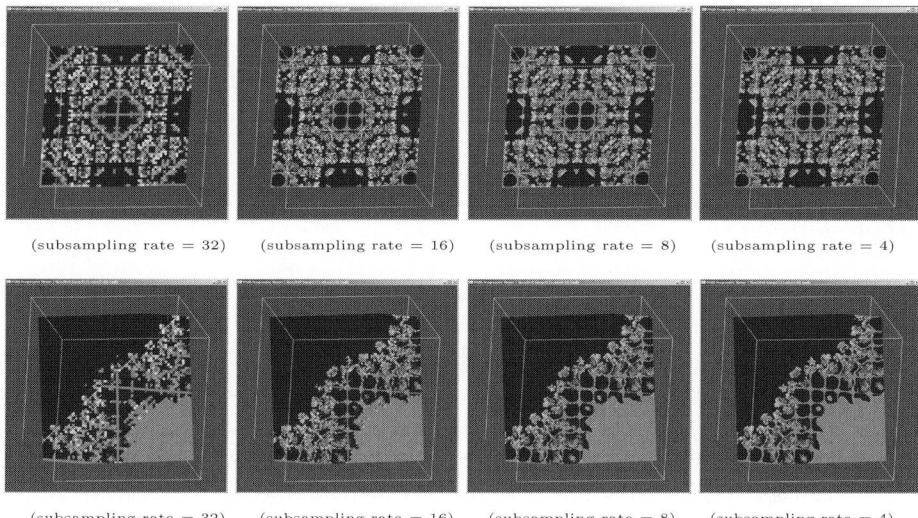

Fig. 11. Progressive refinement of two slices of the PPM dataset. (top row) Slice orthogonal to the x axis. (bottom row) Slice at an arbitrary orientation. (See Color Plate 23 on page 363.)

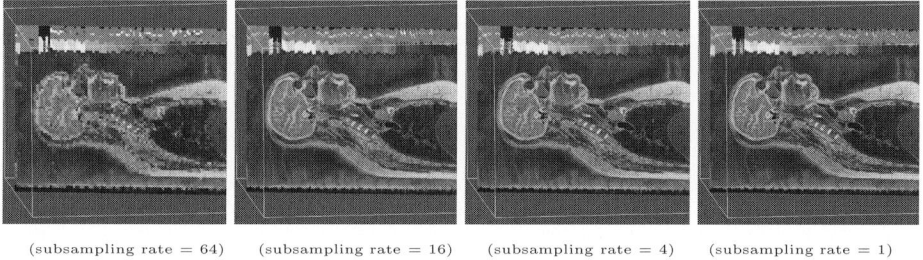

Fig. 12. Progressive refinement of one slice of the visible human dataset. Note how the inclination of the slice allows to show at the same time the nose and the eye. This view cannot be obtained using only orthogonal slice. (See Color Plate 24 on page 363.)

work. For example we can translate and rotate planar slices of an 8k cubed grid achieving half-second interaction rates. In the near future this scheme will be used as the basis for out-of-core volume visualization, computation of isocontours and navigation of large terrains.

Future directions being considered include integration with wavelet compression schemes, the extension to general rectangular grids, distributed memory implementations and application to non-rectilinear hierarchies.

8 Acknowledgments

This work was performed under the auspices of the U.S. Department of Energy by University of California, Lawrence Livermore National Laboratory under Contract W-7405-Eng-48.

References

1. James Abello and Jeffrey Scott Vitter, editors. *External Memory Algorithms and Visualization*. DIMACS Series in Discrete Mathematics and Theoretical Computer Science. American Mathematical Society Press, Providence, RI, 1999.
2. Lars Arge and Peter Bro Miltersen. On showing lower bounds for external-memory computational geometry problems. In James Abello and Jeffrey Scott Vitter, editors, *External Memory Algorithms and Visualization*, DIMACS Series in Discrete Mathematics and Theoretical Computer Science. American Mathematical Society Press, Providence, RI, 1999.
3. T. Asano, D. Ranjan, T. Roos, and E. Welzl. Space filling curves and their use in the design of geometric data structures. *Lecture Notes in Computer Science*, 911:36–44, 1995.
4. C. L. Bajaj, V. Pascucci, D. Thompson, and X. Y. Zhang. Parallel accelerated isocontouring for out-of-core visualization. In Stephan N. Spencer, editor, *Proceedings of the 1999 IEEE Parallel Visualization and Graphics Symposium (PVGS'99)*, pages 97–104, N.Y., October 25–26 1999. ACM Siggraph.
5. L. Balmelli, J. Kovačević, and M. Vetterli. Quadtree for embedded surface visualization: Constraints and efficient data structures. In *IEEE International Conference on Image Processing (ICIP)*, pages 487–491, Kobe Japan, October 1999.
6. Y. Bandou and S.-I. Kamata. An address generator for a 3-dimensional pseudo-hilbert scan in a cuboid region. In *International Conference on Image Processing, ICIP99*, volume I, 1999.
7. Y. Bandou and S.-I. Kamata. An address generator for an n-dimensional pseudo-hilbert scan in a hyper-rectangular parallelepiped region. In *International Conference on Image Processing, ICIP 2000*, 2000. to appear.
8. Yi-Jen Chiang and Cláudio T. Silva. I/O optimal isosurface extraction. In Roni Yagel and Hans Hagen, editors, *IEEE Visualization 97*, pages 293–300. IEEE, November 1997.
9. M. T. Goodrich, J.-J. Tsay, D. E. Vengroff, and J. S. Vitter. External-memory computational geometry. In *Proceedings of the 34th Annual IEEE Symposium on Foundations of Computer Science (FOCS '93)*, Palo Alto, CA, November 1993.
10. M. Griebel and G. W. Zumbusch. Parallel multigrid in an adaptive pde solver based on hashing and space-filling curves. 25:827:843, 1999.
11. D. Hilbert. Über die stetige Abbildung einer Linie auf ein Flächenstück. *Mathematische Annalen*, 38:459–460, 1891.
12. S.-I. Kamata and Y. Bandou. An address generator of a pseudo-hilbert scan in a rectangle region. In *International Conference on Image Processing, ICIP97*, volume I, pages 707–714, 1997.

13. J. K. Lawder. *The Application of Space-filling Curves to the Storage and Retrieval of Multi-Dimensional Data*. PhD thesis, School of Computer Science and Information Systems, Birkbeck College, University of London, 2000.
14. J. K. Lawder and P. J. H. King. Using space-filling curves for multi-dimensional indexing. In Brian Lings and Keith Jeffery, editors, *proceedings of the 17th British National Conference on Databases (BNCOD 17)*, volume 1832 of *Lecture Notes in Computer Science*, pages 20–35. Springer Verlag, July 2000.
15. Y. Matias, E. Segal, and J. S. Vitter. Efficient bundle sorting. In *Proceedings of the 11th Annual SIAM/ACM Symposium on Discrete Algorithms (SODA '00)*, January 2000.
16. A. Mirin. Performance of large-scale scientific applications on the ibm asci blue-pacific system. In *Ninth SIAM Conf. of Parallel Processing for Scientific Computing*, Philadelphia, Mar 1999. SIAM. CD-ROM.
17. B. Moon, H. Jagadish, C. Faloutsos, and J. Saltz. Analysis of the clustering properties of hilbert spacefilling curve. *IEEE Transactions on knowledge and data engeneering*, 13(1):124–141, 2001.
18. R. Niedermeier, K. Reinhardt, and P. Sanders. Towards optimal locality in meshindexings, 1997.
19. Rolf Niedermeier and Peter Sanders. On the manhattan-distance between points on space-filling mesh-indexings. Technical Report iratr-1996-18, Universität Karlsruhe, Informatik für Ingenieure und Naturwissenschaftler, 1996.
20. M. Parashar, J.C. Browne, C. Edwards, and K. Klimkowski. A common data management infrastructure for adaptive algorithms for pde solutions. In *SuperComputing 97*, 1997.
21. Hans Sagan. *Space-Filling Curves*. Springer-Verlag, New York, NY, 1994.
22. J. S. Vitter. External memory algorithms and data structures: Dealing with massive data. *ACM Computing Surveys*, March 2000.

Mesh Fairing Based on Harmonic Mean Curvature Surfaces

Robert Schneider[1], Leif Kobbelt[2], and Hans-Peter Seidel[1]

[1] Max-Planck Institute for Computer Sciences, D-66123 Saarbrücken, Germany
[2] RWTH Aachen, Institute for Computergraphics and Multimedia,
  D-52056 Aachen, Germany

Abstract. We introduce a new surface class, the harmonic mean curvature surfaces (HMCSs). Interestingly, this surface class doesn't seem to appear in the geometric modeling literature for fairing purposes up to now. We show that this surface class is a very interesting alternative to minimal energy surfaces (MESs), which are up to now most frequently used to create high quality surfaces for G^1 boundary constraints, as long as no isolated vertices have to be interpolated. We further present a new iterative HMCS construction algorithm for triangular meshes, where the new vertex positions are determined by a curvature smoothing step that alternates between local and global smoothing operations in combination with a local curvature linearization technique. To speed up the construction algorithm, we use a generic multigrid approach for meshes of arbitrary connectivity.

1 Introduction

While surface representations based on smoothly joining polynomial segments dominated the geometric modeling world for a long time, during the last years, triangle meshes have become more and more popular. The reason for this popularity is that in many cases surfaces nowadays naturally arise in a form where the geometric as well as the topological information is given in a discrete form. Examples here are the output that results from laser-range scanner data or the meshes created by iso-surface extraction of volume data. This development brought up the desire to be able to manipulate meshes with analogous techniques as the ones that are available for their smooth piecewise polynomial counterparts. Here, a very important technique one is often confronted with in the field of geometric modeling is the creation or manipulation of manifolds such that the result looks aesthetic to the human eye. In the technical literature this is known as the surface fairing problem.

There are two major setups where mesh fairing problems arise. In the first case one is interested in smoothing out the noise of an existing mesh with the constraint to preserve the overall global shape. A typical field of application is noise removal of surfaces resulting from scanning techniques or from iso-surface extraction. A simple method that can be used to solve such problems is Laplacian smoothing. More involved schemes were presented e.g. by Taubin [36], Desbrun et al. [8] or Guskov et al. [16]. However, in this paper our goal is not to develop techniques for mesh noise removal, but to contribute to the second setup. Here, fair freeform surfaces based on meshes have to be created from scratch. This is achieved by constructing a fair mesh that satisfies prescribed interpolation constraints. Typical fields of application where such problems arise are

- N-sided hole filling;
- Surface blending;
- Multiresolution modeling.

To solve such problems, very fast mesh fairing algorithms for meshes of arbitrary connectivity have been proposed by Taubin [36] and by Kobbelt et al. [21]. Taubin's approach is based on the idea to combine a signal processing fairing technique that can be used to smooth out the noise of a mesh with a second step that prevents surface shrinkage. Applying such smoothing steps until an equilibrium is reached, his algorithm can be used to create fair free-form surfaces. The mesh fairing algorithm of Kobbelt et al. is based on a different idea. Their approach can be interpreted as the discrete analogue to the partial differential equation (PDE) method that was developed by Bloor and Wilson [1,2] to create fair spline surfaces.

Another discrete method that can be applied in the second setup is the discrete Coons Patch approach introduced by Farin and Hansford [10]. Here, the resulting meshes approximate the same PDE as the approaches of Kobbelt et al. and of Bloor and Wilson.

These algorithms are fast and are only based on simple linear techniques. However, this speed and simplicity does not come for free. These methods are very well suited for interactive mesh fairing purposes, but are not suited to create meshes that satisfy a high quality fairness criteria. Both algorithms have to prescribe a local mesh parameterization strategy in advance and the shape of the resulting mesh highly depends on this chosen strategy. Furthermore, the connectivity of the vertices as well as the resolution of the mesh strongly influences the shape.

To create high quality surfaces one has to use fairing techniques that are only based on intrinsic surface properties, i.e., properties that depend on the geometry alone. However, intrinsic fairing is a nonlinear problem, and while the simple fairing operators are highly efficient and in general mathematically well understood, the analysis in the intrinsic case is much more difficult and theoretical existence proofs are usually not available even for very popular intrinsic schemes.

A standard surface class that is used to solve such an intrinsic problem are minimal energy surfaces (MESs). Algorithms for the construction of MES meshes were e.g. developed by Hsu et al. [19] and Welch and Witkin [38]. In this paper we introduce a new surface class, the harmonic mean curvature surfaces (HMCSs). Interestingly, this surface class doesn't seem to appear in the geometric modeling literature for fairing purposes up to now. We show that this surface class is a very interesting alternative to MESs, which are up to now most frequently used to create high quality surfaces for G^1 boundary constraints, as long as no isolated vertices have to be interpolated. We further present an iterative HMCS construction algorithm for triangular meshes, where the new vertex positions are determined by a curvature smoothing step that alternates between local and global smoothing operations in combination with a local curvature linearization technique. This construction algorithm is excellently suited for the creation of HMCS surfaces.

2 Problem formulation and definitions

In this paper, all discrete surfaces are triangle meshes in a three dimensional space. For each vertex q_i of a mesh Q, let $N(q_i) = \{q_j | \{i,j\} \in \mathcal{E}\}$ be the set of vertices q_j adjacent to q_i. The set $D(q_i) = N(q_i) \cup \{q_i\}$ forms the so called 1-disk of the vertex q_i. The number of vertices in a neighborhood $N(q_i)$ defines the function $valence(q_i) = |N(q_i)|$.

Since we are interested in triangular meshes and not in surfaces that are differentiable in the classical meaning, we have to take two fairness aspects into account. The *outer fairness* criterion determines the shape of the mesh, while the *inner fairness* criterion determines the distribution of the vertices within the surface and the shape of the individual triangle faces. In our case we therefore can formulate the problem: For a given collection of boundary polygons, let P be the set of all boundary vertices p_i. We now have to find a triangular mesh Q with vertices q_i, such that $P \subset Q$, the boundary conditions are satisfied, the mesh forms an aesthetic surface and the prescribed inner fairness condition is satisfied.

Let us partition the vertices of Q into two classes, denoting the set of all boundary vertices with $V_B(Q) = P$ and the set of all vertices in the interior of Q with $V_I(Q)$. The vertices in $V_I(Q)$ are also denoted as free vertices in the following.

3 Important intrinsic surfaces

In this section we present some popular approaches that appear in the literature for the construction of high quality surfaces that solve our boundary problem. As we will see, these approaches are all based on intrinsic energy functional minimization.

3.1 Minimal energy surfaces (MESs)

MESs are the surface analogue to minimal energy curves (MECs)

$$\int \kappa^2 \, ds,$$

where κ is the curvature and s the arc length, and are perhaps the most popular surface type that is used to create fair surfaces. They are defined as surfaces that satisfy the boundary conditions and minimize the intrinsic functional

$$\int_A \kappa_1^2 + \kappa_2^2 \, dA, \tag{1}$$

where κ_1 and κ_2 are the principal curvatures. This functional has a physical interpretation, since it appears when measuring the bending energy of a thin membrane without surface tension. Typically, the boundary condition is specified by G^1 boundary conditions, i.e., curve and tangent plane information along the boundary.

There is a tight connection between MESs and the so called Willmore energy of a surface. The Willmore energy of a surface is defined as

$$\int_A H^2 \, dA, \tag{2}$$

where H denotes the mean curvature. Since $H = \frac{1}{2}(\kappa_1 + \kappa_2)$, we can also express the Willmore energy as

$$\frac{1}{4} \int_A \kappa_1^2 + \kappa_2^2 + 2K \, dA.$$

Exploiting the Gauß-Bonnett theorem for surfaces that tells us that the integral of the Gaussian curvature $\int_A K \, dA$ only depends on the surface topology for a fixed boundary (see e.g. do Carmo [6]), one arrives at the well-known fact that MESs can also be defined as a minimum of the Willmore energy functional.

While MECs are only of limited use for the solution of curve fairing problems, this is completely different for MESs. Although it is to the authors knowledge still not proven that there always exists a MES for a given boundary problem, numerical tests suggest that a MES solution usually can be found for practical applications.

Applying variational calculus onto the minimization problem based on (1) or (2), we arrive at the Euler-Lagrange equation (see Giaquinta and Hildebrandt [12])

$$\Delta_B H + 2H(H^2 - K) = 0, \tag{3}$$

where Δ_B denotes the Laplace-Beltrami operator. Construction algorithms for meshes were presented by Hsu et al. [19] and Welch and Witkin [38]. For smooth tensor-product spline surfaces, a very efficient construction algorithm for MESs was developed by Greiner [15].

MESs do not reproduce all surfaces with a constant mean curvature H. While they reproduce spheres and minimal surfaces with $H = 0$, they for example do not reproduce cylinders. Cylinders have a nonzero constant mean curvature and zero Gaussian curvature, so they can not be a solution of the Euler-Lagrange equation (3).

3.2 Elastic surfaces

It is also possible to extend elastic curves

$$\int \left(C_1 \frac{\kappa^2}{2} + C_2 \right) ds,$$

with $C_1, C_2 > 0$, to the surface case; such surfaces play an important role in physics and other sciences. Here, the surfaces satisfy the prescribed boundary constraints and minimize the functional

$$\int_A \left(C_1 H^2 + C_2 \right) dA,$$

with $C_1 > 0$. The parameter C_2 introduces a tension energy that is added to the bending energy.

The Euler-Lagrange equation resulting from the elastic surface energy formulation becomes (see Hsu et al. [19])

$$\Delta_B H + 2H(H^2 - (K - \frac{C_2}{C_1})) = 0,$$

as we can see we arrive again at the Euler-Lagrange equation of MESs for $C_2 = 0$.

3.3 Minimal variation surfaces (MVSs)

Besides MECs and elastic curves, another curve type that was extended to surfaces are minimal variation curves (MVCs)

$$\int \left(\frac{d\kappa}{ds} \right)^2 ds.$$

For MVCs, this extension was done by Moreton and Séquin (see [25,24,26]) and the resulting MVSs were defined as surfaces that satisfy the boundary constraints and minimize the functional

$$\int_A \left(\frac{d\kappa_1}{d\boldsymbol{e}_1} \right)^2 + \left(\frac{d\kappa_2}{d\boldsymbol{e}_2} \right)^2 dA, \qquad (4)$$

where κ_1 and κ_2 are the principal curvatures and $\boldsymbol{e}_1$ and $\boldsymbol{e}_2$ are the corresponding principal curvature directions.

Compared to MESs, a MVS exhibits higher-order continuity and usually looks more aesthetic. They further naturally reproduce cyclides and therefore it is possible to reproduce e.g. spheres, cylinders, cones and tori. It is also possible to prescribe a wider range of boundary conditions.

The MVSs, however, have a crucial drawback; their complexity. Measuring the curvature variation along the principal directions is expensive. The construction algorithm of Moreton and Séquin produces surfaces of excellent quality, but the construction time is enormous (see Moreton and Séquin [25]).

To our knowledge, no one has had enough patience to derive the Euler-Lagrange equation of MVSs so far.

Remark The extension of MVCs that was given by Moreton and Séquin is not the only possible extension. In our opinion, it is not clear why the principal curvature directions are preferred against other directions. Maybe one should seize the suggestion of Greiner [15], who does not use the functional (4) to extend MVCs but creates a functional that punishes all third order derivatives. This allows to develop a surface construction algorithm for splines that is considerably faster than the one presented by Moreton and Séquin.

Interpreting the gradient operator as surface analogue to the derivative operator for arc-length parameterized curves and the mean curvature as analogue to the planar curvature, another simple functional that extends MVCs is for example

$$\int_A |\text{grad} H|^2 \, dA, \qquad (5)$$

which further simplifies the Greiner functional given in [15].

4 Harmonic Mean Curvature Surfaces

In the following, let us study the following problem: Find a surface that satisfies prescribed G^1 boundary constraints and which in the interior has a harmonic mean curvature distribution, i.e.,

$$\Delta_B H = 0. \qquad (6)$$

Let us name a solution to this problem a *harmonic mean curvature surface* (HMCS). The G^1 boundary conditions can be prescribed along the border curves since the PDE is of fourth order. Before we show how to construct such surfaces, let us first summarize important properties of HMCSs.

4.1 Properties of harmonic mean curvature surfaces

HMCSs can be interpreted as surface extension of planar clothoid splines, which have a vanishing second derivative of the curvature $\kappa'' = 0$ using arc-length parameterization. The Laplace-Beltrami operator Δ_B is the direct analogue to the second derivative operator for curves based on arc length parameterization and the mean curvature is an intrinsic surface measure that comes next to the curvature of a planar curve. This connection is reflected in the properties of HMCSs:

- They are parameterization independent.
- They satisfy a simple intrinsic PDE.
- They include the class of constant mean curvature surfaces (CMCSs), including spheres, cylinders and minimal surfaces.
- They have no local extrema of the mean curvature in the interior.

- They do not depend on the orientation of the normal vector field of the surface.
- They are invariant with respect to translation, rotation, reflection and scaling operations.

When comparing clothoid splines with MECs, one notices that clothoid splines reproduce circles, while MECs do not. If we now compare MESs and HMCSs we see that analogously only HMCSs reproduce all constant mean curvature surfaces. Looking at the Euler-Lagrange equation of MESs, we see that they can reproduce spheres and minimal surfaces with $H = 0$, but they can for example not reproduce a cylinder. We later give an example that shows what surfaces are created by MESs for cylinder boundary conditions. Therefore, the approach to fair surfaces based on HMCSs has even some advantages compared to the traditionally used MESs, although its PDE is a simplification of the Euler-Lagrange equation of MESs. However, the situation changes when isolated vertex information has to be interpolated in the interior of the surface. Here, HMCSs shouldn't be applied for fairing purposes without modifications of the construction algorithm. MESs seem to be much better suited to handle such a problem, we come back to this topic later in Section 7.6. It should also be emphasized that the existence advantage of clothoid splines - to exist also for constellations where a MEC cannot be found - is no longer available in the surface case, since MESs in practical applications usually do exist.

MESs were defined by minimizing an energy functional that punishes large curvature values. Using variational calculus, one can derive a PDE that describes the solution. This leads to the question if there exists an energy functional, whose Euler-Lagrange equation is satisfied for HMCSs. This is indeed possible, such an energy functional was given by Taylor [37], but we don't need this rather complicated functional for our purposes.

Remark Interestingly, we do not know any CAGD literature where HMCSs are used to solve our type of fairing problem. For fourth order problems with prescribed G^1 boundary conditions usually MESs are used in practice. However, there is a connection of HMCSs with another problem that has been studied intensively; the surface diffusion flow. This is a curvature flow where the force function F is specified by

$$F(H, K) = -\Delta_B H$$

and can be interpreted as a diffusion process that takes place within the surface. Such a surface flow was first proposed by Mullins [27] to analyze the dynamics of crystal motion, but meanwhile this flow has also been studied by various other authors (see e.g. Taylor [37], Chopp and Sethian [5]). As we can see, a HMCS is an equilibrium of the surface diffusion flow. Therefore, we could construct HMCSs using a surface diffusion flow, but such a surface flow would have various drawbacks (see Chopp and Sethian [5]).

5 Discretization of the mean curvature

5.1 Our discretization requirements

There are various strategies available to define discrete curvatures on a mesh. For meshes of arbitrary connectivity, such methods were developed e.g. by Moreton and Séquin [25], Welch and Witkin [38], Taubin [35] or Desbrun et al. [7].

To be applicable for our purposes, we need an algorithm that satisfies the following properties:

1. It has to produce accurate results.
2. The discretization has to be a continuous function of the vertices.
3. I should be able to handle vertices that have a complete neighborhood as well as boundary vertices.
4. The mean curvature discretization should enable a linearization technique.

The first property is necessary, since we are interested in a high quality surface construction algorithm and hence need accurate discrete curvatures. The second one is essential for convergence and stability. Since we have to discretize the curvature for inner vertices as well as for boundary vertices, the third condition avoids having mean curvature discretization strategies of different accuracy. Finally the fourth condition is needed in our mesh construction algorithm.

The discretization method of Moreton and Séquin satisfies all of the requirements mentioned above and is therefore used in the following to discretize the curvatures. It requires that a tangent plane is known at every vertex. At interior vertices where no normal vector is known in advance, we define the normal vector to be the normalized sum of the vector crossproducts of the incident triangle faces, in order to minimize square root operations. The problem we encountered here is that the discretization at a vertex q_i requires a neighborhood whose valence is at least 5 to be useful for our purposes. Let us now explain Moreton and Séquin's method in detail and also our modifications for vertices of valence 3 or 4.

5.2 Moreton and Séquin's curvature discretization algorithm

The idea of Moreton and Séquin's approach they presented in [25] is to use the fact that the normal curvature distribution cannot be arbitrary, but is determined by Euler's formula.

Let $\boldsymbol{b}_x$ and $\boldsymbol{b}_y$ be an arbitrary orthonormal basis of the plane defined by the normal $\boldsymbol{n}$. To each vertex $q_j \in N(q_i)$ we can assign a unit direction vector $\boldsymbol{t}_j$ by projecting q_j orthogonally into the plane and scaling this projection to

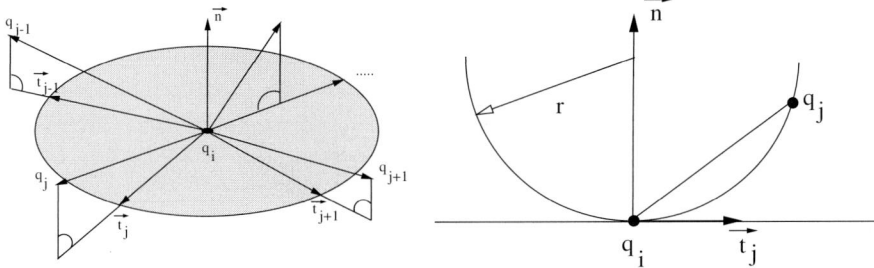

Fig. 1. Left: Projecting the neighborhood of q_i onto the plane defined by $\boldsymbol{n}$ and normalizing the results we get the normal curvature directions $\boldsymbol{t}_j$. Right: The normal curvature along the direction $\boldsymbol{t}_j$ is discretized by interpolating q_i and q_j with a circle and using the inverse of the circle radius r as normal curvature. The center of the circle lies on the line defined by q_i and $\boldsymbol{n}$.

unit length. For each q_j we can now estimate a normal curvature $\tilde{\kappa}_j$ as the inverse of the circle radius defined by q_i, q_j and t_j (Fig. 1)

$$\tilde{\kappa}_j = 2 \frac{\langle q_j - q_i \mid \boldsymbol{n} \rangle}{\langle q_j - q_i \mid q_j - q_i \rangle}. \qquad (7)$$

Using Euler's theorem, we can express the normal curvature κ_n for a direction $\boldsymbol{t}$ by the principal curvatures κ_1 and κ_2 and the principal curvature directions $\boldsymbol{e}_1$ and $\boldsymbol{e}_2$. Let t_x and t_y be the coordinates of $\boldsymbol{t}$ in the basis $\boldsymbol{b}_x, \boldsymbol{b}_y$ and let e_x and e_y be the coordinates of $\boldsymbol{e}_1$, then the normal curvature can be expressed as

$$\kappa_n = \begin{pmatrix} t_x \\ t_y \end{pmatrix}^t \cdot M \cdot \begin{pmatrix} t_x \\ t_y \end{pmatrix}, \qquad (8)$$

with

$$M = \begin{bmatrix} e_x & e_y \\ -e_y & e_x \end{bmatrix} \cdot \begin{bmatrix} \kappa_1 & 0 \\ 0 & \kappa_2 \end{bmatrix} \cdot \begin{bmatrix} e_x & e_y \\ -e_y & e_x \end{bmatrix}^{-1}.$$

The idea of Moreton and Séquin is to use the normal curvatures $\tilde{\kappa}_j$ to create a linear system and to find estimates for the unknown principal curvature values by determining the least squares solution. Let $t_{j,x}$ and $t_{j,y}$ denote the coordinates of $\boldsymbol{t}_j$ and let m be the valence of q_i, then we get by evaluating (8)

$$A\boldsymbol{x} = \boldsymbol{b}$$

where

$$A = \begin{bmatrix} t_{1,x}^2 & t_{1,x}t_{1,y} & t_{1,y}^2 \\ t_{2,x}^2 & t_{2,x}t_{2,y} & t_{2,y}^2 \\ \cdot & \cdot & \cdot \\ t_{m,x}^2 & t_{m,x}t_{m,y} & t_{m,y}^2 \end{bmatrix}, \quad \boldsymbol{b} = \begin{bmatrix} \tilde{\kappa}_1 \\ \tilde{\kappa}_2 \\ \cdot \\ \tilde{\kappa}_m \end{bmatrix}$$

and
$$\boldsymbol{x} = \begin{pmatrix} x_0 \\ x_1 \\ x_2 \end{pmatrix} = \begin{pmatrix} e_x^2 \kappa_1 + e_y^2 \kappa_2 \\ 2 e_x e_y (\kappa_1 - \kappa_2) \\ e_x^2 \kappa_2 + e_y^2 \kappa_1 \end{pmatrix}.$$

Since $x_0 + x_2 = \kappa_1 + \kappa_2$ the discrete mean curvature $\tilde{H}$ is determined by

$$\tilde{H} = \frac{1}{2}(x_0 + x_2). \tag{9}$$

In our case the most efficient method to solve the least squares problem is to use the normal equations approach (see Golub and Van Loan [13]), since this mainly involves to calculate the inverse of a symmetric 3×3 matrix. Using this approach, the least squares solution $\boldsymbol{x}$ can be expressed as

$$\boldsymbol{x} = (A^t A)^{-1} A^t \boldsymbol{b}. \tag{10}$$

The cases when the matrix $A^t A$ becomes singular can be detected by a simple criterion:

Lemma 1. *The Matrix $A^t A$ is singular if and only if all points $(t_{i,x}, t_{i,y})$ are intersection points of the unit circle with two straight lines through the origin.*

Proof: Since singularity of $A^t A$ is equivalent with $\operatorname{\mathbf{rank}} A < 3$, singularity occurs iff any 3 row vectors of A are linear dependent. Therefore, let us first study what happens if three row vectors are linear dependent by looking at a matrix of type

$$\begin{bmatrix} t_{1,x}^2 & t_{1,x} t_{1,y} & t_{1,y}^2 \\ t_{2,x}^2 & t_{2,x} t_{2,y} & t_{2,y}^2 \\ t_{3,x}^2 & t_{3,x} t_{3,y} & t_{3,y}^2 \end{bmatrix}$$

with $t_{i,x}^2 + t_{i,y}^2 = 1$. This matrix is singular, if there are coefficients a_i which are not all zero such that

$$a_1 \begin{pmatrix} t_{1,x}^2 \\ t_{2,x}^2 \\ t_{3,x}^2 \end{pmatrix} + a_2 \begin{pmatrix} t_{1,x} t_{1,y} \\ t_{2,x} t_{2,y} \\ t_{3,x} t_{3,y} \end{pmatrix} + a_3 \begin{pmatrix} t_{1,y}^2 \\ t_{2,y}^2 \\ t_{3,y}^2 \end{pmatrix} = \begin{pmatrix} 0 \\ 0 \\ 0 \end{pmatrix},$$

but this means each point $(t_{i,x}, t_{i,y})$ lies on a curve that satisfies

$$a_1 x^2 + a_2 xy + a_3 y^2 = 0.$$

Depending on the coefficients a_i this equation characterizes a single point (the origin) or two lines through the origin. This shows that three row vectors of the matrix A are linear dependent if and only if the according three points lie on two lines through the origin.

If the matrix $A^t A$ is singular, every three row vectors of A have to satisfy this property. This can only be satisfied if all points $(t_{i,x}, t_{i,y})$ are intersection points of the unit circle with two straight lines through the origin. Otherwise we could find three points that lie on three different lines through the origin and the matrix $A^t A$ cannot be singular. □

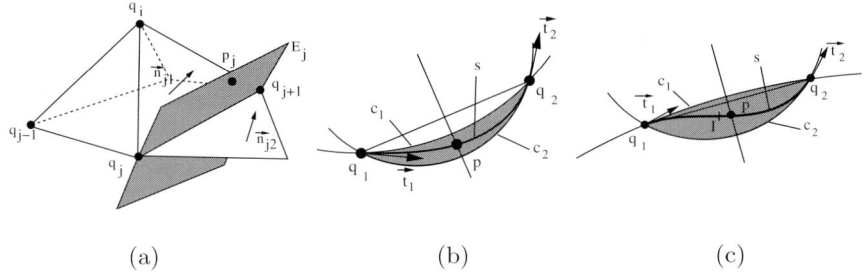

Fig. 2. a) At vertices q_i with valence 3 or 4, new vertices p_j are introduced between all $q_j, q_{j+1} \in N(q_i)$ that share a common edge. The p_j lie on a plane E_j determined by q_j, q_{j+1} and the vector $\boldsymbol{n}_{j1} + \boldsymbol{n}_{j2}$, where $\boldsymbol{n}_{j1}$ and $\boldsymbol{n}_{j2}$ are the triangle normals of the faces adjacent to the edge $q_j q_{j+1}$. b) Approximation of a spiral (planar curve with monotone nonzero curvature) by the area enclosed by two arcs. c) This approximation also produces reasonable results for planar curves with monotone curvature and an inflection point (denoted as I).

5.3 Singularity handling

Perhaps the most obvious strategy to avoid singularities of $A^t A$ is to simply check whether the criterion presented in Lemma 1 is satisfied and to apply a special method only to such cases near singularity. Such an approach is straightforward, but it can lead to instabilities. A small perturbation of one vertex can trigger a special case handling and thus it could switch the mean curvature discretization method, making the process discontinuous.

An elegant solution to this discontinuity problem is to exploit the connection between a possible singularity of $A^t A$ and the vertex valence. Assuming that all points $(t_{i,x}, t_{i,y})$ are distinct, which is automatically satisfied if the mesh approximates a smooth surface, a simple consequence of Lemma 1 is that the matrix $A^t A$ can only become singular, if the valence of q_i is 3 or 4.

For all vertices of valence 3 or 4 we increase the data quantity that serves as input for our algorithm. Instead of enlarging the vertex neighborhood – which lacks symmetry if the valence of the neighbor vertices varies largely – we increase the local input data by estimating new vertices p_j between adjacent $q_j \in N(q_i)$. The p_j and q_j then serve as input for the mean curvature discretization as described in Section 5.2, making the problem well posed.

Simply setting $p_j = (q_j + q_{j+1})/2$ would be fast and convenient, but such an approach distorts the resulting mean curvature discretization considerably and should not be used if high quality results have to be generated. A scheme that has proven to be adequate in our numerical experiments is based on the idea to determine p_j by sampling a planar curve with monotone curvature, that interpolates the vertices and normals at q_j and q_{j+1} (Fig. 2 a). Since it is not obvious what type of spiral (or pair of spirals, if the curve has an inflection point) is most promising and the computation of such an interpolating curve

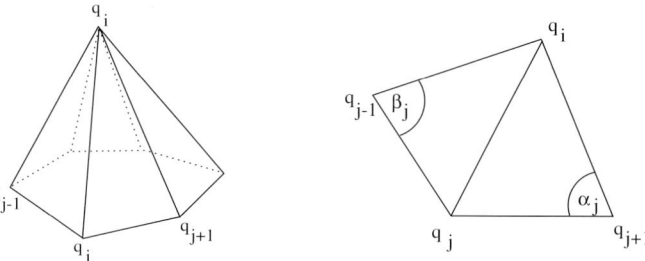

Fig. 3. The Laplace-Beltrami operator at the vertex q_i can be discretized using $D(q_i)$.

can be expensive, we exploit a nice approximation property of spirals (see Marciniak and Putz [23]):

Given two planar points q_1 and q_2 with tangent vectors $\boldsymbol{t}_1$ and $\boldsymbol{t}_2$, let s be an arbitrary spiral that satisfies this G^1 interpolation problem. Let c_1 be the circle defined by q_1, q_2 and $\boldsymbol{t}_1$ and c_2 be the circle defined by q_1, q_2 and $\boldsymbol{t}_2$ (such that $\boldsymbol{t}_1$ and $\boldsymbol{t}_2$ are circle tangents). Then the spiral s can be approximated by the area enclosed by the circular arcs of c_1 and c_2 between q_1 and q_2, if the tangent angle between $\boldsymbol{t}_1$ and $\boldsymbol{t}_2$ does not change exceedingly (Fig. 2 b). This argument is also reasonable for planar curves with monotone curvature and an inflection point (Fig. 2 c).

The tangent vectors $\boldsymbol{t}_1$ and $\boldsymbol{t}_2$ are computed by intersecting the tangent planes at the vertices q_j and q_{j+1} (defined by their normal vectors) with the plane E_j (Fig. 2 a), where we choose in each case the direction that has the smaller angle to the vector $q_{j+1} - q_j$. After intersecting the enclosed area with the perpendicular bisector of q_1 and q_2, we set $p_j = p$ to be the center point of the intersection interval. Each circle has two intersection points with the perpendicular bisector, here we choose only that intersection point that is closer to the line defined by q_1 and q_2.

6 Discretization of the Laplace-Beltrami operator

The discretization relies on the fact that there is a tight connection between the Laplace-Beltrami operator Δ_B and the mean curvature normal of a surface. Desbrun et al. [7] also give an improved discretization that uses a more detailed area calculation based on voronoi regions, but the following discretization is sufficient for our needs. Let $f : \mathbb{R}^2 \to \mathbb{R}^3$ be a parameterization of a surface, then it is well known that the following equation holds

$$\Delta_B f = 2H\boldsymbol{n}.$$

Exploiting this relation, a nice discretization of Δ_B follows directly from the mean curvature flow approach for arbitrary meshes that was presented by

Desbrun et al. [8]. They showed that the mean curvature normal at a vertex q_i of a triangular mesh M can be discretized using its 1-neighborhood by

$$Hn = \frac{3}{4A} \sum_{q_j \in N(q_i)} (\cot \alpha_j + \cot \beta_j)(q_j - q_i), \tag{11}$$

where A is the sum of the triangle areas of the 1-disk at q_i and α_j and β_j are the triangle angles as shown in Fig. 3. This leads to the Δ_B discretization

$$\Delta_B f = \frac{3}{2A} \sum_{q_j \in N(q_i)} (\cot \alpha_j + \cot \beta_j)(q_j - q_i).$$

In this paper, the Δ_B operator is always applied onto the mean curvature function H of a surface, instead of the parameterization f, so for our purposes we need the discretization

$$\Delta_B \tilde{H}_i = \frac{3}{2A} \sum_{q_j \in N(q_i)} (\cot \alpha_j + \cot \beta_j)(\tilde{H}_j - \tilde{H}_i). \tag{12}$$

We later are interested in meshes that are a discrete solution of the equation $\Delta_B \tilde{H} = 0$ at a vertex q_i. Exploiting that in this special case a scaling factor does not influence the result, at every inner vertex

$$\sum_{q_j \in N(q_i)} (\cot \alpha_j + \cot \beta_j)(\tilde{H}_i - \tilde{H}_j) = 0 \tag{13}$$

has then to be satisfied. If this equation is satisfied at all free vertices $q_i \in V_I(Q)$ and if we further know all mean curvature values for the boundary vertices $V_B(Q)$, this discretization of $\Delta_B H = 0$ leads us to a sparse linear system in the unknown discrete mean curvatures $\tilde{H}_i$ at all vertices $q_i \in V_I(Q)$, whose matrix S has the coefficients

$$S_{ii} = \sum_{q_j \in N(q_i)} (\cot \alpha_j + \cot \beta_j), \tag{14}$$

$$S_{ij} = \begin{cases} -(\cot \alpha_j + \cot \beta_j) & : \quad q_j \in N(q_i) \cap V_I(Q) \\ 0 & : \quad \text{otherwise.} \end{cases} \tag{15}$$

The matrix S is symmetric and, as long as no triangle areas of the mesh M vanish, positive definite. To see this, we note that S also appears in a paper by Pinkall and Polthier [30], where a stable construction algorithm for discrete minimal surfaces based on the idea to minimize the discrete Dirichlet energy of a mesh was presented. Hence, for an elegant proof of the mathematical structure of this matrix we can refer to this paper. It should be mentioned that S further appears in a paper about piecewise linear harmonic functions by Duchamp et al. [9], where the discretization of harmonic maps was developed independently of Pinkall and Polthier.

7 Constructing discrete harmonic mean curvature surfaces

In this section we present a construction algorithm that is excellently suited to create discrete HMCSs. The method that seems to be closest to our approach is the intrinsic curvature flow, since in both cases the vertices are moved along the normal vectors. However, the strategies for the new position of the vertices are completely different. Our new algorithm determines the new positions by a curvature smoothing step that alternates between local and global smoothing operations in combination with a local curvature linearization technique, while curvature flow methods locally discretize a flow equation.

7.1 Discrete harmonic mean curvature surfaces

Using the discretization of the mean curvature and the Laplace-Beltrami operator we presented in the Sections 5 and 6, we can now formulate the following

Definition 1. A mesh Q with $P \subset Q$ is called a discrete harmonic mean curvature surface, if the following conditions are satisfied

1. Each free vertex q_i of Q satisfies a specified inner fairness criterion.
2. The discrete mean curvature is a discrete harmonic function:
 $\Delta_B \tilde{H}_i = 0$, whenever $q_i \in V_I(Q)$.

The input data for our algorithm consists of vertices and unit normals that form the G^1 boundary condition and an initial mesh Q^0 that contains the boundary vertices $P \subset Q^0$. The construction algorithm creates a mesh sequence $Q^k, (k = 0, 1, 2, \ldots)$ by iteratively updating all free vertices, until the outer and inner fairness conditions are sufficiently satisfied. Although in our implementation we used a multigrid scheme to construct a solution, let us first assume that only the position of the vertices changes, while the connectivity remains fixed.

Our construction algorithm avoids dealing with explicit fourth order terms by factorizing the fourth order problem into two sub-problems of second order, which are considerably easier to handle than the original problem. Let $\hat{H}^k$ be the discrete mean curvature function of the k-th mesh Q^k in the sequence. The first sub-problem transforms this current scalar curvature function into a new scalar function $\hat{H}^k$. This function is created such that it is closer to a harmonic function than $\tilde{H}^k$, unless $\tilde{H}^k$ is already a solution. At this point only a function is modified, it does not change the current polygon. This is done in the second sub-problem. Here, all vertices of the current mesh Q^k are updated such that the resulting mesh Q^{k+1} has a mean curvature function that is closer to the modified function $\hat{H}^k$ than the curvature function of its predecessor. Expressed in two formulas, this factorization of $Q^k \to Q^{k+1}$ becomes

$$\left.\begin{array}{l} I.\ \Delta_B \hat{H}_i^k \approx 0 \\ II.\ \tilde{H}_i^{k+1} \approx \hat{H}_i^k \end{array}\right\} \quad \forall\, q_i^k \in V_I(Q^k).$$

If the discrete HMCS definition is satisfied, the sequence has reached an equilibrium.

7.2 Inner mesh fairness

To control how the vertices are distributed on the surface of the solution, for every inner vertex $q_i \in V_I$ we assign a generalized discrete Laplacian Δ that determines the local parameterization of the surface. Assigning a scalar weight λ_{ij} to every $q_j \in N(q_i)$ with the constraint $\sum_j \lambda_{ij} = 1$, the discrete Laplacian Δ is defined as

$$\Delta(q_i) = -q_i + \sum_{q_j \in N(q_i)} \lambda_{ij} q_j.$$

A mesh Q is said to satisfy the inner fairness condition, if all $\Delta(q_i)$ have vanishing tangential components, i.e., there are scalar values t_i such that $\Delta(q_i) = t_i n_i$ for all $q_i \in V_I(Q)$.

In our examples we used two different generalized Laplacians Δ. If we assume a local uniform parameterization, we arrive at a Laplacian with weights $\lambda_{ij} = \frac{1}{m}$, where $m = valence(q_i)$ is the valence of the vertex q_i. This Laplacian leads to meshes that have a regular vertex distribution on the surface. However, in some important cases such a kind of mesh parameterization is not wanted. For example, if textured meshes have to be faired, one would like to create a solution that minimizes local distortions to the original mesh Q^0. Another context where local distortions should be minimized is multiresolution modeling of meshes as proposed by Kobbelt et al. [21]. Here, minimal local distortions could relieve the local frame coding scheme.

In theory, local distortions would be minimized if the mesh is conformal to the original mesh Q^0. For meshes, it is usually not possible to get a map that is conformal in the sense that angles are preserved, but we can use the fact that the discretization of $\Delta_B f = 0$ can be interpreted as a discrete harmonic map (see Pinkall and Polthier [30] and Duchamp et al. [9]) and thus approximates a continuous conformal map. Therefore, to minimize local distortions, we use the weights resulting from the discretization of the Laplace-Beltrami operator of the original mesh Q^0. The weights λ_{ij} at a vertex q_i are then given by

$$\lambda_{ij} = \frac{\cot \alpha_j + \cot \beta_j}{\sum_{q_j \in N(q_i)} (\cot \alpha_j + \cot \beta_j)},$$

where α_j and β_j are the triangle angles as shown in Fig. 3.

Remark The drawback of such a harmonic inner fairness strategy is that weights can become negative and this can lead to mesh overlappings in some cases. A planar example where this occurs can be found in Duchamp et al. [9]. Therefore, Floater [11] developed a mesh parameterization that leads to convex combinations and nevertheless minimizes local distortions. This shape-preserving parameterization should be preferred as inner fairness strategy in practical applications. However, even with this strategy, it is still possible to produce overlappings, e.g., for non-convex planar boundary conditions.

7.3 Derivation of the new mean curvature distribution

Let us now formulate the curvature smoothing method. A Jacobi type smoothing step that reproduces HMCSs is given by

$$\hat{H}_i^k = \frac{1}{\sum_{q_j \in N(q_i)}(\cot \alpha_j + \cot \beta_j)} \sum_{q_j \in N(q_i)} (\cot \alpha_j + \cot \beta_j)\tilde{H}_j^k,$$

where the angles are determined as described in Section 6. That this smoothing reproduces a HMCS is a simple consequence of equation (13). Of course, we can also update with a Gauß-Seidel type strategy by exploiting the previously calculated values of $\hat{H}^k$.

If we apply the smoothing step n times on the current mean curvature function $\tilde{H}^k$, we increase the support that influences the new mean curvature value $\hat{H}_i^k$. In the limit $n \to \infty$, the mean curvature function $\hat{H}^k$ is determined only by the values of $\tilde{H}^k$ at the boundary vertices. We then arrive at a Dirichlet problem, where the unknown scalar mean curvature values at the free vertices are determined by a nonsingular linear system with a symmetric and positive definite matrix S whose coefficients are defined in (14) and (15). Solving the resulting nonsingular linear system yields scalar values $\hat{H}_i^k$ at all inner vertices $q_i \in V_I(Q^k)$, that represent a discrete harmonic function.

While discrete harmonic functions do not always have to share all the mathematical properties of their continuous counterparts, e.g., the convex hull property (see Pinkall and Polthier [30]), they will nevertheless approximate continuous harmonic functions. This means our scalar values $\hat{H}^k$ will approximate a function that does not have local extrema and whose maximal values occur at the boundary, so $\hat{H}^k$ will behave well and can be approximately bounded by the current mean curvature values at the boundary vertices. However, it is not necessary to solve the Dirichlet problem exactly. When we have determined the linear system, we apply some iteration steps of an iterative linear solver, using the current mean curvature values $\tilde{H}_i^k$ as starting values.

We noticed that we can considerably improve the convergence rate by using $n \to \infty$ smoothing steps instead of $n = 1$. However, we also noticed that instabilities can occur, when the whole smoothing process only depends on the boundary mean curvature values. We therefore decided to update the vertices using a hybrid approach where we alternate between $n = 1$ and $n \to \infty$ steps. In practice, the step $n = 1$ is based on Gauß-Seidel smoothing and the step $n \to \infty$ is approximated by iterating the linear system using conjugate gradients (see Golub and Van Loan [13]). Convergence of the conjugate gradient iteration is guaranteed, since the matrix S is positive definite.

7.4 Vertex updates

The aim of the update step is to produce a new mesh Q^{k+1} whose discrete mean curvature values $\tilde{H}(q_i^{k+1})$ at the vertices $q_i^{k+1} \in V_I(Q^{k+1})$ are closer to the calculated $\hat{H}^k$ values than those of the previous mesh Q^k. This means, we have to update each inner vertex q_i^k such that $\tilde{H}(q_i^{k+1}) \approx \hat{H}_i^k$, which is again a second order problem. In order to be able to separate between inner and outer fairness, we only allow the vertex to move along the surface normal vector.

We integrate the inner fairness condition we explained in Section 7.2 into the construction algorithm by including it into the update step. Instead of updating $q_i^k \to q_i^{k+1}$ along the line $q_i^k + t\boldsymbol{n}_i$, we update along the line $\tilde{q}_i^k + t\boldsymbol{n}_i$, where $\tilde{q}_i^k$ is the orthogonal projection of $\sum \lambda_{ij} q_j$ onto the plane defined by q_i^k and $\boldsymbol{n}_i$

$$\tilde{q}_i^k = \sum_{q_j \in N(q_i)} \lambda_{ij} q_j + \langle \Delta(q_i) | \boldsymbol{n}_i \rangle \boldsymbol{n}_i.$$

Summarizing the inner and outer fairness condition, the new position of q_i^k is determined by the two equations

$$q_i^{k+1} = \tilde{q}_i^k + t\boldsymbol{n}_i \qquad (16)$$

and

$$\tilde{H}(q_i^{k+1}) = \hat{H}_i^k. \qquad (17)$$

If we take a look at the mean curvature discretization algorithm presented in Section 5.2, we find that the matrix A does not change if we move q_i along the normal vector. Therefore, only the right side of equation (10) is influenced by such a motion and for the normal curvatures (7) we obtain

$$\tilde{\kappa}_j = 2 \frac{\langle q_j - \tilde{q}_i - t\boldsymbol{n} | \boldsymbol{n} \rangle}{\langle q_j - \tilde{q}_i - t\boldsymbol{n} | q_j - \tilde{q}_i - t\boldsymbol{n} \rangle}.$$

This normal curvature discretization is nonlinear in t, but since the position of q_i will not change much during the update step, we assume that the distance between q_i and q_j remains constant. With this linearization technique we get

$$\tilde{\kappa}_j \approx 2 \frac{\langle q_j - \tilde{q}_i | \boldsymbol{n} \rangle}{\langle q_j - \tilde{q}_i | q_j - \tilde{q}_i \rangle} - 2t \frac{1}{\langle q_j - \tilde{q}_i | q_j - \tilde{q}_i \rangle}.$$

Using this assumption, the discretization of $\tilde{H}(q_i^{k+1})$ determined by (9) and (10), becomes a linear function in t. Solving this linear equation for t, the new position q_i^{k+1} is determined by (16) and approximately solves (17).

Here, we noticed that in some rare constellations it is possible that vertices can alternate between two positions. We therefore update the vertices using

$$q_i^{k+1} = \tilde{q}_i^k + \lambda t \boldsymbol{n}_i \tag{18}$$

with a scalar factor $0 < \lambda < 1$. The examples we present in this paper were constructed using the factor $\lambda = 0.9$.

7.5 Multigrid approach

In order to improve the speed of the construction algorithm we apply multigrid techniques. For our surface problem we have to use a generic multigrid scheme that is able to handle meshes that do not have a subdivision connectivity structure.

A generic multigrid fairing scheme for meshes of arbitrary connectivity was presented by Kobbelt et al. [21]. Here, one first has to build a progressive mesh representation as proposed by Hoppe [18] using half-edge collapses (see Kobbelt et al. [20]). To speed up the calculations, one has to create mesh hierarchies using prescribed resolutions of the progressive mesh representation. However, instead of reducing a mesh while trying to keep the details, one is more interested in creating a mesh whose smallest edge length is maximal while avoiding distorted triangles (long triangles with small inner circle). The number of hierarchy levels can be specified by the user. A reasonable strategy to fix the number of vertices per hierarchy is exponential growth (see Kobbelt and Vorsatz [22]).

Our multigrid implementation follows this generic approach. We start with the construction of a discrete solution on the coarsest level of the progressive mesh representation and then each solution on a coarse level serves as starting point for the iteration algorithm on the next finer hierarchy level. Between two hierarchy levels we need a prolongation operator that introduces new vertices using the vertex split information of the progressive mesh. When adding a new vertex q_i, we have to take care that the outer fairness is not destroyed at that position. This is achieved in three steps, where the first two steps are similar to the prolongation operator used by Guskov et al. [16]:

- First we update the mesh topology and introduce q_i at the position given by its inner fairness criterion

$$q_i = \sum_{q_j \in N(q_i)} \lambda_{ij} q_j$$

- In some cases the first step is not enough to avoid triangle distortions, therefore, in the second step we further update the complete 1-ring of q_i. This means we solve the local linear problem $\Delta q_l = 0$ for all $q_l \in D(q_i)$.

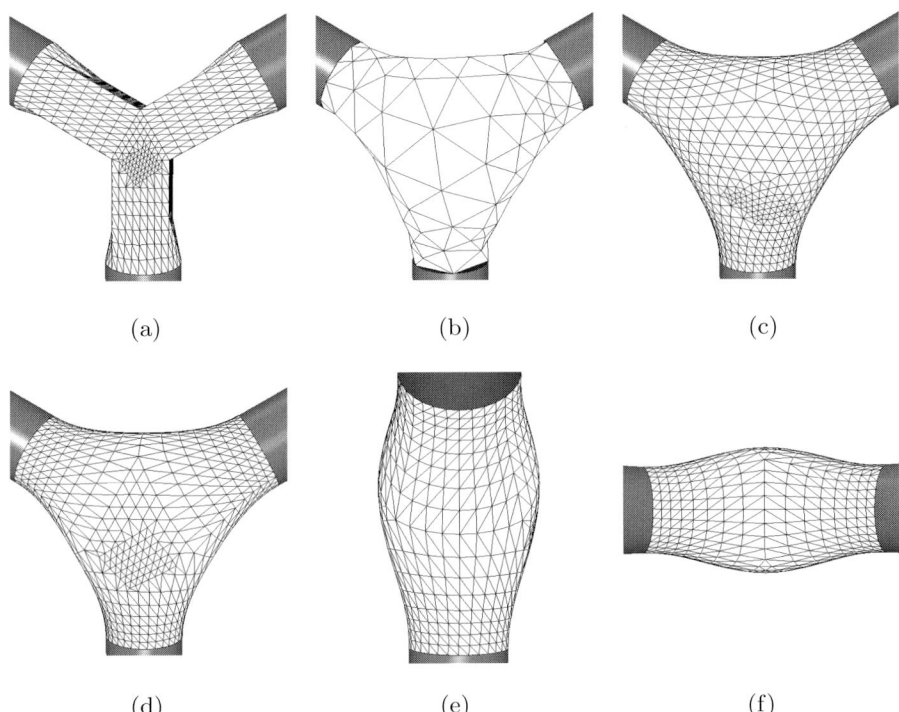

Fig. 4. Our new fairing approach applied on the initial mesh shown in picture (a). In (b) and (c) we optimized the inner fairness with respect to a local uniform parameterization. In (d), (e) and (f) the mesh is discrete conformal to the original mesh in (a).

- Since the second step disturbs the outer fairness, we finally solve $\Delta_B \tilde{H}_l = 0$ for all $q_l \in D(q_i)$ by applying the construction algorithm locally on the 1-disk $D(q_i)$.

Compared to a multigrid approach for a parameter dependent fairing scheme, in our case we have the advantage that the geometry resulting from a coarse mesh already approximates the shape of the smooth surface that is implicitly defined by the PDE very well, increasing the mesh size mainly improves the smoothness of the approximation (see Fig. 4 and 8).

Our mean curvature discretization (Section 5) as well as the update step (Section 7.4) assume that the mesh is not a noisy surface. Therefore, before starting the multigrid algorithm we first construct the linear solution of the problem $\Delta^2 f = 0$ at the coarsest level, using the Laplacian defined by the chosen inner fairness criterion. Later at each hierarchy level our mesh is already presmoothed by the multigrid fairing concept.

In some cases when more boundary curves have to be blended, the initial mesh at the coarsest level that results from the linear biharmonic equation

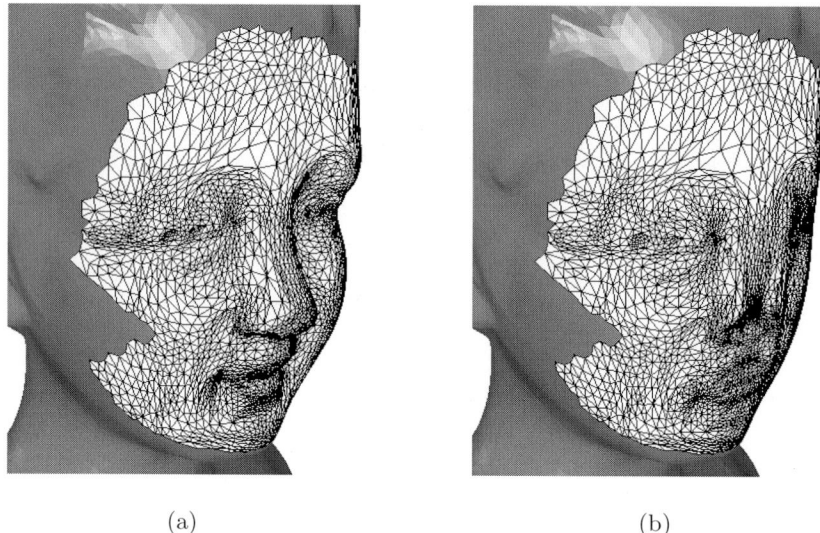

(a)   (b)

Fig. 5. Here, we smoothed the face of a bust model shown in (a). The triangulation of the solution shown in (b) is discrete conformal to the initial mesh.

can have a poor shape. For meshes with an uniform inner fairness, we here improved the mesh fairing algorithm based on discretizing the biharmonic equation. For sparse meshes with very irregular structure, this scheme is much better suited to create an initial mesh (see Schneider et al. [34]).

Remark In this paper we assumed that the final mesh has a prescribed vertex connectivity. If this is not the case, edge flipping should be applied to avoid long triangles. New vertices may be introduced where necessary or removed where the vertex density becomes too high. Strategies for such techniques can be found in Hsu et al. [19] and Welch and Witkin [38].

7.6 Examples

In the example shown in Fig. 4 we determined the HMCS solution of a three-cylinder blend problem. We can see that the chosen inner fairness strategy and the mesh structure have only marginal influence on the shape. The influence of the mesh size is also very small.

As can be seen in Fig. 5, the boundary does not have to be smooth. Here, we used the discrete conformal inner fairness condition during the fairing process. As we can see we can still recognize the face structure of the original mesh on the faired mesh. Such an inner fairness condition is especially interesting for fairing of textured meshes or for multiresolution modeling purposes. The conformal inner fairness strategy is only slightly more time consuming to

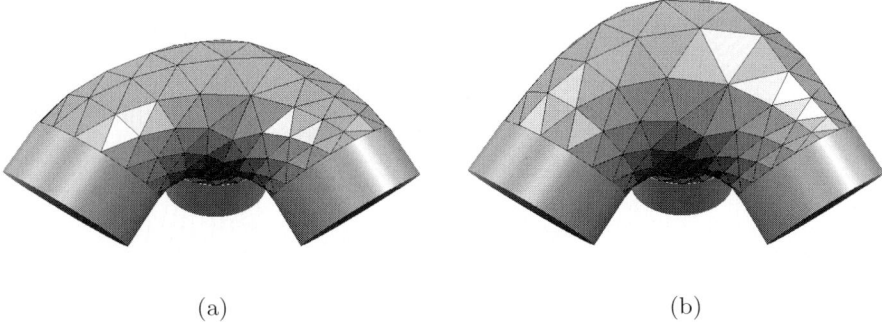

Fig. 6. Three-cylinder blend with sloped cylinders. Picture (a) shows the discrete HMCS solution while (b) shows the according discrete MES solution.

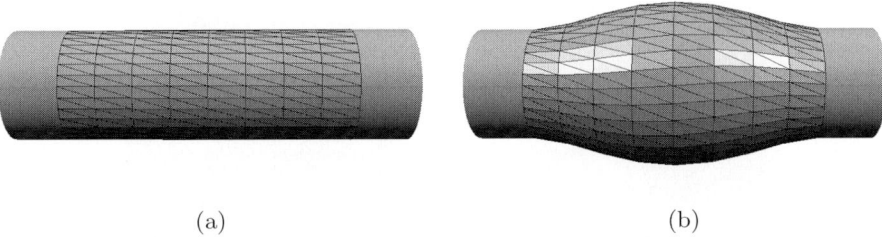

Fig. 7. The two-cylinder blend fairing problem. Picture (a) shows a discrete HMCS solution, while the discrete MES solution is shown in (b).

construct than the uniform parameterization, but requires much more memory. Here, the original mesh has to be reconstructed from the progressive mesh representation parallel to the faired mesh, to be able to update the necessary coefficients of the Laplacian, which also have to be stored in memory.

In the Figs. 6 and 7 we compared HMCS and MES solutions. As we can see MESs do look different when compared to HMCSs. They usually seem to produce surfaces that have larger area and include a larger volume. While in some cases this effect may be wanted, in other cases this effect is definitely disturbing (Fig. 7). In the example shown in Fig. 7 we can also see that the discrete HMCS indeed reproduces a cylinder, while the MES does not. Summarizing the results, one can see that HMCSs are excellently suited for solving our fairing problem without isolated vertex interpolation constraints in the interior. As we will see in the next section, this is no longer true if such constraints have to be satisfied.

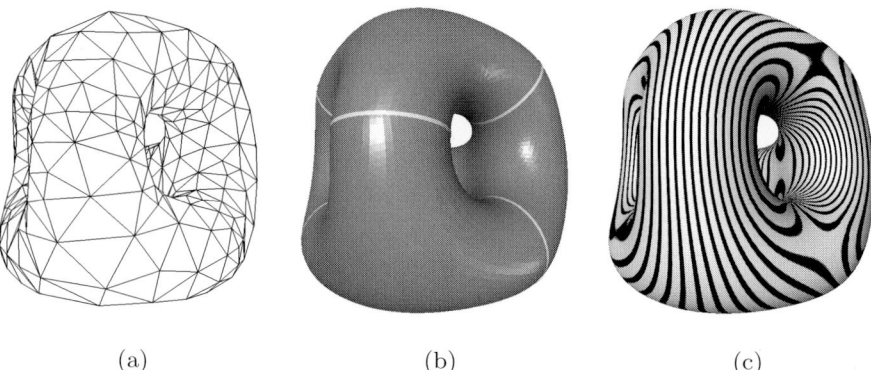

Fig. 8. 6 circles are used to define a tetra-thing. (a) and (b) show the solution of a mesh with 500 resp. 13308 vertices. (c) shows the reflection lines of the mesh (b).

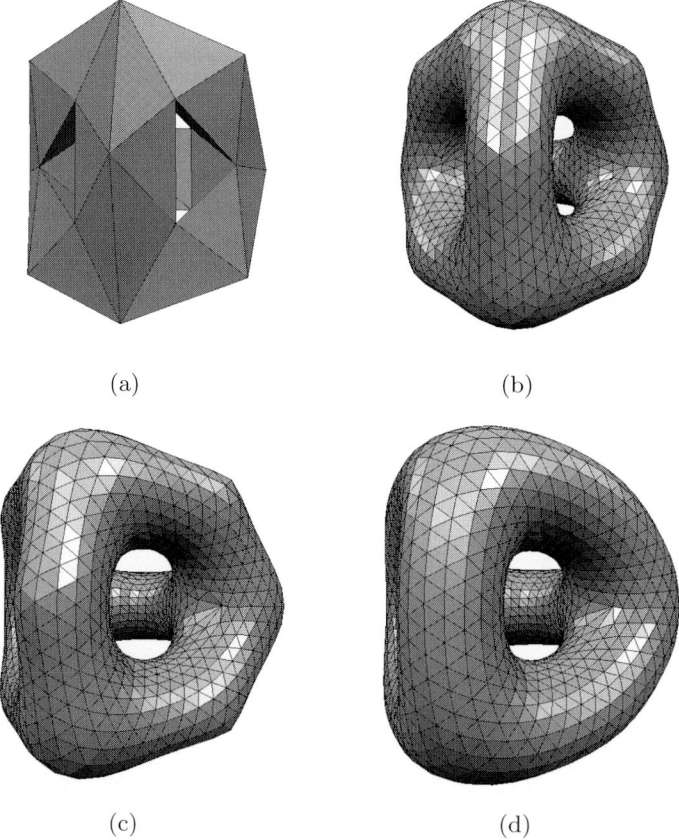

Fig. 9. An example for closed polyhedron interpolation with isolated vertices. Picture (b) and (c) show the HMCS solution and picture (d) the MES solution.

8 Interpolation of curves and isolated vertices in the interior

So far we only considered the surface fairing problem, where no additional interpolation constraints appear in the interior and all conditions were specified at the boundary. Let us in the following show how to handle the general case, which also includes the situation, where there are no boundary conditions at all.

Adding additional isolated vertices or polygonal curves into the interior as interpolation constraints is very easy. At such isolated vertices resp. at the vertices of the polygonal curve we simply have to exclude these vertices from the curvature smoothing and the update process. An example where we used curves as interpolation constraints to define a surface is shown in Fig. 8. Here, six circles were used to define a tetra-thing surface using HMCSs. As we can see such curve constraints are no problem for a surface fairing method based on HMCSs.

However, the situation is completely different for isolated vertex interpolation. Here, at isolated vertices, the mean curvature function of a HMCS can develop spikes. In practice this fact often leads to surfaces that no longer look aesthetic (see Fig. 9(a)–(c)).

In Fig. 9 we also compared our HMCS solution with the according MES solution. As we can see the MES fairing approach is much better suited for isolated vertex interpolation than the HMCS scheme in this example.

9 Conclusion

As we have shown in this paper, our HMCSs can be seen as a very interesting alternative to MESs and are a new surface fairing approach in its own right. Comparing HMCSs and MESs we saw that MESs often tend to create surfaces that include a larger volume than HMCSs what is not always wanted and does not always lead to nicer solutions. However, the situation changes completely, if isolated points have to be interpolated as an additional fairing constraint. Without modifications, HMCSs are not well suited in this case, because of the large mean curvature values that can appear at the isolated interpolation points. Here, the MES approach in our example seemed to be much better suited to handle such vertex constraints.

References

1. M. I. G. Bloor and M. J. Wilson. Generating blend surfaces using partial differential equations. *Computer-Aided Design*, 21:165–171, 1989.
2. M. I. G. Bloor and M. J. Wilson. Using partial differential equations to generate free-form surfaces. *Computer-Aided Design*, 22:202–212, 1990.

3. K. A. Brakke. The surface evolver. *Experimental Mathematics*, 1(2):141–165, 1992.
4. H. G. Burchard, J. A. Ayers, W. H. Frey, and N. S. Sapidis. Approximation with aesthetic constraints. In N. S. Sapidis, editor, *Designing Fair Curves and Surfaces*, pages 3–28. SIAM, Philadelphia, 1994.
5. D. L. Chopp and J. A. Sethian. Motion by intrinsic laplacian of curvature. *Interfaces and Free Boundaries*, 1:1–18, 1999.
6. M. P. do Carmo. *Differential Geometry of Curves and Surfaces*. Prentice-Hall, Inc Englewood Cliffs, New Jersey, 1993.
7. M. Desbrun, M. Meyer, P. Schröder, and A. H. Barr. Discrete differential-geometry operators in nD. submitted for publication.
8. M. Desbrun, M. Meyer, P. Schröder, and A. H. Barr. Implicit fairing of irregular meshes using diffusion and curvature flow. In *SIGGRAPH '99 Conference Proceedings*, pages 317–324, 1999.
9. T. Duchamp, A. Certain, T. DeRose, and W. Stuetzle. Hierarchical computation of pl harmonic embeddings. Technical report, University of Washington, 1997.
10. G. Farin, and D. Hansford. Discrete Coons patches. *Computer Aided Geometric Design*, 16:691–700, 1999.
11. M. S. Floater. Parametrization and smooth approximation of surface triangulations. *Computer Aided Geometric Design*, 14:231–250, 1997.
12. M. Giaquinta and S. Hildebrandt. *Calculus of Variations : The Langrangian Formalism, Vol 1 (Grundlehren Der Mathematischen Wissenschaften, Vol 310)*. Springer, New York, 1996.
13. G. H. Golub and C. F Van Loan. *Matrix Computations*. Johns Hopkins University Press, Baltimore, 1989.
14. G. Greiner. Blending surfaces with minimal curvature. In *Proc. Dagstuhl Workshop Graphics and Robotics*, pages 163–174, 1994.
15. G. Greiner. Variational design and fairing of spline surfaces. *Computer Graphics Forum*, 13(3):143–154, 1994.
16. I. Guskov, W. Sweldens, and P. Schröder. Multiresolution signal processing for meshes. In *SIGGRAPH '99 Conference Proceedings*, pages 325–334, 1999.
17. W. Hackbusch. *Iterative Lösung grosser schwachbesetzter Gleichungssysteme*. Teubner Verlag, Stuttgart, 1993.
18. H. Hoppe. Progressive meshes. In *SIGGRAPH '96 Conference Proceedings*, pages 99–108, 1996.
19. L. Hsu, R. Kusner, and J. Sullivan. Minimizing the squared mean curvature integral for surfaces in space forms. *Experimental Mathematics*, 1(3):191–207, 1992.
20. L. Kobbelt, S. Campagna, and H.-P. Seidel. A general framework for mesh decimation. In *Proceedings of the Graphics Interface conference*, pages 43–50, 1998.
21. L. Kobbelt, S. Campagna, J. Vorsatz, and H.-P. Seidel. Interactive Multi-Resolution Modeling on Arbitrary Meshes. In *SIGGRAPH '98 Conference Proceedings*, pages 105–114, 1998.
22. L. Kobbelt and J. Vorsatz. Multiresolution hierarchies on unstructured triangle meshes. *Computational Geometry*, 14(1-3):5–24, 1999.
23. K. Marciniak and B. Putz. Approximation of spirals by piecewise curves of fewest circular arc segments. *Computer-Aided Design*, 16:87–90, 1984.

24. H. P. Moreton. *Minimum Curvature Variation Curves, Networks and Surfaces for Fair Free-Form Shape Design*. PhD thesis, University of California at Berkley, 1992.
25. H. P. Moreton and C. H. Séquin. Functional optimization for fair surface design. In *SIGGRAPH '92 Conference Proceedings*, pages 167–176, 1992.
26. H. P. Moreton and C. H. Séquin. Scale-invariant minimum-cost curves: Fair and robust design implements. *Computer Graphics Forum*, 12(3):473–484, 1993.
27. W. W. Mullins. Theory of thermal grooving. *J. Appl. Phys.*, 28:333–339, 1957.
28. B. Oberknapp and K. Polthier. An algorithm for discrete mean curvature surfaces. In H. C. Hege and K. Polthier, editors, *Mathematics of Surfaces VIII*. Springer Verlag, 1997.
29. A. Polden. *Curves and Surfaces of Least Total Curvature and Fourth-Order Flows*. PhD thesis, University of Tübingen, Dept. of Mathematics, 1996.
30. U. Pinkall and K. Polthier. Computing discrete minimal surfaces and their conjugates. *Experimental Mathematics*, 2(1):15–36, 1993.
31. J.A. Roulier, T. Rando, and B. Piper. Fairness and monotone curvature. In C. K. Chui, editor, *Approximation Theory and Functional Analysis*, pages 177–199. Academic Press, Boston, 1990.
32. R. Schneider and L. Kobbelt. Discrete fairing of curves and surfaces based on linear curvature distribution. In *Curve and Surface Design - Saint-Malo '99 Proceedings*, pages 371–380, 1999.
33. R. Schneider and L. Kobbelt. Geometric fairing of irregular meshes for free-form surface design. *Computer Aided Geometric Design*, 18(4):359–379, 2000.
34. R. Schneider, L. Kobbelt, and H.-P. Seidel. Improved Bi-Laplacian mesh fairing. Mathematical Methods for Curves and Surfaces OSLO 2000, pages 445–454, 2001.
35. G. Taubin. Estimating the tensor of curvature of a surface from a polyhedral approximation. In *ICCV '95 Proceedings*, pages 902–907, 1995.
36. G. Taubin. A signal processing approach to fair surface design. In *SIGGRAPH '95 Conference Proceedings*, pages 351–358, 1995.
37. J. E. Taylor. Surface motion due to crystalline surface energy gradient flows. In *Elliptic and Parabolic Methods in Geometry*, pages 145–162. A K Peters, Ltd., Wellesley, 1996.
38. W. Welch and A. Witkin. Free-form shape design using triangulated surfaces. In *SIGGRAPH '94 Conference Proceedings*, pages 247–256, 1994.

Shape Feature Extraction

Georgios Stylianou and Gerald Farin

Arizona State University, Tempe AZ 85287

Abstract. In this paper we present a method for automatic extraction of shape features, called crest lines. Shape features are important because they provide an alternative to describing an object, using its most important characteristics and reduce the amount of information stored. The algorithm is comprised of a curvature approximation technique, crest point classification and a crest lines tracing algorithm.

1 Introduction

Three-Dimensional object comparison and matching are very important techniques with applications in robotics and automation, medicine, solid modelling, geometric modelling. It provides the ability to compare or match different objects and automatically identify their degree of similarity based on some distance criteria. Object matching attempts to find the transformation that best matches two given objects. In contrast, when doing object comparison a distance criterion is defined and used to compare different objects and return their distance. This method is usually used to rank objects based on their distance to a given object.

Various methods have been developed in the past years that either use directly the object representation or transform the object to another domain before applying the matching algorithm. Hu [7] calculates all the possible rotations and translations of a scene object from a model object, represented as a connected graph of edges, and uses them to construct a three-dimensional Hough space. Johnson *et al* [8] use spin images of a model object to recognize a scene object. Shum *et al* [9] deform a sphere to the 3D closed surface, genus zero, and use curvature distribution, to define a similarity measure, for object comparison. Zhang and Hebert [10] perform surface matching after reducing the problem to 2D image-matching problem using harmonic maps. Declerck *et al* [11] extract shape features and use them to deform a source surface to a target surface. Most of these methods match objects and are limited within the domain of application. A notable exception is Shum *et al* that use a distance criterion for comparison but is applicable only to surfaces without holes.

In this work, we extract shape features called crest lines that we will use in a later work to construct a feature based object comparison method. Crest lines are 3D lines on a surface that provide us with a satisfactory geometrical representation of important physical properties such as ridge lines and valleys in the case of aerial images [14], or anatomical features in the case of medical images [2,15]. They have been used, so far, for object registration [13], growth simulation [12] and automatic retrieval of anatomical structures [11].

We primarily focus on objects represented as triangulated meshes. With the advent of laser digitizers, we can easily acquire topologically different objects represented by triangulated meshes. Also, triangulations can represent surfaces exhibiting arbitrary topology, whereas B-spline surfaces can only be joined together with some effort in order to represent arbitrary topology.

2 Related work

We look at similar crest lines extraction algorithms and examine their advantages and disadvantages.

Lengagne et al [1], extract crest lines from 3D triangulations. Their algorithm tries to identify the crest points using the definition of a crest point as given in section 3.1. It makes an estimation of the directional derivative of the largest curvature in its direction at every vertex of the mesh using finite differences. According to the paper, the derivative is estimated as $k_1(\mathbf{V}_1) - k_1(\mathbf{V})$, where $\mathbf{V}_1$ is a vertex on the star($\mathbf{V}$) and $\mathbf{V}_1$ is selected such that $\mathbf{V}\mathbf{V}_1.\mathbf{t}_1$ is maximized. $\mathbf{t}_1$ is the direction of the largest curvature k_1 on vertex $\mathbf{V}$. Since at every vertex the derivative is usually positive or negative, a crest point is defined as the zero crossings of the derivative along the edges of the mesh. The problem with this method is that it makes a very rough estimation of the derivative, which fails especially when we have irregular and noisy triangulations. This method, also accumulates error when finding the zero crossings because the positive and negative vertices are not labelled consistently. We show a failing example in Figure 1.

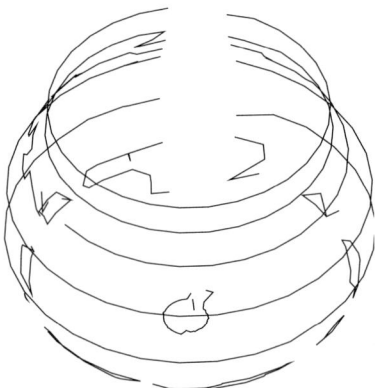

Fig. 1. Crest lines of vase1 as in [1]

Guéziec [3] developed an algorithm to extract crest lines from B-spline parametrized objects. Although, such a parametrization is very accurate because we can calculate the exact curvature, this method is very expensive and most importantly one B-spline surface cannot easily parametrize surfaces of arbitrary topology.

Khaneja et al [2] implemented a dynamic programming algorithm to extract crest lines from triangulations that is stated to have noise immunization. This method is semi-automatic. The problem arises naturally because it requires the presence of someone (usually an expert in the field) to define start-end points for every crest line to be calculated; thus this method fails when there is no expert or the expert fails to identify topologically correct points. Also, it is very slow because of the necessity to manually define start-end points. Therefore, it cannot applied to a large number of surfaces efficiently.

We propose a method that solves the problems appearing in [1], it is automatic in addition to [2] and can be applied in any surface as long as it is represented by a triangulated mesh in contrast to [3].

This article is organized as follows. In Section 4 we briefly describe the underlying theory of our method. Sections 5, 6 present the method and the experimental results, respectively.

3 Background

In this section, we briefly give the basic mathematical framework used to develop our scheme. These include crest lines and normal curvature definition and a local parametrization technique.

3.1 Crest lines

Crest lines are local shape features of a surface. They are defined as the set of of all points satisfying

$$D_{t_1} k_1(u,v) = 0 \qquad (1)$$

where D_{t_1} is the directional derivative in direction t_1, k_1 is the largest principal curvature and t_1 is its domain direction on a point (of the surface) with domain coordinates (u,v).

They are shape features with the main characteristic of using local information to yield a global description of the surface. Crest lines are local shape features by definition. But when all the crest lines are viewed together, they describe the surface. i.e. the viewer can realize just by looking at them, what kind of surface they represent. For instance, on figures 4 and 5, the viewer can vaguely understand that these objects are some kind of vases just by looking at the crest lines, even though he cannot identify their exact shape.

By this means they also provide a global description of an object. Unfortunately, some object exist that do not have any crest lines. These are object having constant or increasing largest curvature.

An example of a crest line is in Figure 2. Because a crest point has maximum largest curvature in its corresponding direction, a crest line naturally follows the direction of the smallest curvature of its composing crest points.

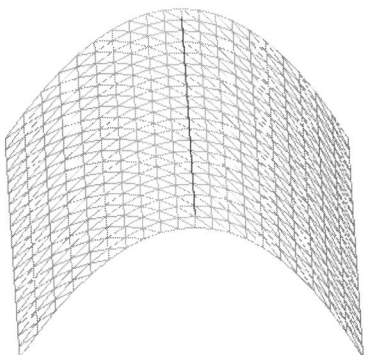

Fig. 2. A crest line example.

3.2 Normal Curvature

Normal curvature [5] measures locally (on a point) the bending of the surface in any direction. Principal curvatures k_1, k_2 are the maximum and minimum values of the normal curvature, in absolute value, in domain directions $\mathbf{t}_1, \mathbf{t}_2$. In the previous section, we defined crest lines with respect to the largest curvature k_1. Here, we rapidly show how principal curvatures can be calculated.

The Weingarten matrix is:

$$W = \begin{bmatrix} \frac{MF-LG}{EG-F^2} & \frac{LF-ME}{EG-F^2} \\ \\ \frac{NF-MG}{EG-F^2} & \frac{MF-NE}{EG-F^2} \end{bmatrix}$$

The eigenvalues and eigenvectors of W are the principal curvatures k_1, k_2 and principal directions $\mathbf{t}_1, \mathbf{t}_2$ respectively.

The coefficients of the first fundamental form E, F, G are:

$$E = \mathbf{x}_u \mathbf{x}_u = \frac{\partial}{\partial u}\mathbf{x}(u,v)\frac{\partial}{\partial u}\mathbf{x}(u,v)$$

$$F = \mathbf{x}_u \mathbf{x}_v = \frac{\partial}{\partial u}\mathbf{x}(u,v)\frac{\partial}{\partial v}\mathbf{x}(u,v)$$

$$G = \mathbf{x}_v \mathbf{x}_v = \frac{\partial}{\partial v}\mathbf{x}(u,v)\frac{\partial}{\partial v}\mathbf{x}(u,v)$$

The coefficients of the second fundamental form L,M,N are:

$$L = \mathbf{n}\mathbf{x}_{uu} = \mathbf{n}(u,v)\frac{\partial^2}{\partial u^2}\mathbf{x}(u,v)$$

$$M = \mathbf{n}\mathbf{x}_{uv} = \mathbf{n}(u,v)\frac{\partial^2}{\partial u \partial v}\mathbf{x}(u,v)$$

$$N = \mathbf{n}\mathbf{x}_{vv} = \mathbf{n}(u,v)\frac{\partial^2}{\partial v^2}\mathbf{x}(u,v)$$

where $\mathbf{x}(u,v)$ is a point on the surface and $\mathbf{n}(u,v)$ is the normal vector on a point with domain coordinates (u,v).

3.3 Parametrization

We use a quadratic bivariate polynomial to locally parametrize the surface. The patch we use has the type

$$\mathbf{x}(\mathbf{u}) = \mathbf{a}u^2 + \mathbf{b}uv + \mathbf{c}v^2 + \mathbf{d}u + \mathbf{e}v + \mathbf{f} \tag{2}$$

where $\mathbf{u} = (u,v)$ is a point on the domain, $\mathbf{x}(\mathbf{u})$ is a point on the surface and $\mathbf{a},\mathbf{b},\mathbf{c},\mathbf{d},\mathbf{e},\mathbf{f}$ are the coefficients of the patch.

3.4 Least Squares fitting

We use the least squares [4] method to fit a polynomial(2) on a point set. Here, we define the problem and describe the domain calculation.

Problem: Given a point set $D = \{\mathbf{p}_i | i = 0,...,L\}$, $L+1$ is the number of the points, we want to find an approximating patch $\mathbf{x}(u,v)$, such that

$$\sum_{i=0}^{L} \|\mathbf{p}_i - \mathbf{x}(u_i,v_i)\|^2 \tag{3}$$

is minimized.

The first step is the domain calculation i.e. the calculation of the parameters $\mathbf{u}_i = (u_i,v_i)$ that correspond to $\mathbf{p}_i$. In other words, we have to calculate the domain parameters $\mathbf{u}_i$ of the prospective range values $\mathbf{p}_i$. The procedure we use is described by Hamann [6]. We briefly give the steps.
In order to approximate the principal curvatures at a point $\mathbf{x}_i$ do

1. Get the neighboring points of $\mathbf{x}_i$.
2. Compute the plane P passing through $\mathbf{x}_i$ and having $\mathbf{n_i}$, the normal vector at $\mathbf{x}_i$, as its normal.
3. Define an orthonormal coordinate system in P with $\mathbf{x}_i$ as its origin and two arbitrary unit vectors in P.
4. Project all the points of D onto the plane P and represent their projections with respect to the local coordinate system in P.

5. The final points are the desired domain points $\mathbf{u_i} = (u_i, v_i)$.

The second step is the patch fitting. First we create the Least Squares System as follows:

$N = 6$, is the number of the coefficients of (2)

$D = [\mathbf{p}_0, ..., \mathbf{p}_L]^T$

$X = [a, b, c, d, e, f]^T$ is the unknown

$$F = \begin{bmatrix} u_0^2 & u_0 v_0 & v_0^2 & u_0 & v_0 & 1 \\ . & . & . & . & . & . \\ . & . & . & . & . & . \\ . & . & . & . & . & . \\ u_L^2 & u_L v_L & v_L^2 & u_L & v_L & 1 \end{bmatrix}$$

The linear system

$$D = FX \qquad (4)$$

is equivalent to minimizing (3). D is the given point set, $L+1$ is the number of the points, F is the domain values of this point set D and X are the unknowns, the coefficients of polynomial (2).

Usually $L > N$; thus system (4) is an overdetermined system. The solution X can be found be solving the linear system $F^T D = F^T F X$. If $L = N$, we solve an interpolation problem.

4 Method

We describe the method we designed for automatic crest line extraction. The method has three steps. The first step approximates the principal curvatures and directions on every vertex of the triangulation, the second step handles crest point classification and the final step traces the crest lines.

4.1 Curvature approximation

The first step is:
For every vertex $\mathbf{v}_i$ of the mesh,

1. Get the star of $\mathbf{v}_i$, denoted by $star(\mathbf{v}_i)$.
2. Fit the quadratic patch (2) on $\mathbf{v}_i$ and $star(\mathbf{v}_i)$.
3. Compute the principal curvatures and their corresponding directions on $\mathbf{v}_i$, as described in the former section.

4.2 Crest point classification

Then we evaluate whether a vertex is a crest point. The local neighborhood of a vertex **v** of a triangulated mesh is its $star(\mathbf{v})$. We can utilize the definition (1) and directly calculate the directional derivative of the largest curvature of every vertex like Lengagne et al [1] or using numerical differences but the gradient is very noise sensitive and it fails to provide consistent values many times, especially on irregularly triangulated meshes. In fact, we have experimented using the numerical differences method and it, often, does not give acceptable results. Consequently, we follow a different approach.

We use the interpretation of definition (1) where it states that *a point is a crest point if its largest curvature is locally maximal in its corresponding direction*. Therefore, after calculating the largest principal curvature k_1 and its corresponding domain direction $\mathbf{t}_1$ on vertex **v**, we use the domain values of the $star(\mathbf{v})$ as follows:

- The domain values of the $star(\mathbf{v})$ are $(\mathbf{u}_1, ..., \mathbf{u}_L)$, L is the number of vertices in the $star(\mathbf{v})$ and $\mathbf{u}_0$ is the domain value of **v**.
- Find the facet that intersects $\mathbf{t}_1$ and get its two vertices $\mathbf{u}_{m1}$ and $\mathbf{u}_{m2}$, $m\{1,2\} \in [1, L]$.
- Do the same in the direction $-\mathbf{t}_1$ and get vertices $\mathbf{u}_{j1}$ and $\mathbf{u}_{j2}$, $j\{1,2\} \in [1, L]$.
- Get the largest curvatures k_1^{m1m2}, k_1^{j1j2} on points $\mathbf{U}_{m1m2}$ and $\mathbf{U}_{j1j2}$ respectively by interpolating along (k_1^{m1}, k_1^{m2}) and (k_1^{j1}, k_1^{j2}).
- If $(k_1^0)^2 - (k_1^{m1m2})^2 > e$ and $(k_1^0)^2 - (k_1^{j1j2})^2 > e$ then **v** is a crest point.

where $e > 0$ controls the level of maximality. It is used to reduce numerical errors or threshold on the number of crest points to keep the most important. Also, k_1^i is the largest curvature on the vertex $\mathbf{u}_i$, $0 \leq i \leq L$.

For example using Figure 3, we interpolate the largest curvatures on vertices **U1**, **U2** to k_1^{12}, and do the same for vertices **U4**, **U5** to get k_1^{45}. If the largest curvature on vertex **U0** satisfies the criterion set then **U0** is a crest point.

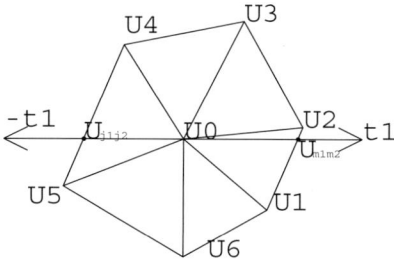

Fig. 3. Crest point classification.

4.3 Crest lines

In the previous section, we have given an algorithm to manage crest point classification. But all we achieved is to identify the crest vertices which are not useful, yet. We still need to design a method to join all these points and create the features of the object. This method is comprised of two parts. First we trace all the crest lines and later we join as many lines as possible, for reasons explained in the subsequent sections.

Let the mesh consist of N vertices. The algorithm is:

1. Initialize linelist LS and set the number of lines $j = 0$.
2. Set the id of the current vertex $i = 0$.
3. if $\mathbf{v}_i$ is not visited, is a crest point and has at most two crest points on its star then
   - call $traceCrestLine(\mathbf{v}_i, ls_j, \text{"first"})$.
   - call $traceCrestLine(\mathbf{v}_i, ls_j, \text{"last"})$.
   - increase j.
4. If $i < N$ increase i and goto step 3.

If a crest point initially has more than two other crest points on its star then it is classified as cross crest point. We do not start tracing the line from this point because it causes problems due to the ambiguity of selecting a path to follow. A cross crest point is a bifurcation point. A crest line sometimes bifurcates to two or more lines. Because we cannot know in advance which line is 'better' to connect to from this point, until we trace it, then we stop at this point and use the second part to join lines.

The procedure $traceCrestLine(\mathbf{v}_i, ls_j)$ is:

1. Mark $\mathbf{v}_i$ as visited and add it to line ls_j.
2. Get all the vertices of $star(\mathbf{v}_i)$ that are crest points and are not visited. Let their number be n.
3. If $n = 1$ then goto step 1 for the new vertex, else exit.

When we call $traceCrestLine$ with the argument "first" it traces the line in one direction and adds every point on the beginning of the line, otherwise with the argument "last", it traces the line in the opposite direction and adds every point on the end of the line. Therefore, we can trace the maximum line segment possible and not just parts of a line, because usually we start tracing a line from a point in that line.

Join crest lines After we trace all the crest lines, we join lines with an identical end point. In many cases, we have more than two lines sharing the same end point. We join the two longest lines in terms of the number of points they contain. Since we want to keep the most important features, we try to create as long lines as possible.

Furthermore, we join all the crest lines, that have endpoints at very small distance. This distance is set to be a multiple of the average edge length

(AEL) of the mesh. This algorithm joins all the lines with endpoints at distance AEL and runs up to distance $\alpha * AEL$. The number α can be defined by the user or automatically. As follows a reasonable value for $\alpha = 2$. At every level given a line, the algorithm finds all the lines on the current distance radius. If it finds more than one it does not join them, unless $\alpha = 0$. When $\alpha = 0$, it joins the two longer lines with respect to the number of points they contain. It runs at every level as many times as needed to join all the lines found and then it proceeds to the next level. The computational complexity at every level is $O(n^2)$ on the number of lines n.

Finally, the algorithm erases from the linelist all the lines that contain less than a predefined number of points, keeping the most important lines (features).

5 Results

We have tested our algorithm on various meshes. First, we used regular triangulated meshes representing different kinds of vases. Figures 4 and 5 show the crest lines superimposed on vase1 and vase2, respectively. The results on those datasets are excellent. We have set $e = 0.001$ and we did not use the line join algorithm. The lines are slightly translated from their actual position, as it is visually observed. This occurs because we trace crest lines over the vertices of the mesh but it does not create a real problem.

Figures 6 and 7 show crest lines on a greek statue, courtecy of the University of Thessaloniki, named Igea. This object has approximately 33500 vertices and 67000 facets. For Figure 6 the settings were $e = 0.001$ and $k = 2$. Also, we kept only the lines of minimum length 10. Before joining we had 3364 lines, after joining 1183 and after deleting smaller lines than length 10 remained 301 lines. For Figure 7 we had the same setting, except that we kept only lines with length 20 or more. Finally remained 134 lines.

Even though, we have very good results with large and complicated meshes such as Igea we still experience problems with very complex meshes that contain a lot of of noise, such as the human brain in Figure 8. This data set has approximately 19900 vertices and 39000 facets. The settings were, also, $e = 0.001$ and $k = 2$. Initially 1421 crest lines and after joining and keeping all the lines with minimum length 10, remained 165 lines. We can see that we still have most of the crest lines, but we do not have the best results.

Most of the results are really nice, especially considering that they are obtained totally automatic. Also, this method is very robust and creates long lines.

6 Conclusions

We have presented a method for automatic crest lines extraction. We used a bivariate polynomial to calculate principal curvatures and directions on every

Fig. 4. Crest lines of vase1.

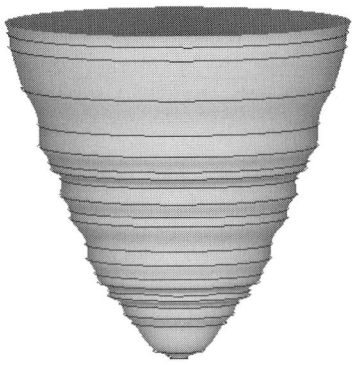

Fig. 5. Crest lines of vase2.

vertex of the mesh and used definition (1) to decide when a vertex is a crest point. Later, we gave an intelligent algorithm to trace the crest lines over the mesh, join them, dispose the small insignificant lines and stored them in a useful data structure for future use.

In future work, we intend to enhance this method to extract crest lines from too complicated and very noisy data sets, make a parallel implementation of this method for extracting crest lines efficiently from triangulated meshes of order of millions of points, and implement a quantitative comparison method using crest lines.

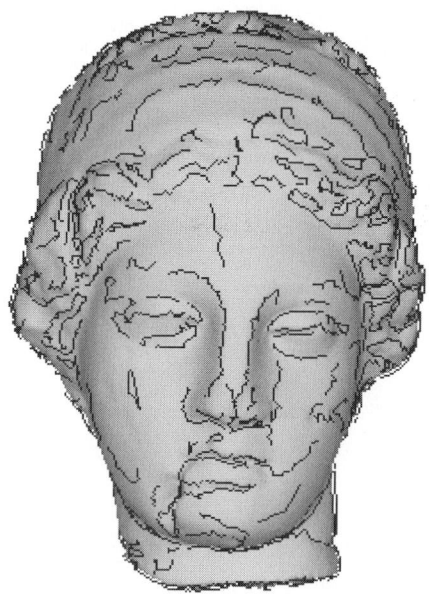

Fig. 6. Crest lines on Igea with minimum line length 10.

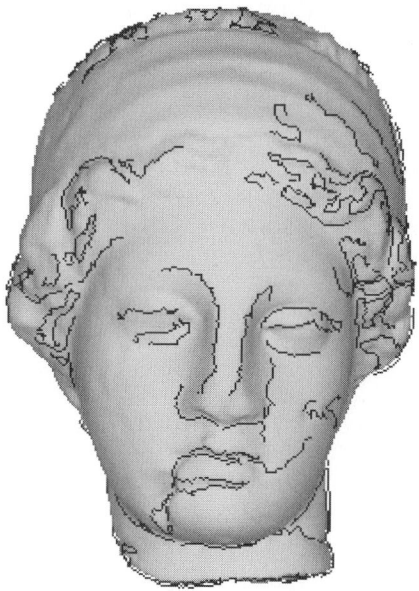

Fig. 7. Crest lines on Igea with minimum line length 20.

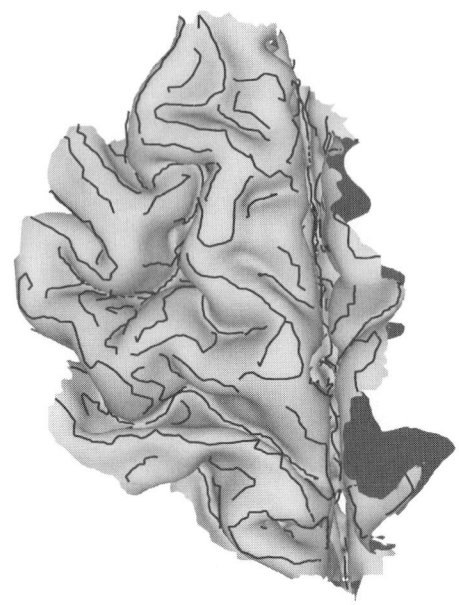

Fig. 8. Crest lines on a part of a human brain.

7 Acknowledgments

This work was supported in part by the Arizona Alzheimer's Disease Research Center.

References

1. R. Lengagne, F. Pascal, O. Monga. Using Crest Lines to Guide Surface Reconstruction from Stereo. *International Conference on Pattern Recognition*,1996.
2. N. Khaneja, M.I. Miller, U. Grenander. Dynamic Programming Generation of Curves on Brain Surfaces. *IEEE Transactions on Pattern Analysis and Machine Intelligence*, vol. 20, 11, 1260-1265, 1998.
3. A. Guéziec. Large Deformable Splines, Crest lines and Matching. *IEEE 4th International Conference on Computer Vision*, 650-657, 1993.
4. G. Farin. *Curves and Surfaces for Computer Aided Geometric Design, Fourth Edition*. Academic Press, Boston, 1997.
5. A. Gray. *Modern Differential Geometry of Curves and Surfaces with Mathematica, Second Edition*. CRC Press, 1998.
6. B. Hamann. Curvature Approximation for Triangulated Surfaces. In *Geometric Modelling*, edited by Farin *et al*, Springer-Verlag, 139-153, 1993.
7. G. Hu. 3-D Object Matching in the Hough Space. *IEEE Int. Conf. on Systems, Man and Cybernetics*, vol. 3, 2718-2723, 1995.
8. A.E. Johnson, M. Hebert. Using Spin Images for Efficient Object Recognition in Cluttered 3D Scenes. *IEEE Transactions on Pattern Analysis and Machine Intelligence*, vol. 21-5, 433-449, 1999.

9. H.Y. Shum, M. Hebert, K. Ikeuchi. On 3D Shape Similarity. *IEEE Conference on Computer Vision and Pattern Recognition*, 526-531, 1996.
10. D. Zhang, M. Hebert. Harmonic Maps and Their Applications in Surface Matching. *IEEE Conference on Computer Vision and Pattern Recognition*, 1999.
11. J. Declerck, G. Subsol, J.P. Thirion, N. Ayache. Automatic Retrieval of Anatomical Structures in 3D Medical Images. In N. Ayache (Ed.), CVRMed'95, vol. 905 of *Lecture Notes in Computer Science*, 153-162, Nice, France, Springer-Verlag.
12. P.R. Andresen, M. Nielsen, S. Kreiborg. 4D Shape-Preserving Modelling of Bone Growth. MICCAI, 1998.
13. A. Guéziec, N. Ayache. Smoothing and Matching of 3D Space Curves. *European Conference on Computer Vision*, 620-629, 1992.
14. O. Monga, N. Armande, P. Montesinos. Thin nets and Crest lines:Application to Satellite Data and Medical Images. *Computer Vision and Image Understanding*, v. 67, n. 3, 285-295, September 1997.
15. J.P. Thirion, A. Gourdon. The 3D Marching Lines Algorithm. *Graphical Models and Image Processing*, 58(6), 503-509, 1996.

Network-based Rendering Techniques for Large-scale Volume Data Sets

Joerg Meyer[1], Ragnar Borg[2], Bernd Hamann[2], Kenneth I. Joy[2], and Arthur J. Olson[3]

[1] NSF-MSU Engineering Research Center (ERC), Department of Computer Science, Mississippi State University, 2 Research Blvd., Starkville, MS 39762-9627, jmeyer@cs.msstate.edu
[2] Center for Image Processing and Integrated Computing (CIPIC), Department of Computer Science, University of California, 1 Shields Avenue, Davis, CA 95616-8562, Ragnar.Borg@proxycom.no, {hamann, joy}@cs.ucdavis.edu
[3] The Scripps Research Institute (TSRI), 10550 North Torrey Pines Road, La Jolla, CA 92037, olson@scripps.edu

Abstract. Large biomedical volumetric data sets are usually stored as file sets, where the files represent a family of cross sections. Interactive rendering of large data sets requires fast access to user-defined parts of the data, because it is virtually impossible to render an entire data set of such an enormous size (several gigabytes) at full resolution, and to transfer such data upon request over the Internet in a reasonable amount of time. Therefore, hierarchical rendering techniques have been introduced to render a region of interest at a relatively higher resolution. Regions rendered at coarser resolutions are provided as context information. We present a dynamic subdivision scheme that incorporates space-subdivision and wavelet compression.

1 Introduction

Real-color volume data sets can be obtained by taking photographs or scanning cross sections of objects. These objects are typically in a frozen state (cryosections). These techniques produce high-resolution image data in real color. The resolution is only limited by the camera or the imaging device, and not so much by principal limitations of the scanning device, because there is no complex matrix transformation required to obtain 2D image data, as it is the case for computed tomography (CT) or magnetic resonance imaging (MRI). Therefore, real-color volume data sets tend to be much larger than CT or MRI data.

A typical setup is a client-server architecture, where a large-scale data set is stored on a powerful server, and the rendering is performed on the client side. In order to make a data set available on a visualization server and transmit data progressively to a rendering client, we need to compactify the data set and break it down into smaller "bricks". The order of transmission and the size or resolution of the bricks is determined and driven by the client application. Our system uses a Windows NT-based server system which is both data repository and content provider for shared rendering applications. The NT system is connected to a Unix file system from where it accesses the data. The client accesses the server via a web-based interface.

The client selects a data set and sends a request to the server. The server analyzes and interprets the request and returns a customized Java applet together with an appropriate representation of the data set. The Java applet is optimized for a specific rendering task. This means that the rendering algorithm is tailored to a particular problem set. This keeps the applet small and avoids additional overhead and considering different cases. The initial data set is also small. It is refined later upon additional requests by the client. Bricks of different sizes and different resolutions may be requested from the server. We present a method that combines dynamic space-subdivision algorithms, such as adaptive octrees for volumes, wavelet-based data representation, and progressive data transmission for hierarchically stored volume data sets [1], [8], [10].

2 Indexing scheme

Previous work on space-subdivision [4], [5], [11] has shown that octrees provide an efficient method to store large volume data sets, as long as the depth of the octree is limited. Otherwise the data structures become so complex that tree traversal causes additional overhead [9]. Wavelet compression has been proven as an efficient method to transform a data set into a multiresolution representation. Unfortunately, it is difficult to extract sub-volumes from a compressed data set. We present a technique which combines both methods in order to optimize performance.

A web-based user interface with local rendering capability on the client side [7] requires hierarchical data representation on the server site. The server transmits a coarse overview representation of the entire data set (context information), which allows the user to select a region of interest. When this region is specified, the client application sends a request to the server, which responds by transmitting sub-volumes of the specified region at increasingly higher resolutions. The client progressively refines the image for the specified region. A prototype implementation is described in [7].

Biomedical imaging data are usually structured as sets of files, which represent a series of 2-D cross sections. By arranging all slices in a linear array, we obtain a 3-D volume. Unfortunately, when accessing the data, in most cases we do not make use of the implicit coherency across single slices. This coherency is only useful for extraction of cross sections perpendicular to the scanning direction, i.e., within a single image plane. Instead, in most cases we need brick-like coherency within sub-volumes. Therefore, we present a new data structure, which uses a combination of delimited octree space-subdivision and wavelet compression techniques to achieve better performance.

In this article, we present an efficient indexing scheme, an adaptive data reduction method, and an efficient compression scheme. All techniques are based on integer arithmetic and are optimized for speed. Binary bit operations allow for memory-efficient storage and access.

We use a standard file system (Unix or FAT32) to store our derived data structures, and we use file names as keys to the database. This allows us to avoid additional overhead, which is typically caused by inserting additional access layers between the application and the underlying storage system. We found that this method provides the fastest method to access data. Our indexing scheme in conjunction with the underlying file system provides the database system (repository) for the server application, which reads the data from the repository and sends it to a remote rendering client upon request. Initially, a low-resolution representation is requested from the repository and rendered on the client side. This coarse representation provides context cues and sufficient information for initial navigation. After a user has specified a sub-volume or region of interest, the client application sends a new request to the server to retrieve a sub-volume at a higher level of detail. When using the data structures described below, this updating procedure typically requires considerably less time compared to the single-slice representation, because a smaller number of files needs to be accessed. The initial step, which requires reading the initial section of every file, i.e., all bricks, can be accelerated by storing an additional file that contains a reduced version of the entire data set.

3 Storage scheme

The file size f for storing the leaves of the octree structure should be a multiple n of the minimum page size p of the file system. The value of p is typically defined as a system constant in `/usr/include/sys/param.h` on Unix systems. The value of n depends on the wavelet compression method described below. If the lowest resolution of the sub-volume requires b bytes, the next level requires a total of $8 \cdot b$ bytes (worst case, uncompressed), and so forth.

We assume that we have a recursion depth of r for the wavelet representation, leading to $8^r \cdot b$ bytes that must fit in f. This implies that

$$f = n \cdot p \geq 8^r \cdot b. \tag{1}$$

Both r and b are user-defined constants. Typical values are $b = 512$, which corresponds to an $8 \times 8 \times 8$ sub-volume, and $r = 3$, which provides four levels of detail over a range between $8^0 \cdot 512 = 512$ and $8^3 \cdot 512 = 262{,}144$ data elements, which is a range of more than 2.7 orders of magnitude.

For optimal performance, and in order to avoid gaps in the allocated files, we can assume that

$$n \cdot p = 8^r \cdot b, \tag{2}$$

thus

$$n = 8^r \cdot \frac{b}{p}. \tag{3}$$

The enormous size of the data sets (see Section 4) requires that the data is subdivided into smaller chunks that can be loaded into memory within a reasonable amount of time [2], [6]. Since we are extracting sub-volumes, it seems natural to break the data up into smaller bricks. This can be done recursively using an octree method [4], [5], [11]. Each octant is subdivided until we reach an empty region that does not need to be subdivided any further, or until we hit the file size limit f, which means that the current leaf fits into a file of the given size.

Each leaf contains a part of the original data set at full resolution. Memory space is reduced by skipping empty regions. Typically, the size of the data set shrinks to about 20% of the original size (see Section 5).

Progressive data transmission and extraction of a region of interest requires to access the data set in a hierarchical fashion. Therefore, it is useful to convert the leaves into a multiresolution representation. This representation must be chosen in a way that the reconstruction can be performed most efficiently with minimal computational effort [12]. Haar wavelets satisfy these requirements. They also have the advantage that they can be implemented easily with integer arithmetic [13]. The lowest resolution (Figure 1, lower-right image) is stored at the beginning of the file, thus avoiding long seek times.

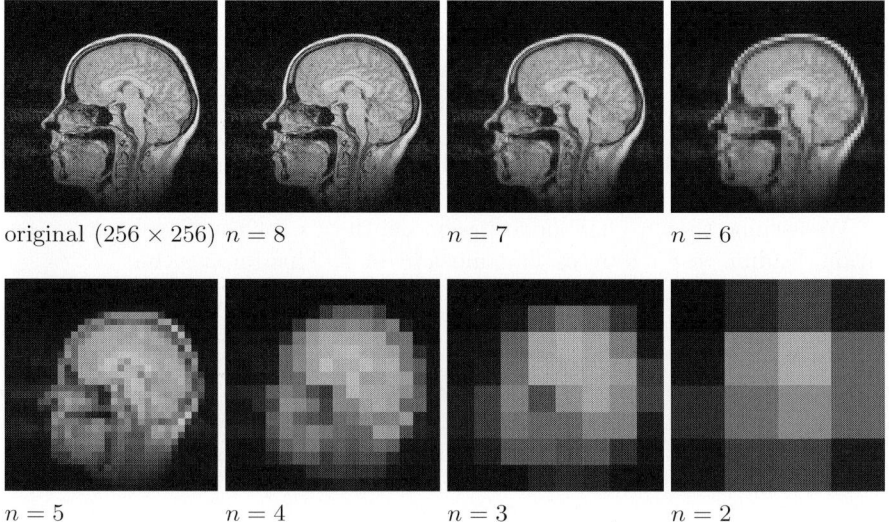

Fig. 1. Haar wavelet compression scheme (2-D case)

For the wavelet representation, we associate each sub-volume with a vector space V that consists of a set of piecewise linear functions. The space V^0 is associated with a constant function that is defined over the domain $[0, 1)$ and describes a single pixel. The space V^i consists of 2^i intervals, with a constant function defined on each of these intervals. All vector spaces are subsets of each other, i.e.,

$$V^i \subset V^{i+1}, i \in \mathbf{N}_0. \tag{4}$$

We choose the following scaling functions as the basis functions for V^i:

$$\phi_{i,j}(x) = \phi(2^i x - j), \; j \in \{0, ..., 2^i - 1\}, \text{ where} \tag{5}$$

$$\phi(x) = \begin{cases} 1, & \text{if } 0 \leq x < 1 \\ 0, & \text{otherwise}. \end{cases}$$

Figure 2 illustrates the scaling functions.

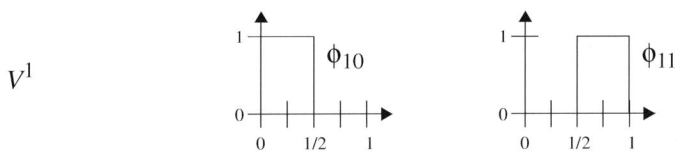

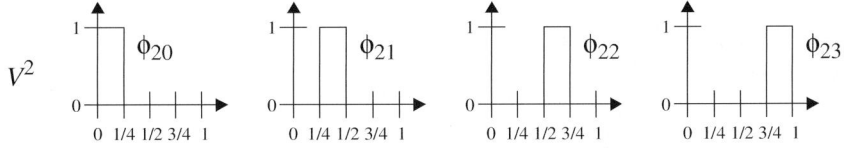

Fig. 2. Wavelet transformation: scaling functions

We define another vector space W^i that comprises all functions of V^{i+1}, and which is orthogonal to all functions in V^i. These basis functions, which span W^i, are the *Haar wavelets*, defined as

$$\psi_{i,j}(x) = \psi(2^i x - j), \; j \in \{0, ..., 2^i - 1\}, \text{ where}$$

$$\psi(x) = \begin{cases} 1, & \text{if } 0 \leq x < \frac{1}{2} \\ -1, & \text{if } \frac{1}{2} \leq x < 1 \\ 0, & \text{otherwise.} \end{cases} \quad (6)$$

Figure 3 illustrates these basis functions.

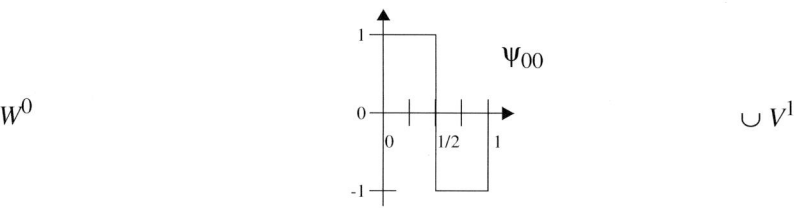

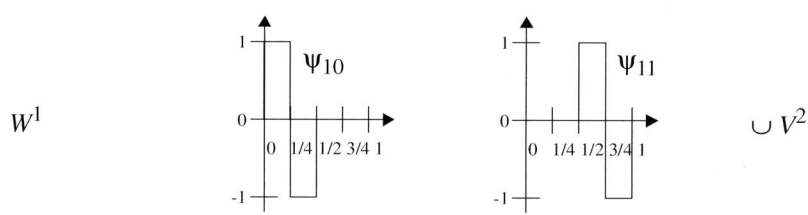

Fig. 3. Haar wavelets

A discrete signal in V^{i+1} can be represented as a linear combination of the basis functions of V^0 and $W^0 \ldots W^i$. We can represent an image I as

$$I(x) = \sum_{i=1}^{n} l_i \cdot \phi_{\frac{n}{2}-1, i-1}(x), \quad (7)$$

where $l_i, i \in \mathbf{N}_0$, is the image data. After the first transformation, we obtain

$$I(x) = \sum_{i=1}^{n/2} l'_i \cdot \phi_{\frac{n}{2}-2, i-1}(x) + \sum_{i=1}^{n/2} c'_i \cdot \psi_{\frac{n}{2}-2, i-1}(x), \quad (8)$$

where the first sum represents the averaged image data, which can also be described as a low-pass filtered image, and the second sum comprises the detail coefficients, representing a high-pass filtered image, or the difference between one of two adjacent pixels and their average value. A detail coefficient describes the symmetric error or deviation of the averaged value from one of the two original values.

It is easy to lift this scheme to the 3-D case. A simple solution would be an enumeration of all grid points, e.g., row-by-row and slice-by-slice, in a linear chain, so that one can still apply the 1-D algorithm given above (standard decomposition). This, of course, would reduce the data set only by a factor of two, instead of $2^3 = 8$. Thus, we apply the algorithm alternatingly to each dimension. Figure 4 shows the method for the 2-D case [12].

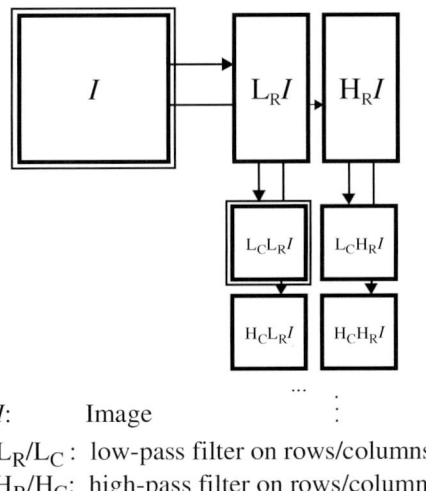

I: Image
L_R/L_C: low-pass filter on rows/columns
H_R/H_C: high-pass filter on rows/columns

Fig. 4. Wavelet compression scheme

First, all rows of the original image I are decomposed into a low-pass-filtered image $L_R I$ (reduced image) and the high-pass-filtered components $H_R I$ (detail coefficients). For the next transformation, the algorithm is applied to the columns, which results in $L_C L_R I$, $H_C L_R I$, $L_C H_R I$, and $H_C H_R I$ (prefix notation). The same technique can be used in three dimensions.

After each cycle, we end up with a reduced image in the upper-left corner. Subsequently, the algorithm is only applied to this quadrant (see Figure 5). The algorithm terminates when the size of this quadrant is one unit length in each dimension.

This method of alternating between dimensions is known as non-standard decomposition (Figure 1 shows the $L_C L_R$ components for an MRI scan at different levels of detail).

A very useful property of the wavelet scheme is the fact that even for lossless compression a volume converted into the wavelet representation requires

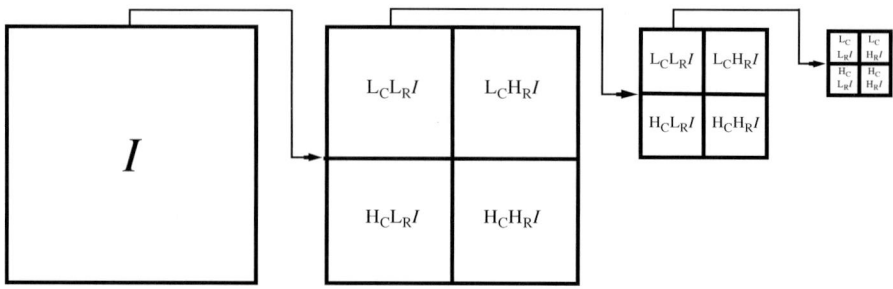

Fig. 5. Wavelet compression: memory management

the exact same amount of memory as the original representation. Since many coefficients are relatively small, the number of different discrete values is also small, provided we use integer arithmetic. Extremely small values can be neglected to obtain even better compression rates (lossy compression). We use a simple run-length encoding (RLE) scheme that turns out to be efficient, especially for small brick sizes b. It also allows for easy decoding. The space requirement (lossless compression) is the same for all subsequent wavelet recursions, i.e., for all levels of detail. The wavelet algorithm terminates when it reaches a pre-defined minimum sub-volume size b. The lower bound is the size of a single voxel.

Each octant can be described by a number [3]. We use the following numbering scheme (see Figure 6): A leaf is uniquely characterized by the octree recursion depth and the octree path. We limit the recursion depth to eight, which allows us to encode the depth in three bits. In order to store the

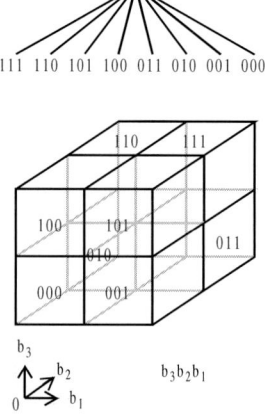

Fig. 6. Numbering scheme

path, we need three bits per recursion step, which gives us 24 bits. Four bits are spent to encode the depth of the wavelet recursion. The remaining bit is a flag that indicates whether the file is empty or not. This prevents us from opening and attempting to read the file and accelerates the computation. The total number of bits is 32 (double word).

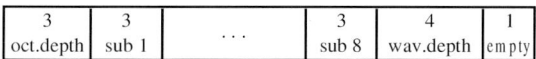

Fig. 7. Tree encoding

Each bit group can be easily converted to an ASCII character by using binary arithmetic, e.g., $(OCT\_DEPTH >> 29) | 0x30$ ('C' notation) would encode the octree depth as an ASCII digit. By appending these characters we can generate a unique file name for each leaf.

In order to retrieve a sub-volume, we have to find the file(s) in which the corresponding parts are stored. We start with the lower-left-front corner and identify the subvoxel by recursive binary subdivision of the bounding box for each direction. Each decision yields one bit of the sub-volume path information. We convert these bits to ASCII characters, using the same macros as above. The first file we are looking for is 7xxxxxxxx??, where the 'x's describe the path, and '?' is a wildcard. If this file does not exist, we keep looking for 6xxxxxxx???, and so forth, until we find an existing leaf. If the file name indicates that the file is empty (last digit), we can skip the file. The file name also indicates how many levels of detail we have available for a particular leaf. This allows us to scale the rendering algorithm. In order to retrieve the rest of the sub-volume, we must repeat this procedure for the neighboring leaves. The number of iterations depends on the recusion depth and therefore on the size of the leaves found. The algorithm terminates when all files have been retrieved and the sub-volume is complete.

4 Results

Our test application focusses on biomedical imaging. We have designed a prototype to support 3-D visualization of a human brain, which allows us to study details by moving tools, such as an arbitrary cutting plane and different-shaped lenses, across the data set. The various data sets are typically between 20 MB and 76 GB in size, which makes them too large to transfer over the Internet in real time. The rendering client operates independently from the size of the data set and requests only as much data as can be displayed and handled by the Java applet.

The web-based user interface combines HTML-form-driven server requests with customized Java applets, which are transmitted by the server to accomplish a particular rendering task.

Our prototype application (Figure 8) features 2-D/3-D preview capability; interactive cutting planes (in a 3-D rendering, with hierarchical isosurface models to provide context information); and a lens paradigm to examine a particular region of interest (variable magnification and lens shape, interactively modifiable region of interest). Complex scenes can be pre-computed on the server side and transmitted as a VRML2 file to the client so that the client can render the scene, and the user can interact with it in real time.

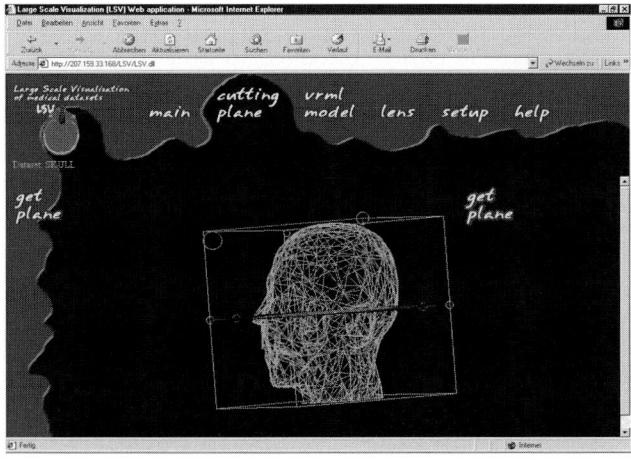

Fig. 8. Prototype application. (See Color Plate 25 on page 364.)

5 Comparison

Our data structure uses considerably less memory than the original data set, even if we choose lossless compression. In our experiments we found that we can save between 62% and 85% of memory for storing a typical biomedical data set (about 14 MB). By selecting appropriate thresholds for the wavelet compression algorithm, we can switch between lossless compression and extremely high compression rates. Computing time is balanced by choosing an appropriate file size (see Section 3).

One of the advantages of our approach is the fact that the computing time does not depend so much on the resolution of the sub-volume, but merely on the size of the sub-volume, i.e., the region of interest. This is due to the fact that the higher-resolution versions (detail coefficients in conjunction with the lower-resolution versions) can be retrieved in almost the same time from disk as the lower-resolution version alone. All levels of detail are stored in the same file, and the content of several files ("bricks"), which make up the sub-volume, usually fits into main memory. Since seek time is much higher than read time for conventional harddisks, the total time for data retrieval primarily depends on the size of the sub-volume, i.e., the number of files that need to be accessed, and not so much on the level of detail.

Figure 9 shows the reduction in the amount of memory required to store a large data set when we use an octree at two different levels. The column on the right represents the original data set. The wavelet decomposition takes about 0.07 sec for a 64^3 data set, and 68 sec for a 1024^3 data set. Reconstruction can be done more efficiently and usually requires about 30% of the time (measurements based on an R12000 processor). For the above data we assume lossless wavelet decomposition. RLE or other (lossy) compression/decompression algorithms take an additional amount of time, but this is negligible compared to data transmission time.

Algorithm	Octree				BSP tree	
	level 1		level 2			
Data type	MRI	CT	MRI	CT	MRI	CT
Pre-processing	56	63	98	97	156	159
Depth	4	4	5	5	12	12
Memory	5.412.610 / 14.548.992	3.996.526 / 14.811.136	3.831.488 / 14.548.992	2.358.442 / 14.811.136	3.299.516 / 14.548.992	2.137.326 / 14.811.136

Fig. 9. Space-subdivision algorithm

6 Conclusions

We have presented an efficient numbering scheme and access method for hierarchical storage of sub-volumes on a regular file system. This method allows us to access a region of interest as a set of bricks at various levels of detail. The simplicity of the method makes it easy to implement. The algorithm is scalable by increasing word length and file name length. Future work will address better wavelet compression schemes, lossy compression techniques, and the integration of time-varying data sets.

We are currently working on the integration, adaptation and evaluation of these tools in the National Partnership for Advanced Computational Infrastructure (NPACI) framework. Future research efforts will include integration of San Diego Supercomputer Center's High-performance Storage System (HPSS) as a data repository to retrieve large-scale data sets, and accessing the data via NPACI's Scalable Visualization Toolkits (also known as VisTools).

7 Acknowledgements

Data sets are courtesy of Arthur W. Toga, UCLA School of Medicine. We thank Kwan-Liu Ma, Ikuko Takanashi, Eric B. Lum, Eric C. LaMar, UC Davis; Elke Moritz, Mississippi State University, and Edward G. Jones, Neuroscience Center, UC Davis.

This work was supported by the U.S. Army Engineer Research and Development Center (DACA42-00-C-0039); the Army Research Office under contract ARO 36598-MA-RIP; the National Science Foundation under contracts ACI 9624034 (CAREER Award), through the Large Scientific and Software Data Set Visualization (LSSDSV) program under contract ACI 9982251, and through the National Partnership for Advanced Computational Infrastructure (NPACI, NSF SPA#00120410), the Office of Naval Research under contract N00014-97-1-0222; the NASA Ames Research Center through an NRA award under contract NAG2-1216; the Lawrence Livermore National Laboratory under ASCI ASAP Level-2 Memorandum Agreement B347878 and under Memorandum Agreement B503159; the Lawrence Berkeley National Laboratory; the Los Alamos National Laboratory; and the North Atlantic Treaty Organization (NATO) under contract CRG.971628. We also acknowledge the support of ALSTOM Schilling Robotics and SGI. We thank the members of the Visualization and Graphics Research Group at the Center for Image Processing and Integrated Computing (CIPIC) at the University of California, Davis.

References

1. Bertram, M., Duchaineau, M. A., Hamann, B. and Joy, K. I. (2000), Bicubic subdivision-surface wavelets for large-scale isosurface representation and visualization, in: Ertl, T., Hamann, B. and Varshney, A., eds., *Visualization 2000*, IEEE Computer Society Press, Los Alamitos, California, 389–396 (presented at: "Visualization 2000," Salt Lake City, Utah).
2. Heiming, Carsten (1998) Raumunterteilung von Volumendaten. Thesis, Department of Computer Science, University of Kaiserslautern, Germany.
3. Hunter, G. M.; Steiglitz, K. (1979) Operations on Images Using Quad Trees. IEEE Trans. Pattern Anal. Mach. Intell., 1(2), 145–154.
4. Jackins, C.; Tanimoto, S. L. (1980) Oct–Trees and Their Use in Representing Three–Dimensional Objects. CGIP, 14(3), 249–270.
5. Meagher, D. (1980) Octree Encoding: A New Technique for the Representation, Manipulation, and Display of Arbitrary 3–D Objects by Computer. Technical Report IPL-TR-80-111, Image Processing Laboratory, Rensselaer Polytechnic Institute, Troy, NY.

6. Meyer, Joerg (1999) Interactive Visualization of Medical and Biological Data Sets. Ph. D. thesis; Shaker Verlag, Germany.
7. Meyer, Joerg, Borg, Ragnar, Hamann, Bernd (2000) VR–based Rendering Techniques for Time–Critical Applications. Proceedings of Scientific Visualization 2000, Schloss Dagstuhl, Germany.
8. Pinnamaneni, Pujita, Saladi, Sagar, Meyer, Joerg (2001) 3–D Haar Wavelet Transformation and Texture–Based 3–D Reconstruction of Biomedical Data Sets, in: Hamza, M. H., ed., Proceedings of *Visualization, Imaging and Image Processing (VIIP 2001)*, ACTA Press, 389–394 (presented at: "Visualization, Imaging and Image Processing," The International Association of Science and Technology for Development (IASTED), Marbella, Spain).
9. Pinskiy, Dmitriy V., Meyer, Joerg, Hamann, Bernd, Joy, Kenneth I., Brugger, Eric, Duchaineau, Mark A. (1999) A Hierarchical Error–controlled Octree Data Structure for Large–scale Visualization. ACM Crossroads, Data Compression, Spring 2000 – 6.3, 26–31.
10. Pinskiy, D. V., Brugger, E. S., Childs, H. R. and Hamann, B. (2001) An octree-based multiresolution approach supporting interactive rendering of very large volume data sets, in: Arabnia, H. R., Erbacher, R. F., He, X., Knight, C., Kovalerchuk, B., Lee, M. M.-O., Mun, Y., Sarfraz, M., Schwing, J. and Tabrizi, M. H. N., eds., Proceedings of *The 2001 International Conference on Imaging Science, Systems, and Technology (CISST 2001)*, Volume 1, Computer Science Research, Education, and Applications Press (CSREA), Athens, Georgia, 16–22 (presented at: "The 2001 International Conference on Imaging Science, Systems, and Technology," Las Vegas, Nevada).
11. Reddy, D.; Rubin, S. (1978) Representation of Three–Dimensional Objects. CMU-CS-78-113, Computer Science Department, Carnegie-Mellon University, Pittsburgh, PA.
12. Schneider, Timna Esther (1997) Multiresolution–Darstellung von 2D–Schichtdaten in der medizinischen Bildverarbeitung. Thesis; Department of Computer Science, University of Kaiserslautern, Germany.
13. Stollnitz, Eric J., DeRose, Anthony D., Salesin, David H. (1996) Wavelets for Computer Graphics: Theory and Applications. Morgan Kaufmann Publishers.

A Data Model for Distributed Multiresolution Multisource Scientific Data

Philip J. Rhodes, R. Daniel Bergeron, and Ted M. Sparr

University of New Hampshire, Durham NH 03824

Abstract. Modern dataset sizes present major obstacles to understanding and interpreting the significant underlying phenomena represented in the data. There is a critical need to support scientists in the process of interactive exploration of these very large data sets. Using multiple resolutions of the data set (multiresolution), the scientist can identify potentially interesting regions with a coarse overview, followed by narrower views at higher resolutions.

Scientific data sets are often multisource coming from different sources. Although it may be infeasible to physically combine multiple datasets into a single comprehensive dataset, the scientist would often like to treat them as a single logical entity. This paper describes formal conceptual models of multiresolution and distributed multisource scientific data along with an implementation of our multisource model. Our goal is to allow a scientist to describe a dataset that combines several multisource multiresolution datasets into a single conceptual entity and to provide efficient and transparent access to the data based on functionality defined by the model.

1 Introduction

New data gathering and data generation tools have created an explosion of data available to scientists. The existence of such large amounts of data opens up a wide range of opportunities for scientific data exploration. Widespread availability of ever increasing compute power provides some hope that we can realize these opportunities. The size and nature of the data, however, do present substantial obstacles.

Multiresolution data consists of several representations of a dataset at various resolutions, ranging from the original resolution to a very coarse overview. The scientist can use the overview to identify significant regions of the data and then examine such regions in greater detail using one of the finer resolutions.

The problems presented by large data size are especially difficult when a scientist wishes to combine a set of related multisource data sets into a single conceptual entity. In order to access such data efficiently, we need to understand the fundamental nature of the data and capitalize on its inherent structure.

This work is supported by the National Science Foundation under grants IIS-0082577 and IIS-9871859

We have developed a formal data model and an implementation of that model that supports a wide range of scientific datasets with a variety of data organizations [16]. The complete model encompasses hierarchies of multiresolution and adaptive resolution data in a distributed environment.

Our model and our implementation is intended to support a concept analogous to a database view. For example, a scientist ought to be able to define a dataset representing points in a three-dimensional space with 4 data attributes, where one of the attributes is defined (for all points) in a single physical file and the other 3 attributes are partitioned into spatial blocks which are distributed to multiple locations across the web. The composite dataset will never be instantiated as a single object, but should appear to the user as a single unified entity.

We first describe the kind of data we wish to model and then give an overview of our formal data model. We present a brief summary of our implementation of the multisource data model and present a specific example describing how a scientist might define and access a composite multisource data set. We conclude with some observations and directions for future development.

1.1 Problem Definition

The research described here focuses on developing database support for this method of scientific research based on interactive exploration of very large distributed multiresolution data. We believe that a major weakness of current scientific database efforts is the lack of a comprehensive model that encapsulates the structure inherent in the data. Such a model should allow a database system to store and access this data efficiently without needing to understand the meaning of the data for the application domain. The most important requirements for a data model for distributed multiresolution data include the following:

- The model must be general-purpose while still able to rigorously encapsulate the most important aspects of the data.
- In addition to describing the multiple resolution levels of a data set, it must be able to describe an *adaptive resolution* level, i.e., a single level of the data set that is itself composed of data that has different resolutions in different regions.
- It must be able to describe both the physical domain in which the scientific phenomenon actually occurs (we call this the *geometry* of the problem) as well as the structure of the data that represents that phenomenon (which we call the *topology* of the data).
- It must incorporate information about how data points in each level of the multiresolution representation relate to data points in the other levels. We have developed notions of *support* and *influence* which encapsulate these relations in an application-independent manner.

- It must allow scientists to assemble data from multiple local or remote sources into a single conceptual entity.

1.2 Summary of Research

The major components of our research include:

MR/AMR Model. We are developing a data model for hierarchical multiresolution (MR) and adaptive multiresolution (AMR) data representations that supports interactive exploration of scientific data. This includes a model of error and error operations that helps keep the experimenter informed of the quality of data at various resolutions. Also, we characterize the kinds of operations that can be performed on MR/AMR data, especially as they relate to data in a distributed computing environment.

Geometry and Topology. Our data model distinguishes between the geometry and topology of a dataset, allowing us to characterize a wide variety of data types. This work should allow us to develop a taxonomy of scientific data that helps exploit regularities in both geometry and topology. We see the topology as a bridge between the scientists geometric data view and the index-oriented view of the underlying database.

Lattice Model. The lattice model is a single-level model of data that incorporates our ideas about geometry and topology, and is an important component of the formal model especially for representing adaptive resolution data. Geometry and topology forms the basis of a lattice class hierarchy that can efficiently represent a variety of scientific data.

Domain Representation. We are developing an efficient way to represent domains, and especially the extent of subdomains within an enclosing domain. The representation should be space efficient and quick to access. This work is particularly important because we represent certain metadata (e.g., extracted features, classifications) as labeled subdomains.

Data Storage. We have developed a model for multisource data and implemented a prototype for evaluation. For array-based data, our approach is spatially coherent, i.e., given a point, we have efficient database access to its geometric neighbors.

Evaluation. The model will be evaluated by implementing a prototype and testing its performance with large datasets. Our goal is a system that is flexible and expressive enough to be of real assistance as the scientist works with the data.

2 Distributed Multiresolution Scientific Data

We envision the creation of a standard multi-level data hierarchy for large scientific data sets . The original data sets would be stored permanently at some repository site or sites. A preprocessing operation would generate a complete multiresolution representation with increasingly coarse representations and the associated localized error representations. All components of this data representation would be available for downloading arbitrary subsets at arbitrary resolutions. A scientist is likely to extract some small coarse component of the hierarchy to store at his or her workstation, may access the next several levels on a data server on the local network, and perhaps access the finest resolution representations over the Internet. In addition, the scientist would like to define virtual datasets composed of sections or attributes taken from other datasets.

This model works because only very small subsets of the total data set will need to be accessed at the higher resolutions. Note that we can assume that the original data is essentially read-only, although we certainly need to be able to dynamically update both the lower resolution representations and the metadata associated with the data set.

In order for this multi-level storage representation to be effective, the scientist needs to have a cumulative local error value associated with the lower resolution representations of that data. The error representation must be an integral component of the data exploration process providing the scientist with critical feedback concerning where the current representation is likely to be accurate and where it isnt. Without such information, the scientist could not trust anything but the finest resolution.

3 Data Model Foundations

Pfaltz et al. [14] identify the major features of scientific data as large size, complex entities and relationships, and volumetric retrieval. Although such characterization is correct, we need a more rigorous definition if we hope to provide effective database support for scientific data. For our purposes, scientific data is a collection of values that represents some natural phenomenon [7] that is a function over a domain [11] which might be time, space, radio frequency, etc. or some multidimensional combination. The *value space* of the function defined over the domain usually consists of the cartesian product of the value ranges of several data attributes. This is equivalent to saying that any point in D has a number of attributes – the value of the data function at that point.

3.1 Dimensional Data

Much scientific data can be meaningfully represented in a continuous n-dimensional data space [2,8]. If a data set consists of *some* attributes that are ordinal, independent, and defined on a continuous value range, the data set contains *dimensional data*, and those attributes are *dimensional*. Each possible combination of dimensions defines a *view* of the data, a notion similar to the view capability found in traditional databases. *Spatial data* is dimensional data that represents an actual physical space. A data set can be dimensional without being spatial but even non-spatial dimensional data can often be visualized as if it were spatial, since humans find this representation familiar and easy to grasp. It may also be convenient to treat a set of attributes as if they are dimensional attributes even though they may not satisfy all the conditions for dimensional data. For example, we might want to treat a set of attributes as independent for exploration purposes with the goal of either validating or disproving that assumption.

Spatial (dimensional) data is often represented as points defined on a wide variety of regular and irregular grid types [2,4,6,18]. The choice of the grid, usually based on a natural organization of the data, impacts the nature of the representation chosen for the data and the specification of algorithms for analyzing it.

3.2 Geometry, Topology, and Neighborhoods

The terminology used in the literature to describe various systems of grids is not standardized. We propose a more comprehensive and consistent framework for describing and defining grids that encompasses most reported grid structures [9,11], including both point and cell data organizations. We separately represent the underlying space in which the grid is defined, which we call the *geometry*, and the point and cell relationships implied by the grid, which we call the *topology*. Thus, the geometry of a data set refers to the space defined by the dimensions; the topology of a data set defines how the points of the grid are connected to each other. A data sets topology is a graph with data points or cells as nodes and arcs between nodes representing a *neighbor* or *adjacency* relationship.

This approach enables database support for application algorithms to process data either geometrically or topologically. In many cases, the topology and/or geometry do not have an explicit representation within the data set because they derive easily from the indexes of an array that stores the data. The array and its index structure compose the *computational space* of a data set. Other more complex geometries and topologies may have a separate representation from their computational spaces.

In some applications the data has no inherent geometry or topology. For example, categorical data is normally not defined in a geometric space and scatter data has no predefined topology. Our data model allows a user to represent and manipulate both kinds of data. However, it may be useful to impose additional structure on the data. For example, we could impose a topology on scatter data either for efficiency of access or to support an alternative conceptual model for the data. Similarly we have shown that topology can be an effective vehicle for imposing a metric space upon categorical data [11,12].

Many scientific applications require selecting the neighborhood of a point [5,12,15]. The neighborhood of a point p consists of points "near" p. Nearness may be defined *geometrically* (e.g., as the set of points within distance d of p in the geometric space) or *topologically* (e.g., as the set of points within n arcs of p in a topological space).

3.3 Error

Most scientific data contains some inherent error. This includes measurement error from sampling or computational error from simulation. Furthermore, operations and analyses may introduce additional error. Our model of scientific data includes *localized error* that is estimated at every point within the domain [3,19].

We expect that the process of developing lower resolution representations will introduce additional error that increases as resolution decreases. Hence we refer to the error as cumulative error. Error information plays an important role in helping a scientist determine the appropriate level of resolution for his or her needs.

3.4 Data Representation

Effective exploration tools for very large data sets are best developed on a rigorous conceptual model of the data. Such a model must be accessible to both the programmer and the user and must be able to adapt to the actual data in a natural and efficient way. We now present a data model that describes scientific data that can be organized into a multiresolution hierarchy. The basic motivation for this model is to support distributed *interactive* data exploration.

A rigorous definition of a *data representation* is the formal basis of our model of the scientists data set. Although the data set represents a phenomenon defined over a continuous domain, D (a *geometric* space with an infinite number of points), the data set is a finite sampling of this space. Consequently, our data representation is defined over a finite set of points $\Delta \subset D$, known as *sample points*, within the domain D [8]. A sampling function f_Δ maps D to a subset Ω of a value space V [10], denoted by

$$f_\Delta : \Delta \to \Omega , \tag{1}$$

with *sampling error* described by a localized error function E_Δ mapping each sample point to an error space E, denoted by $E_\Delta : \Delta \to E$. Formally, we define a data representation R to be a quadruple:

$$R = \langle \Delta, \Omega, f_\Delta, E_\Delta \rangle \tag{2}$$

where Δ is a set of sample points in D that are sampled using the sampling function with an error function E_Δ, and Ω is the range of f_{Delta}. (By convention, we use dot notation to refer to the components of a tuple. Thus, $R.\Delta$ identifies the sample points of the representation R.)

3.5 The Lattice Representation

Although the *data representation* definition is comprehensive enough to encompass most kinds of scientific data, it only represents the actual data and does not incorporate any notion of how the different data elements might be related in a grid structure. We incorporate the grid definitions into our data model by adopting and extending the *lattice* model [1]. A lattice includes a topology, τ, as well as a geometry [11]. A lattice L_k^n has n topological dimensions that define a topological space, and k attributes for a point located in that space. A 0-dimensional lattice is simply an unordered set, a 1-dimensional lattice is an ordered list, a 2-dimensional lattice lies in a plane, and so on. The lattice geometry need not have the same dimensionality as the lattice topology. For example, a 2D lattice can be mapped to a curvilinear surface that exists in three-dimensional space. Formally, a lattice L consists of a data representation R and a topology τ; that is, $L = \langle R, \tau \rangle$. Operations on lattices include *value transformations* (e.g., normalization), *geometric domain transformations* (including affine transformations like scaling, translation, and rotation), and *topological transformations* such as mesh simplification. Operations like extension and projection can be either geometric or topological transformations (or both) depending on which components are altered.

The separation of geometry and topology in our lattice model allows us to represent both cell-based and point based datasets in a unified fashion. Either a cell-based or point-based topology can be imposed on the same dataset, allowing the user to easily switch between these two views. Similarly, data read into the system as a collection of cells can be converted into a point-based representation, and *vice versa*.

3.6 Simple Data Model

Our notions of *data representation* and *lattice* are sufficient to represent a gridded scientific data set, but they do not provide a representation for the phenomenon that the data set is intended to model. We now define a simple *data model* which uses the lattice to approximate the phenomenon in the domain, as well as the error. Formally, an *application data model* M consists of a lattice, and functions f_D and E_C to approximate the data value and its associated error at every point in the domain; i.e.,

$$M = \langle L, f_D, E_C \rangle. \tag{3}$$

The *approximating function*, f_D is normally based on the sampling function and returns a value that approximates the phenomenon in the domain; i.e., $f_D : D \to V$. *Interpolating functions* are approximating functions that satisfy the condition: $\forall d \in \Delta, f_D(d) = f_\Delta(d)$. That is, the interpolating function and the sampling function agree at each point in the sample domain Δ.

A possible definition of the initial E_C, representing error over the entire domain, could be f_{E_Δ}, which uses the approximating function to find values E_D from the values of E_Δ, the original sampling error function. For application data models derived from other application data models, E_C is the *cumulative* error including both sampling error and the error introduced by the derivation process.

4 Multiresolution Data Model

Although the basic data model described above represents a very wide range of basic data sets, it is not adequate as a model for multiresolution data. A multiresolution (MR) model allows a researcher to view data using resolutions ranging from very coarse to very fine (the original data). Using a coarse resolution can vastly reduce the size of the data that needs to be stored, manipulated, and displayed. It also serves as an overview of the entire dataset, allowing the researcher to pick out regions of interest without examining the original data directly. Once an interesting region has been identified, the researcher may examine it at finer resolutions, perhaps even accessing the original data. "Drilling down" allows the researcher to examine only data of interest at fine resolution, minimizing processing and display costs.

4.1 Multiresolution Data Representation

The MR representation offers a tradeoff between detail and efficiency. Incorporating multiple resolution capability into the data model allows the database system to provide direct support for managing and using data at the resolution most appropriate to the immediate task.

A *reducing operator* transforms one data model into another data model, where the new representation is smaller than the old [2]. This reduction introduces additional associated localized error which must be modeled. An MR hierarchy is formed by repeated applications of reducing operations. The process is repeated a number of times until the size of the data has been reduced sufficiently or until further reductions would introduce too much error. Certain classes of wavelet functions form an ideal basis for reducing functions because of their localized error characteristics [19], but our model is also appropriate for very different kinds of data reduction techniques such as triangle mesh simplification [2].

4.2 Adaptive Multiresolution

An *adaptive resolution (AR) representation* allows resolution to vary within a single lattice. The resolution near a point may depend on the behavior of the sampling function, on local error, or on the nature of the domain in the neighborhood of the point. A reducing operator that behaves differently over parts of D can define an *adaptive multiresolution representation* (AMR), which is an MR hierarchy in which each layer is an AR representation. For example, it can reduce resolution in areas with lowest error when forming the next level. It might also try to preserve resolution in areas of rapid value change and reduce resolution in less volatile areas. Because an AMR contains multiple resolutions within each level, it has the potential to achieve a representation with the same accuracy as MR using less storage. Alternatively, for a given amount of memory, it can retain increased detail and accuracy in important regions of the domain.

4.3 Support and Influence

Our model for MR is very general. In practice, most MR hierarchies are defined entirely by operations on the sampling set, and they often place further restrictions on a reducing function such as requiring spatial coherence. Typically, any neighboring set of sample points S_j in λ_i should map to a neighboring set of sample points S_k in λ_{i+1}. S_j forms the *support* for S_k as shown in Fig. 1. For any point p there is a set of points in the next level that claim p as part of their support. We call this set of points the *influence* of p (see Fig. 1). By building the notions of *influence* and *support* explicitly into the data model (and into the database support system), we can provide a framework for better implicit support for efficient data distribution and distributed computation.

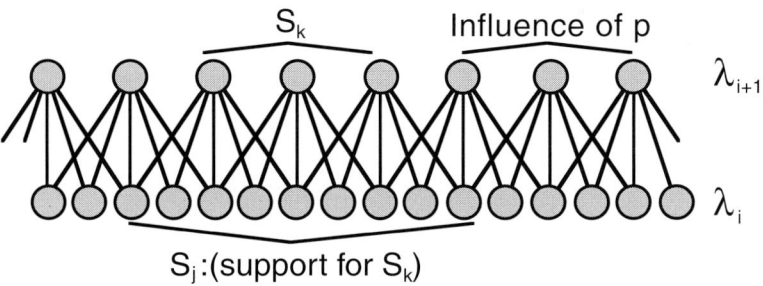

Fig. 1. Support and Influence

5 Taxonomy of Geometry and Topology

Since our representation of data must be efficient, it is worthwhile to categorize the way that data points lie in the geometry, and how they are connected in the topology. Our classification is motivated by the desire to exploit patterns within the spacing of the sample points, so the data can be represented efficiently. The taxonomy must be able to represent both cell and point based grids and transformations between them such as Delaunay and Voronoi techniques. Our software design for the lattice representation follows from this taxonomy. We are particularly interested in how much information must be stored in order to describe the geometry and topology of the dataset.

5.1 Periodic Tilings and Data

The study of tilings (tessellations) has some relevance to our research since topologies often define a tiling. A review of this field can be found in [17].

If a tiling is *periodic*, then it is possible to duplicate the tiling, translate it some distance, and place it down again so that it matches exactly with the original copy. That is, the tiling consists of a number of translated repetitions of some pattern of tiles. An important and related property of periodic tilings is that there exists a subset of the space S that can be repeatedly copied and translated throughout the space to complete the tiling. A minimal subset of this kind is called a *fundamental domain* or *generating region*.

A *regular tiling* is a periodic tiling made up of identical regular polygons [17]. The three tilings shown in Fig. 2 are the only regular tilings for 2D space.

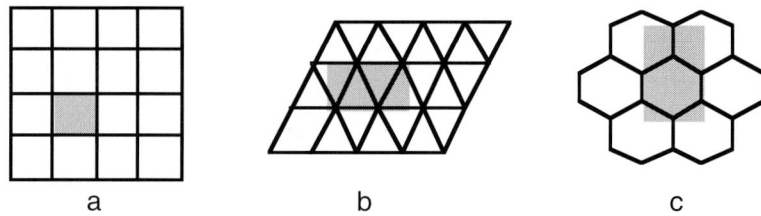

Fig. 2. The fundamental domains of the 2D regular tilings

This approach does not explicitly distinguish between the geometry and topology components of a grid. The definition of a tiling includes aspects of topology but most concepts are geometry specific. We are adapting these ideas to our work with geometry and topology.

We use the notion of the *supercell* to represent periodic sampling topologies. As shown in Fig. 3, a supercell represents a generating region for the topology, allowing the entire topology to be conceptually represented by a grid of repeated supercells while only storing a single supercell definition. If we can find where in the grid a point lies, we can very easily form a search key from the position in the grid (i.e. supercell identifier), and the position of the point within the supercell (i.e. point identifier). Such a technique promises a quick way to access a point's data given its geometric position.

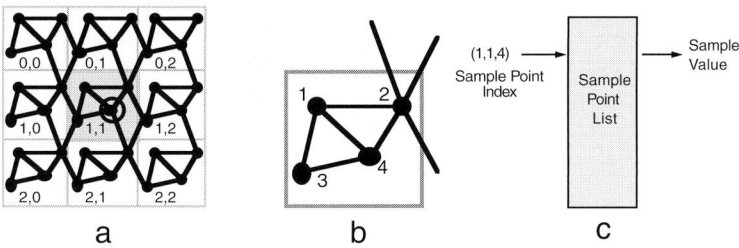

Fig. 3. A possible supercell implementation

5.2 Regular Data

Usually, for scientific data to be considered regular, it must lie within a mesh of squares or cubes, perhaps displaced by a shear operation [2]. By this definition, data points arranged in a hexagonal fashion, as in case c of Fig. 2, would not be considered regular, though the first two would. However, this may only be the case because of the ubiquity of array storage. Researchers tend to think of regular data as any data that can be stored very easily in an array.

It should be possible to develop a rigorous definition of regular data. Besides being regular in the mathematical sense, the patterns in Fig. 2 have an interesting property: if we store the vertices (sample points) in an array, it is possible to map a point's array indices to its locations in both the geometry and topology without using any other information. Since this property is of immediate interest to designers of scientific databases, it might serve well as a definition of regular scientific data.

5.3 Classifying Irregular Data

With irregular data, it is not possible to map array indices to a location in geometric space without using extra information, if at all. Of course, arrays may still be used to merely store the data points. Figure 4 shows examples of irregular data. In Fig. 4a, there is no way to map indices to a geometric location without referring to the spacing between the rows and columns, which varies for each row and column. Therefore, the mapping between indices and geometry must take this spacing as another parameter, i.e. as "extra information". The situation in Fig. 4b is even worse. Here, there is no pattern whatsoever to the position of the sample points, so a simple mapping from indices to geometry is out of the question.

We further classify irregular data according to how much information is required to represent the pattern of sample points within the domain. In Fig. 4a, points are lined up in rows and columns, so we only need to store

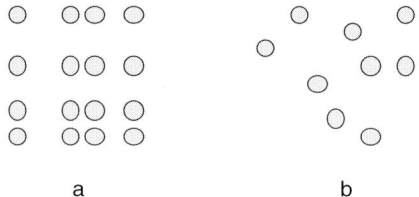

Fig. 4. Irregular data

the spacings for each row and column. The space required to store this information is proportional to the number of rows plus the number of columns. In Fig. 4b, there is no pattern whatsoever to the sample points, so we must store coordinates for each individual point. Here, the space required is proportional to the number of points. Of course, it is possible to have datasets that combine these attributes, behaving in different ways along different dimensions.

6 Datasources

Lattices provide the scientist with a conceptual view of his or her data that should be consistent with the operations that need to be applied to the data. In principle, this conceptual view will be reflected in the organization of the physical data. In practice, however, this is often not feasible. The scientist may need different views of the same data and the data may be too large to replicate and reorganize to match each desired view. In general, multisource data and distributed computing require sophisticated ways of dividing large files into smaller pieces while maintaining a simple view of the distributed data.

6.1 Mapping Lattices to Data

A lattice is able to map locations in the geometry to locations in the topology. It remains to map topological locations to offsets in file or network streams. A *datasource* provides the lattice with a single, unified view of multisource data. This simplifies the mapping from topological locations to file and network stream offsets.

Some datasources are directly associated with a local file or remote source, and are known as *physical* datasources. Other datasources are *composite*, meaning they are made up of more than one component datasource. For example, a datasource that performs an attribute join would be composite. It is possible to perform very complex operations by combining several datasources together in a tree structure, with the *root datasource* at the top of the tree providing the lattice with a simplified view of the data.

6.2 Datasource Model

A datasource can be modeled as an n-dimensional array containing δ, a subset of the set of lattice sample points Δ. We think of arrays as an *index space* d paired with a collection of associated data values. An index space can be expressed as the cross product of several indices, each defined as a finite subset of the integers:

$$I_1 \times I_2 \times \cdots I_n \tag{4}$$

where each I_k is an integer in the range $[a_k \ldots b_k]$. The dimensionality of a datasource's index space may or may not match the dimensionality of the lattice domain D. If these dimensionalities do match, then the neighborhood relationships present in the lattice may be reflected in the adjacencies present in the underlying storage. In other cases, there is no simple pattern in the distribution of Δ in D, so the lattice topology must map points from D into the underlying index space.

6.3 Physical Datasource

A *physical datasource* is a simple datasource that is directly connected to a file or network stream that contains sample points. While the stream is in reality a one dimensional entity, a physical datasource may have an index space that is n-dimensional. In this case, the physical datasource is responsible for mapping the n-dimensional view to the underlying data. This mapping can be expressed as a function m that maps an index space to a single index I_f used to access the actual data block.

$$m : I_1 \times I_2 \times \cdots I_n \to I_f \tag{5}$$

For example, consider a file of three dimensional array data. Perhaps the layers are stored in order of increasing z value, and each layer is stored in row major order. The physical datasource is responsible for mapping the x, y, and z values of its data space to a file index, which is really an offset from the beginning of the file. Of course, a single file may contain many attributes, so a physical datasource should be capable of returning all values corresponding to a location in the index space.

6.4 Blocked Datasource

A *blocked datasource* is a composite datasource in which the index spaces of the component datasources are joined together in contiguous, non-overlapping fashion to form a single index space[1].

$$d = \bigcup_{i=1\ldots k} d_i \text{ where } \forall i, j : i \neq j \to (d_i \cap d_j = \varnothing) \tag{6}$$

---

[1] As defined here, this is an outer natural join. It is possible to relax the contiguity and non-overlap constraints, but this is beyond the scope of this paper's goals.

For example, consider two datasources *ds1* and *ds2* that might represent two contiguous satellite image files, as shown in Fig. 5. Their index spaces can be joined together in the fashion shown by *ds3*, a blocked datasource, producing a single index space that can be manipulated as a single entity. Of course, a blocked datasource can have an arbitrary number of component datasources, allowing large amounts of data to be viewed as a single entity, but stored and accessed in a distributed fashion.

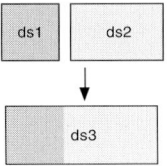

Fig. 5. Two datasources joined by a blocked datasource

6.5 Attribute Join Datasource

An *attribute join datasource* is a composite datasource for which each point in δ is composed of attributes taken from two or more component datasources. If A is the attribute set of an attribute join datasource, then we say:

$$A = \bigcup_{i=1...n} A_i \qquad (7)$$

where A_i are the attribute sets of the component datasources. For example, suppose *ds1* is a datasource with attributes {*salinity, pH, oxygen*} and *ds2* is a datasource with attributes {*temperature, depth*}. If these two datasources are combined by an attribute join datasource *ds3*, then each point in the index space of *ds3* has attributes {*salinity, pH, oxygen, temperature, depth*}. Such an operation is particularly useful when data has been organized into separate files, perhaps because it was gathered by different instruments.

7 Datasource Implementation

In this section we describe the primary elements of our prototype implementation and give a simple example of how the system processes multisource data for both rectilinear and unstructured lattices.

7.1 Implementation Components

Figure 6 depicts the key components of the implementation of our model. Lattice methods provide the functionality by which scientific applications interact with lattice data. Each lattice has associated Geometry and Topology objects that map sample points of the lattice to a computational space in the form of an N-dimensional array implemented as a DataSource object. We use an array because physical storage is easily conceptualized as an array. The mapping may be simple as when the sample grid forms a regular pattern over the domain so that each sample point can be naturally associated with a unique array element. The mapping is more complex for unstructured sampling patterns.

The DataSource transforms computational space to low level file or network URL requests. When all data appear in one file or one internet data server, the DataSource may be implemented directly as a PhysicalDataSource which maps an N dimensional data identity (index position) onto a 1 dimensional storage device. Our DataSource is designed to be flexible enough to integrate data from multiple physical data sources. We allow both spatial and attribute joins. The BlockedDataSource is composed of multiple sources for separate regions of the index space. The AttributeJoinDataSource combines attributes from multiple datasources. Blocked and attribute join data sources can be combined and nested to reflect multiple files and/or network URLs.

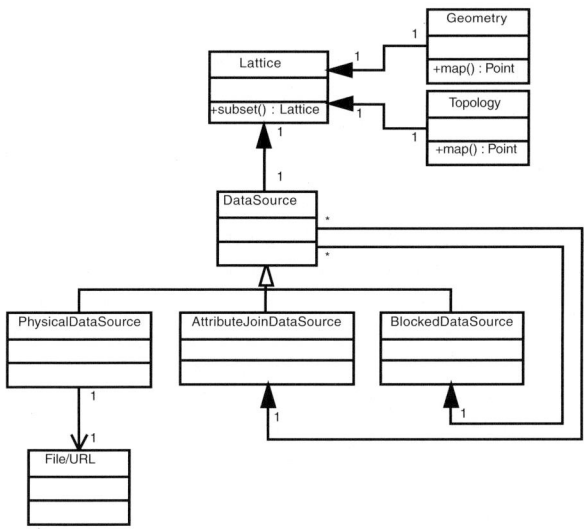

Fig. 6. A UML representation of key components of our system

7.2 An Example Scenario

In this section we discuss two running examples to illustrate the concepts of the lattice, datasource, and the relationship between them. Our first example involves rectilinear data in which there is a simple mapping between the location of a point in the lattice geometry and its location in the datasource index space. The second example involves unstructured data for which there is no such simple mapping, forcing the lattice topology to play a greater role.

A Rectilinear Lattice Consider the two dimensional lattice in the diagram below. The sample points are obviously placed in a regular fashion within the geometry. It is most likely that the experimenter will choose to use a rectilinear topology for such data as shown in Fig. 7b, although it should be understood that a different choice could be made. Lets suppose that each point has attributes {*time, carbon, nitrogen, oxygen*}.

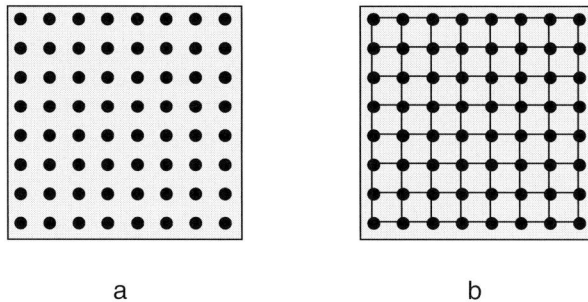

a   b

Fig. 7. A 2D Rectilinear Lattice: **(a)** geometry only, and **(b)** with topology imposed

The query `lattice.datum(p)` asks for a value corresponding to a point p in the domain, which is perhaps the simplest kind of lattice query. The lattice topology is responsible for associating a point in the geometry with a location in a datasource index space. Since we know there is a simple mapping from geometric location to index space, it is easy to check whether p is a sample point, or sufficiently close to one that it can be treated as such. In this case, the Topology computes the datasource index for the point and then issues the query `ds.datum(new TwoDIndex(i,j))` to the root DataSource, which then returns the correct value for the sample point.

If the query point does not correspond to a sample point, the lattice approximating function must be used to generate an approximate value. To do so it must be given the values of nearby sample points. Because the topology closely reflects the Geometry, it is quite efficient to use the lattice topology to select the four sample points that surround p. Their indices are then placed into a list, after which the query `ds.datumList(indexList)` is issued to the root DataSource. The values returned can then be given to the approximating function to return a computed value.

An Unstructured Lattice The process for an unstructured lattice is similar overall, except that there is no simple mapping between the location of

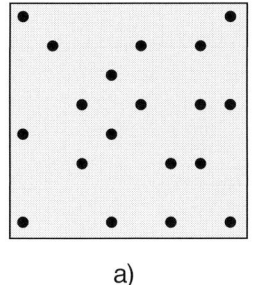 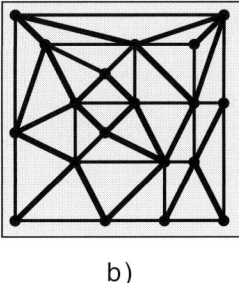

Fig. 8. A 2D Unstructured Lattice: **(a)** geometry only, and **(b)** with topology imposed

sample points in the geometric space and their location in the datasource index space. A 1D datasource is often appropriate for unstructured data, effectively treating the data as a list of sample points. The lack of a simple mapping forces the topology to do more work when computing the datasource indices for sample points near the query point in the geometric space. In fact, a multidimensional access method [5] such as a quadtree could be used to allow efficient search through the geometric space for nodes in the topology graph that correspond to nearby sample points. These nodes will contain the datasource index of the sample point they represent. Once nearby sample points have been found, it is relatively easy to navigate the Topology graph to find the set of points nearest the query point, and then send an array of indices to the root datasource to have the corresponding values retrieved. As with the rectilinear example, the query `ds.datumList(indexList)` will be sent to the root datasource after the indexes of neighbor points have been put into the indexList.

An Example Datasource with Rectilinear Data Now that we have seen how a lattice translates geometric queries into datasource queries, we examine how datasource queries are satisfied. Suppose the data consists of one attribute that is stored in a single file organized as a 2D array and the other three attributes are stored in four files that represent the four quadrants of the 2D array. This structure is shown by the DataSource tree in Fig. 9. The index spaces of the DataSources in this example are always two dimensional. The four physical DataSources in the lowest level of this tree are associated with network streams communicating with remote machines. Lets assume that their attribute sets are all {*carbon, nitrogen, oxygen*}. The BlockedDataSource in the middle level joins the index spaces of its components together, but shares their attribute set. The PhysicalDataSource in the second level is associated with a local file containing the single attribute {*time*}. The two DataSources on the second level have identical index spaces

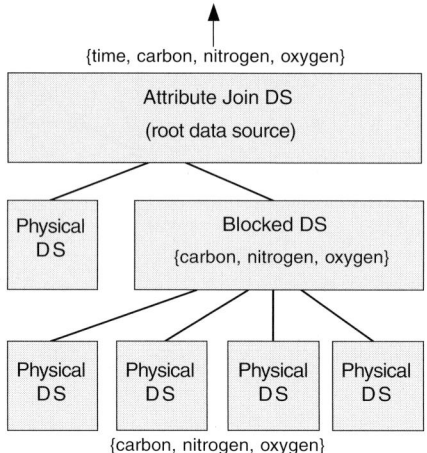

Fig. 9. An example DataSource tree

and are combined using an AttributeJoinDataSource. This topmost DataSource is the root DataSource that communicates with the Lattice itself.

Let's consider the query `ds.datum(new TwoDIndex(i,j))` given to the root DataSource by the Lattice. The parameter is an index for the root DataSource. Since this DataSource is an attribute join, it sends the same query to both of its component DataSources. No transformation is necessary because the component DataSources have the same index space as the root DataSource. Of course, the PhysicalDataSource in the second level maps the two dimensional index to a one dimensional file offset and then return values for its attribute. The BlockedDataSource has more work to do. It must decide which of its component DataSources contains the index and then transform it into the index space for that DataSource. After retrieving the result from the chosen physical DataSource, the BlockedDataSource sends the three attributes up to the root DataSource, which joins them with the one already retrieved from the physical DataSource. Finally, the root DataSource returns the four attributes to the Lattice.

An Example DataSource with Unstructured Data For unstructured data, the principal difference at the datasource level is that the index spaces of the various DataSources do not match the dimensionality of the Lattice. Since unstructured data is often stored as a list of points, the datasource index space may well be simply one dimensional.

To continue our unstructured data example, let's assume that the DataSource tree is the same as the one depicted in Fig. 9. The root DataSource still receives the query `ds.datum(index)` from the Lattice, but now the index is one dimensional. The BlockedDataSource picks which PhysicalDataSource in the bottom level to query based on the index value, retrieving {*carbon, nitrogen, oxygen*} from the appropriate component. The PhysicalDataSource in the second level retrieves {*time*}, and finally the root DataSource joins the four attributes together. Notice that except for the dimensionality of the index space, the tree works in a fashion which is identical to the rectilinear case.

8 Performance Issues

Systems that handle very large datasets should be designed to minimize redundant data access and to take advantage of any structure in patterns of access. It is also important to avoid unnecessary duplication of data.

Our prototype system allows us to experiment with and evaluate techniques for minimizing access costs. We are particularly interested in finding ways to reduce overhead associated with distributed multisource data.

8.1 Lazy Evaluation

The concept of a datasource fits naturally with lazy evaluation. This is especially apparent if we think of a tree of datasources as a description of a data file that does not physically exist. When the tree is built, no data is actually processed until the Lattice begins asking for sample points. Even at this point, the file that the Lattice sees is only conceptual. Sample points are assembled from component datasources but the entire file is never materialized.

We have already added a small enhancement to our prototype system by allowing datasource `datum()` queries to take an argument that specifies which attributes should be retrieved. In practice, this information is sent down to the physical datasources so that only desired attributes are read and processed by the rest of the datasource tree. Though conceptually simple, this enhancement could greatly reduce the volume of data that is processed in situations where each sample point has a large number of attributes.

We also have defined a subset operation to both the lattice and datasource levels. At the lattice level it is useful to specify a new lattice as a subset of an existing one. However, it is not practical to duplicate the lattice sample points; it is much more efficient to implement a new *subset datasource* that presents the new lattice with a subset of the original data without ever materializing an actual subset of the original. Of course, this scheme relies on the fact that much scientific data is read-only. More complex mechanisms are needed to support multiple lattices writing to the same data.

8.2 Caching and Prefetching

Future versions of our prototype system will incorporate caches as a primary tool for avoiding redundant access to data.

When data is accessed in a regular pattern, it is possible to predict which elements will be required in the near future. Processes that iterate over a lattice in a predictable fashion can be greatly accelerated by prefetching data that is likely to be needed. Prediction may be straightforward with rectilinear data since iteration will often proceed through rows and columns in a regular fashion [13]. Effective prefetching of unstructured data is harder but can be done using a topological or geometric *neighborhood*.

In fact, a *flexible neighborhood query* can serve as the basis for prefetching for both structured and unstructured data. With this kind of query, a lattice would ask its root DataSource to return two sets of points, a *primary* set and a *secondary* set (such as a neighborhood). The DataSource must return the primary set and may return some or all of the secondary set if it is convenient. For example, a DataSource might return a partial result consisting of the primaries and readily available secondaries. Other secondaries may later arrive and can be cached for future queries.

The Topology and Geometry provide valuable information for identifying appropriate secondary data points based on the nature of the processing being done.

9 Conclusion

We have developed a formal data model for describing distributed multiresolution multisource scientific data sets. The lattice provides the logical view of data as it appears to the scientist. The DataSource is the principal mechanism for an efficient implementation of multisource datasets. Lattices also include an explicit definition of topology which allows us to represent a wide variety of grid structures. The topology uses the datasource representation to map data points to physical locations. The entire model serves as the basis for a scientific data management support environment that can provide nearly transparent access to distributed multisource data.

We have an initial implementation of a prototype system to support the principal features of our data model. This implementation allows a scientist to describe a single dataset that represents a unified view of data that is physically stored in multiple datasets.

Although our current prototype is designed to include caching and prefetching, these features have not yet been implemented. Similarly, our current support for unstructured data is minimal and we need to integrate multiresolution and adaptive resolution functionality into the prototype.

References

1. R. Daniel Bergeron, Georges G. Grinstein, "A Reference Model for the Visualization of Multi-Dimensional Data", Eurographics '89, Elsevier Science Publishers, North Holland, 1989

2. Paolo Cignoni, Claudio Montani, Enrico Puppo, Roberto Scopigno, "Multiresolution Representation and Visualization of Volume Data", IEEE Trans. on Visualization and Computer Graphics, Volume 3, No. 4, IEEE, Los Alamitos, CA, 1997
3. P. Cignoni, C. Rocchini and R. Scopigno, "Metro: Measuring Error on Simplified Surfaces", Computer Graphics Forum, Vol. 17, No. 2, Blackwell Publishers, Oxford, UK, 1998
4. Mark de Berg, Katrin T.G. Dobrindt, "On Levels of Detail in Terrains", Graphical Models and Image Processing 60:1–12, Academic Press, 1998
5. Volker Gaede, Oliver Günther, "Multidimensional Access Methods", ACM Computing Surveys, Vol. 30, No. 2, ACM, New York, 1998
6. R.B. Haber, B. Lucas, N. Collins, "A Data Model for Scientific Visualization with Provisions for Regular and Irregular Grids", Proceedings of IEEE Visualization 91, San Diego, CA, 1991
7. W.L. Hibbard, C.R. Dyer, and B.E. Paul, "A Lattice Model for Data Display", Proceedings of IEEE Visualization '94, IEEE, Washington, DC, 1994
8. W.L. Hibbard, D.T. Kao, and Andreas Wierse, "Database Issues for Data Visualization: Scientific Data Modeling", Database Issues for Data Visualization, Proc. IEEE Visualization '95 Workshop, LNCS 1183, Springer, 1995
9. D.T. Kao, R. Daniel Bergeron, Ted M. Sparr, "An Extended Schema Model for Scientific Data", Database Issues for Data Visualization, Proceedings of the IEEE Visualization '93 Workshop (LNCS 871), Springer, Berlin, 1993
10. D.T. Kao, M.J. Cullinane, R.D. Bergeron, T.M. Sparr, "Semantics and Mathematics of Scientific Data Sampling", in Wierse, Grinstein and Lang (Eds.), Database Issues for Data Visualization, LNCS 1183, Springer-Verlag, Berlin, 1996
11. D.T. Kao, "A Metric-Based Scientific Data Model for Knowledge Discovery", Ph.D. Thesis, University of New Hampshire, Durham, 1997
12. D.T. Kao, R.D. Bergeron, T.M. Sparr, "Efficient Proximity Search in Multivariate Data", 10th International Conference on Scientific and Statistical Database Management, Capri, Italy, 1999
13. Heng Ma, "Remote Transformation and Lattice Manipulation", Masters Thesis, University of New Hampshire, Durham, NH, 1992
14. John L. Pfaltz, Russell F. Haddleton, James C. French, "Scalable, Parallel, Scientific Databases", Proceedings 10th International Conference on Scientific and Statistical Database Management, IEEE, Los Alamitos, CA, 1998
15. Richard J. Resnick, Matthew O. Ward, and Elke A. Rundensteiner, "FEDA Framework for Iterative Data Selection in Exploratory Visualization", Proceedings of Tenth International Conference on Scientific and Statistical Databases, IEEE Computer Society Press, Los Alamitos, CA, 1998
16. Philip J. Rhodes, R. Daniel Bergeron, Ted M. Sparr, "A Data Model For Distributed Scientific Visualizations", Proceedings of the Scientific Visualization Conference:DAGSTUHL 2000 (To be published)
17. Doris Schattschneider, Marjorie Senechal, "Tilings", Handbook of Discrete and Computational Geometry, CRC Press, Boca Raton, 1997
18. D. Speray, S. Kennon, "Volume Probes: Interactive Data Exploration on Arbitrary Grids", Computer Graphics, Vol. 24, No. 5, ACM, 1990
19. Pak Chung Wong, R. Daniel Bergeron, "Authenticity Analysis of Wavelet Approximations in Visualization", Proceedings of IEEE Visualization '95, IEEE Computer Society Press, Los Alamitos, CA, 1995

Adaptive Subdivision Schemes for Triangular Meshes

Ashish Amresh, Gerald Farin, and Anshuman Razdan

Arizona State University, Tempe AZ 85287-5106
Email: amresh@asu.edu Phone: (480) 965 7830 Fax: (480) 965 2751.

Abstract. Of late we have seen an increase in the use of subdivision techniques for both modeling and animation. They have given rise to a new surface called the subdivision surface which has many advantages over traditional Non Uniform Rational B-spline (NURB) surfaces. Subdivision surfaces easily address the issues related to multiresolution, refinement, scalability and representation of meshes. Many schemes have been introduced that take a coarse mesh and refine it using subdivision. They can be mainly classified as Approximating – in which the original coarse mesh is not preserved, or Interpolating – wherein the subdivision forces the refined mesh to pass through the original points of the coarse mesh. The schemes used for triangular meshes are chiefly the Loop scheme, which is approximating in nature and the Modified Butterfly scheme which is interpolating. Subdivision schemes are cost intensive at higher levels of subdivision. In this paper we introduce two methods of adaptive subdivision for triangular meshes that make use of the Loop scheme or the Modified Butterfly scheme to get approximating or interpolating results respectively. The results are obtained at a lower cost when compared with those obtained by regular subdivision schemes. The first method uses the angles between the normal of a face and the normals of its adjacent faces to develop an adaptive method of subdivision. The other method relies on user input, i.e. the user specifies which parts of the mesh should be subdivided. This process can be automated by segmentation techniques, e.g. watershed segmentation, to get the areas in the mesh that need to be subdivided. We compare our methods for various triangular meshes and present our results.

1 Introduction

When Catmull and Clark [1] and Doo and Sabin [2] published their papers little did they expect that subdivision would be used so extensively as it is being used today for the purposes of modeling and animation. It has been used to a large extent in movie production, commercial modelers such as MAYA 3.0, LIGHTWAVE 6.0 and game development engines.

The basic idea behind subdivision can be traced as far back as to the late 40s and early 50s when G. de Rham used *corner cutting* to describe smooth curves. In recent times the applications of subdivision surfaces has grown in the field of computer graphics and computer aided geometric design(CAGD) mainly because it easily addresses the issues raised by multiresolution techniques to address the challenges raised for modeling complex geometry. The subdivision schemes introduced by Catmull and Clark [1] and Doo and Sabin [2] set the tone for other schemes to follow and schemes like Loop [6], Butterfly [3] and Modified Butterfly [14], Kobbelt [4] have become popular. These schemes are chiefly classified as either approximating, where the original vertices are not retained at newer levels of subdivision, or interpolating, where subdivision makes sure that the original vertices are carried over to the next level of subdivision. The Doo-Sabin, Cattmull-Clark and Loop schemes are approximating and Butterfly, Modified Butterfly and Kobbelt schemes are interpolating.

It is seen that all the schemes provide a process of global refinement at every level of subdivision. This can lead to a heavy computational load at higher levels of subdivision. For example, it is observed that in the Loop scheme every level of subdivision increases the triangular count by four. It is also observed that for most surfaces there are regions that become reasonably smooth after few levels of subdivision and only certain areas of the surface where there is a high curvature change need high subdivision levels to make it smooth. It therefore is not ideal to have a global subdivision scheme being applied at every level. Adaptive Subdivision aims at providing a local subdivision rule that governs weather or not a given face in a mesh needs to be subdivided at the next level of subdivision.

2 Existing Methods

Mueller [8] proposed an adaptive process for Catmull-Clark and Doo-Sabin subdivision schemes. In his method the adaptive refinement is controlled by the vertices at every level of subdivision. The approximation is carried by an error calculated at every vertex of the original mesh before it is subdivided. This error is the distance between the vertices of the original mesh and their limit point. All the vertices that lie in the error range are labeled differently and special rules are applied for subdividing a polygon when it contains one or more of these labeled vertices.

Xu and Kondo [12] devised an adaptive subdivision scheme based on the Doo-Sabin scheme. In their method the adaptive refinement is controlled by the faces of the original mesh. Faces are labeled as *alive* or *dead* if they have to be subdivided or not. The labeling is based on the angle between the normal vectors of adjacent faces and a tolerance limit for this angle is set. If a face satisfies the set tolerance then it is labeled as dead and further refinements are stopped for that face.

Kobbelt has developed adaptive refinement for both his Kobbelt scheme and newly introduced $\sqrt{3}$ subdivision [5]. His refinement strategy ia also centered around the faces. In both the schemes adaptive refinement presents a face cracking problem (discussed later in the paper). He solution is to use a combination of mesh balancing and the Y-technique for his Kobbelt scheme. For his $\sqrt{3}$ subdivision he uses a combination of dyadic refinement, mesh balancing and gap fixing by temporary triangle fans. This process is well known in the finite element community under the name red-green triangulation [11].

Zorin et al. have developed adaptive refinement strategies in [15], where they have additional constraints that require a certain number of vertices in the neighborhood of those vertices calculated by adaptive subdivision to be present. Their methods have been implemented on the Loop scheme.

We observe that essentially the adaptive strategies can be developed in two ways, one by classifying which vertices need to be subdivided (vertex spilt operation at the next level) or two by identifying those faces that should be subdivided (face split at the next level).

3 The Loop Scheme

The Loop scheme is a simple approximating face-split scheme for triangular meshes proposed by Charles Loop [6].The scheme is based on the triangular splines [10], which produces C^2-continuous surfaces over regular meshes. A regular mesh is a mesh which has no extraordinary vertices, i.e. vertices whose number of nieghbors does not equal six. It also means that the vertex has a valance of six. The Loop scheme produces surfaces that are C^2-continuous everywhere except at extraordinary vertices, where they are C^1-continuous. A boundary vertex is regular or even if it has a valance of three and is extraordinary for any other valance. The masks for the Loop scheme are shown in Figure 1. For boundaries special rules are used. These rules produce a cubic spline curve along the boundary. The curve only depends on control points on the boundary. The scheme works as follows

- for every original vertex a new vertex(odd vertex) is calculated by calculating β from equation (1), where k is the number of adjacent vertices for the vertex, and finding the suitable coefficients for the adjacent control points as shown in Figure 4.
- for every edge in the original mesh a new vertex(even vertex) is calculated by using the mask shown in Figure 1.
- every triangle in the original mesh gives rise to six new vertices, three from original vertices and three from original edges, these six vertices are joined to give four new triangles.

In Figure 1, k is the no of adjacent vertices for a given vertex and β can be chosen as

$$\beta = 1/k \left(5/8 - (3/8 + 1/4 cos 2\pi/k)^2\right) \tag{1}$$

The value for β [6] was so found that the resulting surface is C^1-continuous at the extraordinary points. For regular vertices the coefficients for calculating the new vertices are obtained by substituting k as six in the mask for even vertices shown in Figure 1.

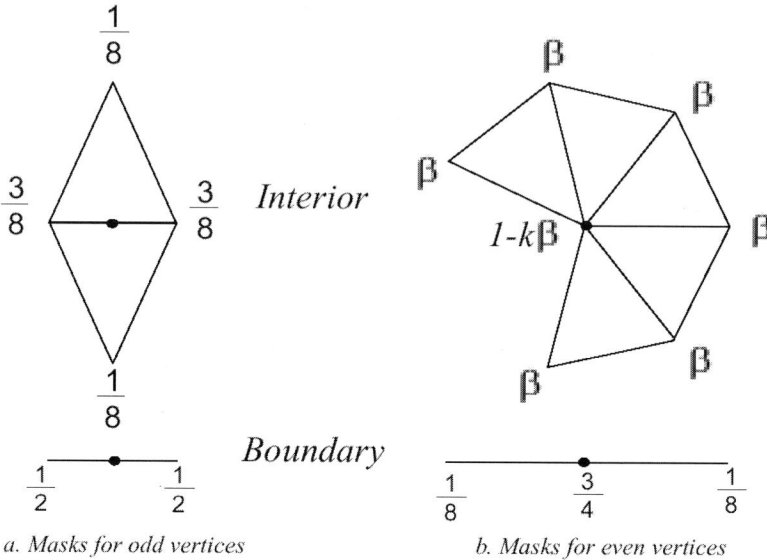

Fig. 1. Masks for Loop scheme

4 Our Methods

We now discuss the methods we have developed for adaptively subdividing meshes generated by the Loop scheme. Our first method is based on identifying which faces are "flat" and proposing suitable mesh refinements based on the properties of the of its neighboring faces. The second method is based on user interaction, where the user can select areas on the mesh where refinements are desired.

4.1 Dihedral Angle Method

Our first method is called the dihedral angle method because it uses the angles between normals of a face with adjoining face normals to determine if the face needs to be subdivided or not. If the angles are within some tolerence limit then we classify the face as flat. This adaptive process introduces cracking between faces and triangle fans are introduced to solve this problem. We take care of the cracking problem in our scheme by a process of refinement that takes into account the nature of the adjacent faces before refining a given face. The process is shown in Figure 2. We introduce an adaptive weight a that controls the tolerence limit for the angles between the normals. Our scheme works as follows:

- The normal for each face is calculated

- For every face the angle between its normal and normals of adjoining faces are calculated
- If all angles lie below a certain threshold then the face is set to be flat
- For every face a degree of flatness, which is the number of adjoining faces that are flat, is set. The maximum value for this degree can be three and minimum will be zero. Based on the degree of flatness, refinement is done as shown in Figure 2.

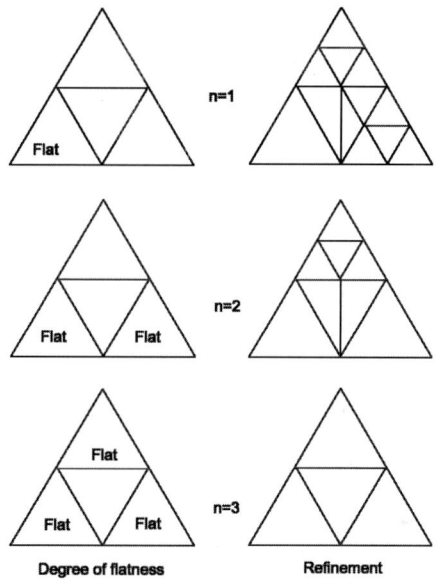

Fig. 2. Mesh refinement based on the degree of flatness

4.2 Analysis of Results

It is seen that many levels of adaptive refinement produce degenerate triangulations and therefore we suggest that after a level of adaptive subdivision a suitable realignment of triangles should be done. Realignment introduces some more computation and if speed becomes important a good strategy could be to alternate between adaptive and normal subdivision methods. We now take a complicated mesh of a cat and compare our adaptive scheme with the normal approximating scheme at two levels of subdivision. Figure 3 shows the comparison of the normal Loop scheme at two levels (left) and using our adaptive scheme and then applying the loop scheme at two levels of subdivision (right).

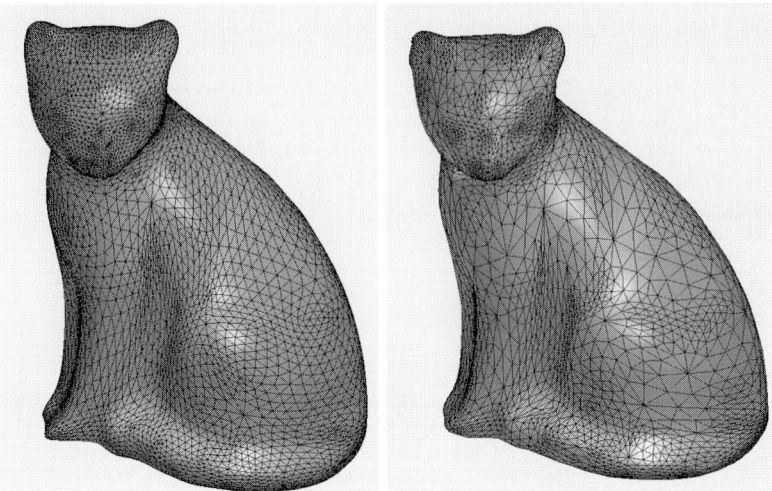

Fig. 3. Using normal angle adaptive Loop subdivision method on a cat data file, the image on the left is obtained using normal subdivision and the one on the right is subdivided adaptively

4.3 Watershed Segmentation Method

The main drawback of the dihedral method is that it acts on the whole mesh and goes through a significant computation to identify the triangle types and find their degree of flatness. We could identify the regions in a mesh that need to be subdivided then these computations can be avoided. We perform this identification using a segmentation process which segments meshes based on their geometry. Watershed segmentation [7] is an effective technique to employ before adaptive subdivision. A brief explanation of the watershed segmentation algorithm is as follows:

- Estimate curvature at each vertex of the mesh.
- The vertices whose curvature is less than the curvatures of their neighbors is labeled as a minimum.
- From every vertex that is not a minimum, a token is sent in the direction of its neighbor with lowest curvature. The token stops when it hits a minimum. The minimum is copied into every vertex in the tokens path, and when the token stops every minima establishes its own region.
- Merging of regions is performed if they satisfy certain conditions of similarity as described in [7].

The adaptive subdivision scheme now works as follows:

- Apply the normal Loop scheme to a coarse mesh until a reasonable resolution is reached. Usually two levels of subdivision are sufficient.

- Identify regions using the watershed segmentation method.
- Find points along the boundary between segmented regions, and mark all triangles that contain these points.
- Decide on a triangle threshold limit, i.e. the number of adjacent triangles to the triangles on the boundary and mark these triangles.
- Subdivide only the marked triangles.

4.4 Analysis of Results

Figure 4 shows the results of adaptively subdividing a simple uniform mesh of a vase data. The areas of refinement happen to be the areas where the curvature changes and these have been identified by the user. The results for a golf head mesh are shown in Figure 5. Watershed segmentation was used to segment regions and this can be viewed as an automated process. The figure also shows a comparison of subdividing the golf head normally by Loop subdivision(left), subdivide it by normal-angle subdivision(middle) and watershed method(right). It can be seen that for data that has significant curvature changes and irregularity the normal angle method is a better way of adaptive subdivision as the whole mesh is filled with irregularities and a process like watershed segmentation would have to compute all the various regions. Suitable applications would be in terrain modeling where we find a lot of irregularity in the data. In cases where the regions can be marked with ease and where change in curvature is consistent a segmentation approach should be applied to get faster results as the computational value of the normal angle method could be a burden. Suitable applications could be meshes used in character animation and industrial design prototypes.

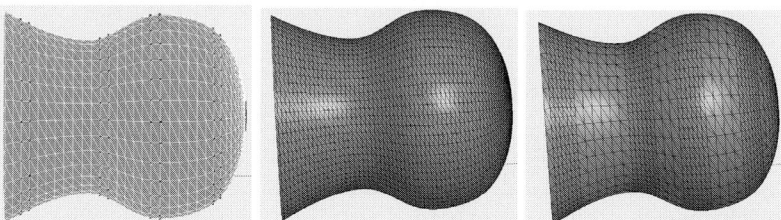

Fig. 4. Using watershed segmentation method for a vase data file, image on the left is the base mesh with points picked, one in the middle uses loop subdivision and the one on the right is subdivided adaptively after Watershed segmentation process.

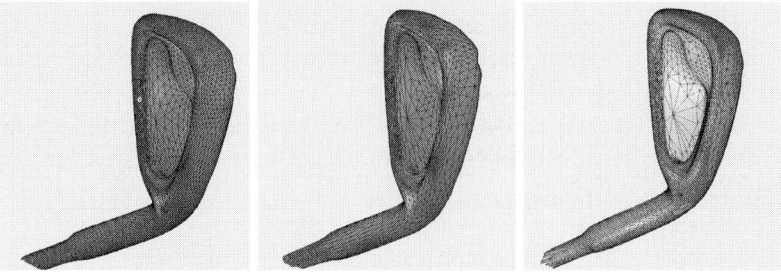

Fig. 5. Comparison of normal subdivision, normal angle adaptive subdivision and watershed adaptive subdivision applied on to a golf head file

5 Computational Analysis

We compare costs of the two schemes in this section. As discussed earlier at every level of the Loop scheme the number of triangles generated increases by a factor of four. For the results obtained by the dihedral scheme, shown in Figure 3, the base mesh of the cat consists of 698 triangles. A second level Loop subdivision produces 9168 triangles as opposed to 3337 triangles produced by dihedral subdivision. For the results obtained by watershed segmentation, shown in Figure 5, the base mesh of the golf head consisted of 33 triangles and a second level Loop subdivision produces 528 triangles, upon which watershed segmentation is used to identify the regions. Continuing with Loop subdivision produces 8448 triangles at the fourth level as opposed to 7173 triangles produced by the watershed segmentation method. It can be seen that that saving in the size of the mesh is considerably high while using the dihedral method but it also must be noted that the dihedral method cannot be applied in succession. We also note that when the mesh is highly undulating, then using a curvature based user selected process like watershed segmentation will not result in considerable savings in mesh size.

6 Future Work and Conclusions

We have presented adaptive schemes for triangular meshes based on the Loop scheme. These could be very well applied to the Modified Butterfly scheme. These schemes could also be extended to subdivision schemes that work on polygonal meshes like Catmull-Clark or Doo-Sabin. Our methods are a shift from the methods proposed in [5], [15] where the additional constraints due to smoothing require extra storage for temporary triangles.

Our first method tries to alternate normal and adaptive subdivision to bring in smoothing properties to the resulting surface. For a reasonable angle tolerance the results obtained are in accordance to the results obtained by normal Loop subdivision. Our second method adds intelligence by picking regions before an adaptive subdivision scheme is applied. It is seen that automating the interaction by a process like watershed segmentation brings in good results. However watershed segmentation requires meshes with a reasonable amount of resolution, but since we need adaptive subdivision for only higher subdivision levels, this is not a problem. Subdivision at lower levels can be achieved with good speed by normal subdivision.

We therefore propose to have a coarse mesh subdivided using the normal Loop scheme for two levels and achieving a reasonable resolution. The watershed algorithm is now run to get the points on the boundary of various segments. The faces lying along the boundary are the only ones subdivided at the next levels of subdivision. The area of applying segmentation along with subdivision has a lot of promise and holds potential for future research.

References

1. E. Catmull and J. Clark. Recursively generated b-spline surfaces on arbitrary topological meshes. *Computer Aided Design*, 10:350–355, 1978.
2. D. Doo and M. Sabin. Behaviour of recursive division surfaces near extraordinary points. *Computer Aided Design*, 10:356–360, 1978.
3. N. Dyn, J. Gregory, and D. Levin. A butterfly subdivision scheme for surface interpolation with tension control. *ACM Trans. Graph.*, 9:160–169, 1990.
4. L. Kobbelt. Interpolatory subdivision on open quadrilateral nets with arbitrary topology. *Proceedings of Eurographics.*, 409–420, 1996.
5. L. Kobbelt. $\sqrt{3}$ Subdivision. *Computer Graphics Proceedings.*, 103–112, SIGGRAPH 2000.
6. C. Loop. Smooth subdivision surfaces based on triangles. *Masters Thesis.*, University of Utah, Dept. of Mathematics, 1987.
7. A. Mangan and R. Whitaker. Partitioning 3D meshes using watershed segmentation. *IEEE Transactions on Visualization and Computer Graphics*, 308-321, oct-dec, 1999.
8. H. Mueller and R. Jaeschke. Adaptive subdivision curves and surfaces *Proceedings of Computer Graphics International '98*, 48–58, 1998.
9. J. Peters and U. Reif. The simplest subdivision scheme for smoothing polyhedra. *ACM Trans. Gr.16(4).*, 420–431, 1997.
10. H. Seidel. Polar forms and triangular B-Splines. *Tutorial notes in Pacific Graphics*, 61–112, 1997.
11. M. Vasilescu and D. Terzopoulos. Adaptive meshes and shells: Irregular triangulation, discontinuities and hierarchical subdivision. *Proceedings of the Computer Vision and Pattern Recognition conference*, 829–832, 1992.
12. Z. Xu and K. Kondo. Adaptive refinements in subdivision surfaces. *Eurographics '99, Short papers and demos*, 239–242, 1999.
13. D. Zorin. Subdivision and multiresolution surface representations. *PhD Thesis.*, Caltech, Pasadena, 1997.
14. D. Zorin, P. Schroder, and W. Sweldens. Interpolating subdivision for meshes with arbitrary topology. *SIGGRAPH '96 Proceedings*, pages 189–192, 1996.
15. D. Zorin, P. Schroder, and W. Sweldens. Interactive multiresolution mesh editing. *SIGGRAPH '97 Proceedings*, pages 259–268, 1997.

Hierarchical Image-based and Polygon-based Rendering for Large-Scale Visualizations

Chu-Fei Chang[1], Zhiyun Li[2], Amitabh Varshney[2], and Qiaode Jeffrey Ge[3]

[1] Department of Applied Mathematics, State University of New York, Stony Brook, NY 11794, USA, chchang@cs.sunysb.edu
[2] Department of Computer Science and UMIACS, University of Maryland, College Park, MD 20742, USA, {zli, varshney}@cs.umd.edu
[3] Department of Mechanical Engineering, State University of New York, Stony Brook, NY 11794, USA, ge@design.eng.sunysb.edu

Abstract. Image-based rendering takes advantage of the bounded display resolution to limit the rendering complexity for very large datasets. However, image-based rendering also suffers from several drawbacks that polygon-based rendering does not. These include the inability to change the illumination and material properties of objects, screen-based querying of object-specific properties in databases, and unrestricted viewer movement without visual artifacts such as visibility gaps. View-dependent rendering has emerged as another solution for hierarchical and interactive rendering of large polygon-based visualization datasets. In this paper we study the relative advantages and disadvantages of these approaches to learn how best to combine these competing techniques towards a hierarchical, robust, and hybrid rendering system for large data visualization.

1 Introduction

As the complexity of 3D graphics datasets has increased, different solutions have been proposed to bridge the growing gap between graphics hardware and the complexity of datasets. Most algorithms which effectively reduce the geometric complexity and overcome hardware limitations fall into the following categories: visibility determination [29,33,7,5,30,23,34], level-of-detail hierarchies [16], and image-based rendering (IBR) [10].

IBR has emerged as a viable alternative to conventional 3D geometric rendering, and has been widely used to navigate in virtual environments. It has two major advantages over the problem of increase in complexity of 3D datasets: (1) The cost of interactively displaying an image is independent of geometric complexity, (2) The display algorithms require minimal computation and deliver real-time performance on workstations and personal computers. Nevertheless, use of IBR raises the following issues:

- Economic and effective sampling of the scene to save memory without visually perceptible artifacts in virtual environments,
- Computing intermediate frames without visual artifacts such as visibility gaps,

- Allowing changes in illumination, and
- Achieving high compression of the IBR samples.

To address some of the above issues we have developed a multi-layer image-based rendering system and a hybrid image- and polygon-based rendering system. We first present a hierarchical, progressive, image-based rendering system. In this system progressive refinement is achieved by displaying a scene at varying resolutions, depending on how much detail of the scene a user can comprehend. Images are stored in a hierarchical manner in a compressed format built on top of the JPEG standard. At run-time, the appropriate level of detail of the image is constructed on-the-fly using real-time decompression, texture mapping, and accumulation buffer. Our hierarchical image compression scheme allows storage of multiple levels in the image hierarchy with minimal storage overhead (typically less than 10%) compared to storing a single set of highest-detail JPEG-encoded images. In addition, our method provides a significant speedup in rendering for interactive sessions (as much as a factor of 6) over a basic image-based rendering system.

We also present a hybrid rendering system that takes advantage of the respective powers of image- and polygon-based rendering for interactive visualization of large-scale datasets. In our approach we sample the scene using image-based rendering ideas. However, instead of storing color values, we store the visible triangles. During pre-processing we analyze per-frame visible triangles and build a compressed data-structure to rapidly access the appropriate visible triangles at run-time. We compare this system with a pure image-based, progressive image-based system (outlined above), and pure polygon-based systems. Our hybrid system provides a rendering performance between a pure polygon-based and a multi-level image-based rendering system discussed above. However, it allows several features unique to polygon-based systems, such as direct querying to the model and changes in lighting and material properties.

2 Previous Work

Visibility determination is a time- and space-consuming task. Good visibility information often takes significant time to compute and space to store. Current applications involving this problem often pre-compute the visibility information and store it for later use to improve the rendering speed. Teller and Sequin [29] divide a building into rooms and compute room-to-room visibility. Yagel and Ray [33] have proposed an algorithm to compute cell-to-cell visibility by applying a subdivision scheme. They also propose a clustering scheme to cluster cells with similar visibility. Coorg and Teller [6,7] use large occluders in the scene to perform occlusion culling for a viewpoint. Cohen-Or et al. [5] use large convex occluders to compute cell-to-object visibility. Wang et al. [30] pre-compute visible sets and simplify the regions where the visible sets are very large. Coorg and Teller [6] compute visibility information by using frame-to-frame incremental coherence. Panne and Steward [23] have presented two algorithms which effectively compress pre-computed visible sets over three different types of models.

Image-based Rendering (IBR) has recently emerged as an alternative to polygon-based rendering. The study of IBR has focused on image morphing and image interpolation for walkthroughs in virtual environments. Several ideas have been proposed to solve these problems including use of textures, environment maps, range images, depth information, movie maps, and so on. Several computer vision techniques have been being used in IBR for solving problems such as disparity maps, optical flows, and epipolar geometry. The techniques in computer graphics and computer vision are merging gradually in the newer applications to IBR.

An image morphing method usually involves two steps. The first step constructs the correspondence (mapping) between images. The second step uses the mapping to interpolate the intermediate images. Chen and Williams [4] have proposed an image morphing method. Their method uses the camera transformation and image range data to determine the pixel-to-pixel correspondence between images. They use a Z-buffer algorithm on pixels to solve the pixel overlap problem and interpolate adjacent pixels to fill holes. Chen [3] has described an image-based rendering system, which is now known as QuickTime VR. He uses 360° cylindrical panoramic images. In this system, a fixed-position camera can roll freely by simply rotating images. The pitch and yaw can be achieved by reprojecting an environment map. To achieve high quality in continuous zooming, this system interpolates the adjacent levels in a hierarchical image representation.

Adelson and Bergen [1] have proposed the plenoptic function concept. They used a plenoptic function to describe the structure of information in the light impinging on an observer. The plenoptic function is parameterized by eye position (V_x, V_y, V_z), azimuth and elevation angles θ and ϕ from any viewable ray to the eye, and a band of wavelengths λ. A view from a given eye position in a given direction is thus formulated as $P(V_x, V_y, V_z, \theta, \phi, \lambda)$. McMillan and Bishop [20] have proposed *plenoptic modeling* for image-based rendering. They have formulated the relative relationships between a pair of cylindrical projections (cylindrical epipolar geometry). They have resolved the visibility problem efficiently by a simple partitioning and enumeration method on cylinders, in a back-to-front ordering. Mark et al. [17] have proposed a post-rendering 3D warping algorithm by first reconstructing the image-space 3D mesh from reference frames and then warping the mesh into the derived frame. Rademacher and Bishop [25] have proposed "multiple-center-of-projection images"(MCOP). An MCOP image consists of a two-dimensional array of pixels and a parameterized set of cameras. A Z-buffer algorithm is used here to solve the visibility problem. To achieve a better quality, blending methods in [24] and [9] can be applied. MCOP images provide connectivity information among adjacent samples and allow different parts of scene to be sampled at different resolutions.

Shade et al. [27] have proposed layered depth images (LDI). In a LDI, each pixel contains a set of depth values along one line of sight sorted in front-to-back order. They have also proposed two rendering algorithms which heavily depend on McMillan's ordering algorithm ([19,18,20]). After pixel ordering, traditional splatting methods are applied to render the warped image. Oliveira and Bishop [22] have proposed an image-based representation for complex three-dimensional objects, the image-based object (IBO). Each IBO is represented by six LDIs sharing a single center of projection (COP). They have proposed a list-priority algorithm, which is based on epipolar geometry and an occlusion compatible order [20] for rendering.

Darsa et al. [9] have constructed image-space triangulation ([8]) from cubical environment maps with depth information. The goal of the triangulation is to establish the 3D geometric information of an environment map as accurately as possible. The triangulation represents a view-dependent simplification of the polygonal scene. Each triangle is assigned a quality which is related to the angle that the normal vector of the triangle makes with the viewing ray. In real time, the warping process involves projecting the triangles of the visible triangulations and texture mapping. A Z-buffer is used for hidden surface removal. Several blending schemes are used to render the intermediate images and improve the quality. This warping highly depends on the hardware texture mapping and transformations.

Levoy and Hanrahan [15] and Gortler et al. [11] have reduced the 5D plenoptic function (without λ) to a 4D Light Field or Lumigraph. They make use of the fact that the radiance does not change along a line unless it is blocked in free space. Light fields or Lumigraphs may be represented as functions of oriented lines. The authors have used *light slabs* as the representations. Both methods use quadralinear basis function to improve the result of interpolations. Sloan et al. [28] have extended the work of Gortler et al. [11] by using hardware texture mapping. Heidrich et al. [13] have improved the image quality of Lumigraph by adding new images to the current Lumigraph. They warp the closest images from the Lumigraph to a new viewpoint that lies on the viewpoint plane and add the new images. The warping step relies on depth information and performs the depth-correction precisely. Schirmacher et al. [26] have presented an adaptive acquisition algorithm by applying the work of Heidrich et al. They predict the error or potential benefit in image quality when adding a new image, and decide which new image should be rendered and added to the Lumigraph. Instead of using two slabs, Camahort et al. [2] have proposed a two-sphere parameterization (2SP) and a sphere-plane parameterization (SPP)for a more uniform sampling.

Nimeroff et al. [21] have effectively re-rendered the scene under various illuminations by linearly interpolating a set of pre-rendered images. Wong et al. [32,31] have proposed an algorithm which allows image-based objects to be displayed under varying illumination. They compute a *bidirectional reflectance distribution function* (BRDF) [14] for each pixel over various illuminations, and store the tabular BRDF data in spherical harmonic domain for reducing the storage.

3 Multi-Level Image-Based Rendering

In this section, we present an image-based rendering system. This system composes a scene in a hierarchical manner to achieve the progressive refinement by using different resolution images. Progressive refinement is achieved by taking advantage of the fact that the human visual system's ability to perceive details is limited when the relative speed of the object to the viewer is high. We first discuss the pre-processing and then the run-time navigation.

3.1 Image Sampling and Collection

Data sampling and collection plays a very important role in an image-based rendering system. It directly affects the storage space and the real-time performance of the system including image quality, rendering speed and user's visual perception. Different sampling strategies can be applied depending on the purpose of the system.

Environment Setting In our system the model is placed at the center of a virtual sphere. The viewer (camera) is positioned on the sphere with the viewing direction toward the origin. The viewer can move around the sphere along longitude and latitude. The camera takes one snapshot every $\Delta\theta$ degree along longitude and $\Delta\phi$ degree along latitude. Due to the symmetry of the sphere, we will have $360/\Delta\theta \times 180/\Delta\phi$ camera positions. The sampling density of camera positions may be adjusted by changing the values of $\Delta\theta$ and $\Delta\phi$. In our implementation, we use $\Delta\theta = \Delta\phi = 5°$ to achieve a reasonably smooth and continuous motion with 2592 images.

3.2 Multi-Level Image Construction Algorithm

The algorithm computes n different levels of resolution of images as the basis for building the system image database. Our algorithm has the following steps:

Step 1: Decide the number n of progressive refinement levels in the system and the resolution of the display window, say $W \times W$, where $W = 2^m$, and $m \leq n$.

Step 2: Dump a Level 0 image, say I_0, at the display window resolution ($W \times W$).

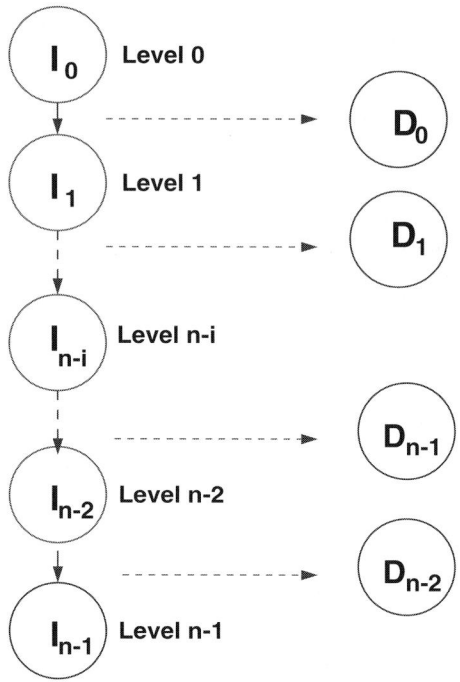

Fig. 1. Multi-Level Image Construction. (See Color Plate 26 on page 364.)

Step 3: Construct Level $i+1$ image (resolution $= W/2^{i+1} \times W/2^{i+1}$), i.e., I_{i+1}. The RGB values of the level $i+1$ image are constructed from the RGB values of the level i image by the following equations:

$$R_{j,k}^{i+1} = min\{R_{2j,2k}^{i}, R_{2j+1,2k}^{i}, R_{2j,2k+1}^{i}, R_{2j+1,2k+1}^{i}\} \qquad (1)$$
$$G_{j,k}^{i+1} = min\{G_{2j,2k}^{i}, G_{2j+1,2k}^{i}, G_{2j,2k+1}^{i}, G_{2j+1,2k+1}^{i}\} \qquad (2)$$
$$B_{j,k}^{i+1} = min\{B_{2j,2k}^{i}, B_{2j+1,2k}^{i}, B_{2j,2k+1}^{i}, B_{2j+1,2k+1}^{i}\} \qquad (3)$$

where $i = 0, 1, \ldots, n-2$. For example, we compute R_{00}^{i+1} as the minimum amongst $\{R_{00}^{i}, R_{10}^{i}, R_{01}^{i}, R_{11}^{i}\}$. We repeat this step until image I_{n-1} is computed.

Step 4: Compute the $W/2^i \times W/2^i$ resolution image I_i from the $W/2^{i+1} \times W/2^{i+1}$ resolution image I_{i+1} as follows. Display I_{i+1} on a $W/2^i \times W/2^i$ resolution window using texture mapping and dump the displayed window image as T_i. Compute image difference D_i as:
$D_i = I_i - T_i, \quad i = 0, 1, \ldots, n-2$
Repeat this step until image difference D_{n-2} is computed, see Figure 1.

Step 5: Store $I_{n-1}, D_{n-2}, D_{n-3}, \ldots, D_0$ in JPEG format as the database images.

This algorithm works well since texture mapping hardware provides speed and antialiasing capabilities through the OpenGL function `glDrawPixels()`. Also, image differences compress better than full images and provide an easy way to generate progressive refinement. For compression and decompression we use the public-domain JPEG software [12] in our implementation. It supports sequential and progressive compression modes, and is reliable, portable, and fast enough for our purposes. The reason we take the minimum value in equations 1–3 is so that we can store all RGB values of D_i as positive values and save a sign bit in storage.

3.3 Progressive Refinement Display Algorithm

Let us define $Tex(I_{n-1})$ as the image generated by mapping texture image I_{n-1} on a $W \times W$ resolution window, and define $Tex(D_i)$ as the image created by mapping texture image D_i on a $W \times W$ resolution window, where $i = 0, 1, \ldots, n-2$. At run time, image I_i is displayed by accumulating images $Tex(I_{n-1}), Tex(D_{n-2}), Tex(D_{n-3}), \ldots, Tex(D_i)$, where $i = 0, 1, \ldots, n-1$.

If $i = n-1$, we only display image $Tex(I_{n-1})$, which has the lowest detail. We add image $Tex(D_i)$ to $Tex(I_{n-1})$, for $i = n-2, n-3, \ldots, 0$, to increase the image details. Image I_0, which is $Tex(I_{n-1}) + \sum_{i=0}^{n-2} Tex(D_i)$, has the highest detail (see Figure 2). Notice that all images are decompressed before texture mapping. The implementation is done using the OpenGL accumulation buffer and texture mapping.

In a real-time environment, progressive refinement can be achieved by displaying different levels of images, depending on how much detail of the scene the user needs to see. If the user moves with high speed, we can simply display lowest detail. As the user speed reduces, we can raise the level of detail of the displayed image in a progressive fashion.

3.4 Our Implementation and Results

In our implementation, we use three different image resolutions, 128×128, 256×256, and 512×512, for progressive refinement. We use sequential mode with quality setting 80 for JPEG compression, which gives us an unnoticeable difference from the highest quality setting of 100. We observed that the composite image quality in our system is not only affected by the lossy JPEG compression, but also by the error from image difference and the geometric error from texture mapping. Table 1 shows the JPEG image reduction from full images I to image differences D. $\sum I_i$ is the sum of storage space required for all the I_i (level i) images. Similarly, $\sum D_i$ is the sum of storage space required for all the D_i (level i) images. The total storage space required is computed as $\sum (I_2 + D_1 + D_0)$. Note that the total memory consumption

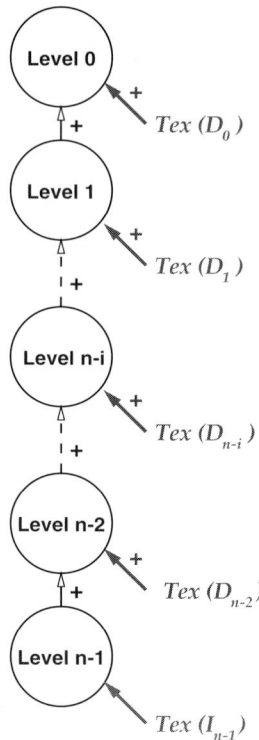

Fig. 2. Hierarchical Accumulation of JPEG Images. (See Color Plate 27 on page 364.)

compares quite favorably to the original (non-progressive) storage space requirements ($\sum I_0$).

Table 2 shows the decompression time, rendering time, and frame rate on different image levels. All numbers in this table are average numbers over different models. I_i, $i = 0, 1, 2$ are full images of 512×512, 256×256, and 128×128 resolutions, respectively. D_i, $i = 0, 1$ are the image differences we discussed in Section 3.2. The *image error* is the root-mean-squared difference between the two images. The errors reported are with respect to the I_0 image.

4 Hybrid Rendering

Image-based rendering is a two-stage process. The first stage is off-line preprocessing that includes sampling of the necessary scene information and setting up data structures, possibly with hierarchy and compression, to reduce access times. The second stage deals with real-time rendering of pre-processed image data which may include image interpolation and warping. Like conventional

Table 1. Storage for I_i, D_i and the total system

Model	Level 2	Level 1		Level 0		Total (MB)
	$\sum I_2$	$\sum I_1 \rightarrow$	$\sum D_1$	$\sum I_0 \rightarrow$	$\sum D_0$	$\sum (I_2 + D_1 + D_0)$
Bunny	10.70	21.39	10.70	53.11	31.34	52.74
Submarine	19.35	40.54	29.63	112.53	70.13	119.11
Enoyl Hydratase	21.29	37.88	21.31	107.34	46.26	88.86
Stanford Dragon	12.19	25.73	21.15	64.70	42.61	75.95
Stanford Buddha	10.70	20.15	14.22	47.22	30.86	55.78

Table 2. Multi-Level Image Rendering Comparison

Image Level	Decompression Time (msec)	Rendering Time (msec)	Speed (fps)	Image Error
$I_2 + D_1 + D_0$	98.6	23.8	7.99	0.435
$I_2 + D_0$	25.4	16.6	23.80	0.077
I_2	6.3	9.4	63.96	0.079
I_1	20.9	10.1	32.31	0.025
I_0	78.9	17.4	10.37	0.0

image-based methods, our hybrid method also has two stages; and the key difference is that, instead of using three- or four-channel color values for each image, we compute the exact visibility of each triangle for each viewpoint, and only the visible (displayed) triangles are stored for each viewpoint. We outline our hybrid system below:

Preprocessing

1. **Initialization**
   1.1 Load polygonal dataset.
   1.2 Encode all primitive ids into RGB values.
   1.3 Set environment parameters such as viewing angle, distance, and resolution.
2. **Dumping process**
   2.1 Dump visible triangles for each camera position.
3. **Compression process**
   3.1 Single frame compression (run-length and Huffman compression).
   3.2 Frame-to-frame compression (intersection and union analysis).

Run-time Navigation

1. **Initialization**
   1.1 Load frame data file.
   1.2 Set frame specific world environment parameters.
   1.3 Construct Huffman trees and locate viewpoints for frames.
2. **Real time**
   2.1 New viewer position.
   2.2 Get the corresponding frame for current viewer node.
   2.3 Decompress the frame.
   2.4 Display frame.

These steps are explained in the following sections.

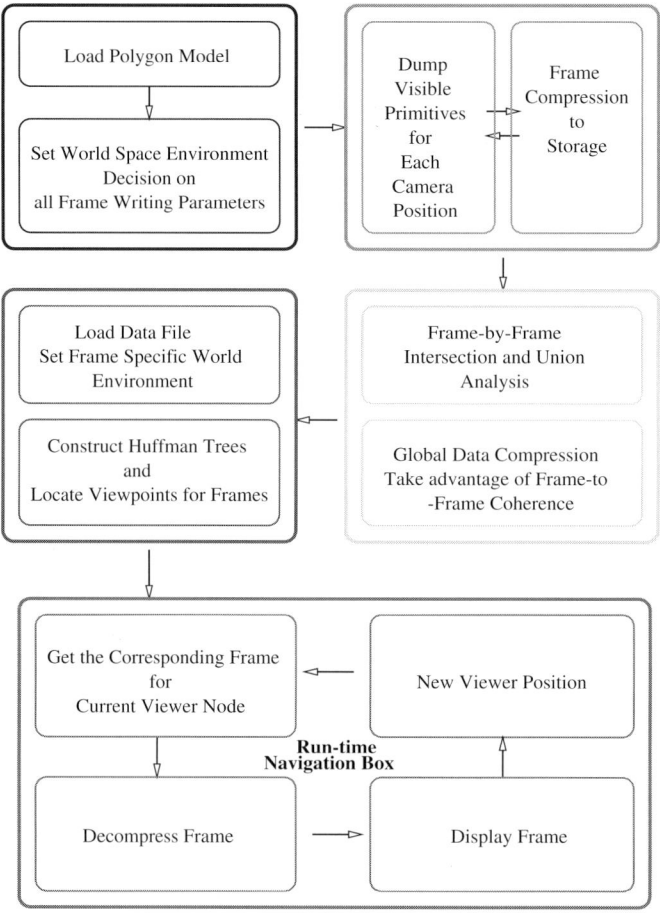

Fig. 3. Hybrid Rendering System Overview

4.1 Preprocessing

We adopt the same environment settings as we did in the JPEG image-based rendering system, see section 3.1.

Encoding Triangle IDs In order to compute the visibility for each triangle, we assign each triangle a unique id when we load the dataset. We then decompose the number, in binary format, into three consecutive bytes and assign them to R, G, and B in order. During the dumping process, we render the whole dataset with the given RGB value for each triangle as its color. Notice here that in order to render all colors correctly, the illumination and antialiasing function in OpenGL should be turned off. We then read the color buffer of this image to get the color for each pixel and compose the R, G, B back to the id. We currently use unsigned char for each single color value, which means, with a one-pass encoding process, we can encode as many as $(2^8)^3 = 16$ million triangles. For larger datasets, multiple-pass encoding processes may be needed. In our method we dump triangles for each camera position (θ, ϕ) by using the dumping process we discussed in Section 3.1 into an occupancy bit-vector, say TriMap(θ, ϕ).

Compression Process Two types of compression are relevant in image-based navigation of virtual environments: single-frame compression and frame-to-frame compression. We have only worked with single frame compression at this stage; the multiple frame compression, which needs more analysis and work, will be dealt with in the future. For representing the visible triangles in a single frame we use an occupancy bit vector (an unsigned char array) in which each bit represents the triangle id corresponding to its position in the vector. The bit is 1 if the triangle is visible in that frame, 0 otherwise.

As the size of 3D datasets increases and the resolution of image space remains fixed, the number of dumped triangles will saturate around the display resolution. In our results, the million triangle Buddha model has on an average only $5 \sim 6\%$ visible triangles for a 512×512 resolution window. It means that most bits in a bit vector would be 0, and consecutive-0-bit-segment cases would occur frequently. This result inspires us to use run-length encoding and Huffman compression.

4.2 Run-time Navigation

At run time the 3D dataset and precomputed information in compressed format is loaded first. The precomputed information not only includes the visible primitives for each frame but also the viewing parameters including viewing angle, distance, and so forth. The run-time viewing parameters should be exactly the same as those used in the dumping process. In the system, each camera position has a frame pointer pointing to the corresponding frame in compressed format. A Huffman tree, which is used for decompression, is also constructed for each frame.

At run time the viewer moves around in a virtual environment following discrete camera positions at $\Delta\theta$, $\Delta\phi$ increments which were used in the dumping process. For a given viewer position, we can locate the corresponding frame by following its frame pointer and decompress the frame by retracing the Huffman tree.

The rendering speed of the system highly depends on the number of visible triangles and the decompression time. In our implementation, the decompression function doesn't have to go through a whole data frame, it stops the decompression loop immediately whenever it detects that all dumped triangles have been found and sends them to the graphics engine. However, the decompression time still depends on the size of the frame (the size of object model) and the number of visible triangles in the frame.

5 Results

We have tested five different polygonal models on SGI Challenge and Onyx2. All models are tested at a window resolution of 512×512 pixels with 2592 images. We describe our results in this section.

Bunny, Dragon, and Buddha are scanned models from range images from the Stanford Computer Graphics Lab. The submarine model is a representation of a notional submarine from the Electric Boat Division of General Dynamics. The E. Hydratase Molecule (Enoyl-CoA Hydratase) is from the Protein Data Bank. All models have vertex coordinates (x, y, z) in floating-point format, and all triangles are represented by their three vertex indices (integers). The submarine dataset provides RGB color values (unsigned char) for each triangle. The E. Hydratase Molecule has a normal vector for each vertex.

Table 3 shows the average compression ratios for all models. The *average dumped tris %* is the average percentage of dumped triangles over all 2592 images.

In Table 4, *P* refers to conventional *Polygonal* rendering, *H* refers to the *Hybrid* rendering system discussed in Section 3, *I* refers to the *Multi-level Image-based* rendering system discussed in Section 2. The *Image Error* is the root-mean square error with respect to the images rendered using the conventional polygonal rendering method . *Dcmprs* is the time for decompression. As can be seen, the *Hybrid* method has consistently low image errors. Multi-level image-based rendering has the highest image error amongst these methods, since JPEG compression, image differences, and texture mapping all contribute to the final image error. For the *Hybrid* method, all visible triangles are stored in a bit vector and compressed by two steps: run-length encoding and Huffman compression. The decompression and rendering speeds are highly dependent on the displayed frame size and the number of dumped tri-

Table 3. Compression Ratios for the Hybrid Method

Model	Avg Dumped Tris %	Run Length Ratio	Huffman Ratio	Total Ratio
Bunny	35.68 %	1.47	1.29	1.82
Submarine	2.83 %	6.27	1.46	9.16
Enoyl Hydratase	11.21 %	3.54	1.24	4.38
Dragon	7.79 %	1.68	1.60	2.69
Buddha	4.04 %	2.32	1.75	4.07

Table 4. Comparison Results for Different Methods

Model	System	Storage MB	Time and Speed			Image Error
			Decompression (msec)	Rendering (msec)	Overall Speed (fps)	
Bunny 69K tris	P	1.54	0.0	61.3	16.39	0.0
	H	12.35	10.8	81.7	10.79	2.02E-4
	I	52.74	83.6	28.2	8.94	2.66E-2
Submarine 376K tris	P	12.85	0.0	3549.0	0.28	0.0
	H	13.32	11.1	118.1	7.73	7.29E-3
	I	119.11	107.8	27.1	7.40	1.42E-1
Enoyl Hydratase 717K tris	P	14.95	0.0	777.4	1.28	0.0
	H	37.93	32.1	177.8	4.76	4.15E-4
	I	88.86	108.6	28.9	7.26	2.69E-2
Dragon 871K tris	P	19.19	0.0	1306.9	0.76	0.0
	H	104.98	87.8	223.45	3.10	4.82E-4
	I	75.95	102.2	28.7	7.63	2.45E-3
Buddha 1087K tris	P	23.92	0.0	1638.3	0.61	0.0
	H	86.56	69.4	191.9	3.82	5.14E-4
	I	55.78	95.9	27.9	8.07	9.60E-2

angles in that frame. The rendering speed on the Submarine is much slower than on the other models because we do the coloring for each rendered triangle. Without coloring, the *Polygonal* method has the average rendering speed about 7 − 9 frames/sec, and the *Hybrid* method is over 10 frame/sec.

The traditional polygon rendering has the best quality amongst all methods and requires least storage space, but it has the lowest rendering speed. The hybrid method which only renders visible triangles has very good ren-

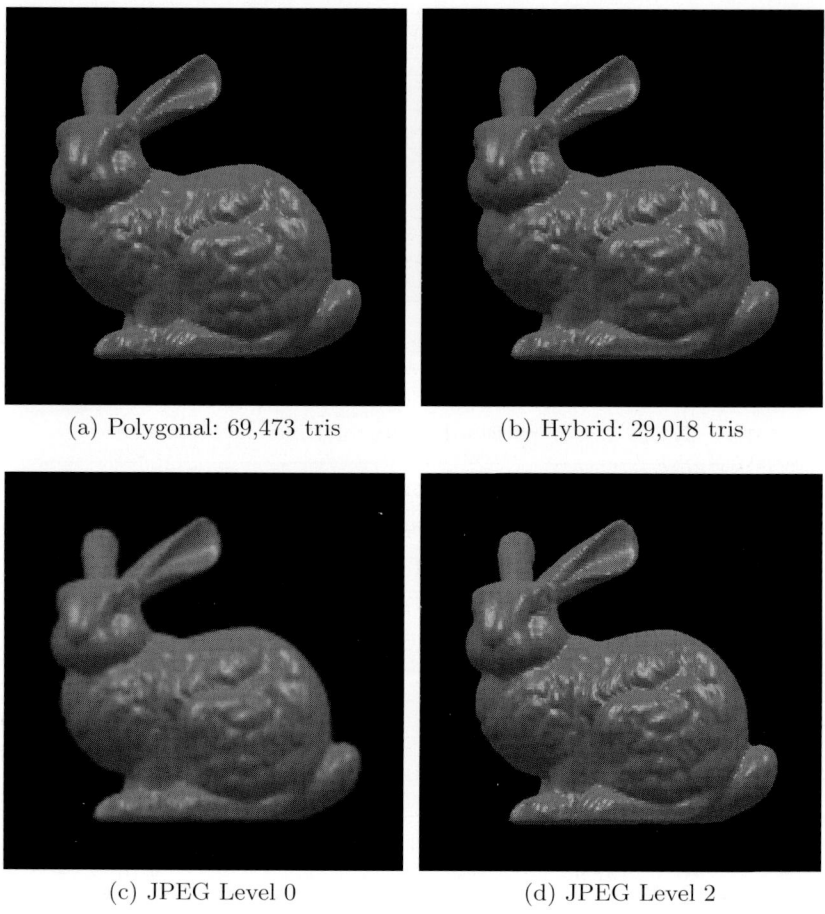

Fig. 4. Different Rendering Methods for the Bunny. (See Color Plate 28 on page 365.)

dering speeds, but needs much more storage space than traditional polygon rendering. Multi-level JPEG provides progressive refinement and has the lowest rendering complexity, but needs a lot of storage space. Figure 4, 5 and 6 show the images displayed by these methods on various models.

6 Conclusions

In this paper we have presented a hybrid method as well as a progressive refinement image-difference-based rendering method for high-complexity rendering. Our hybrid method takes advantage of both conventional polygon-

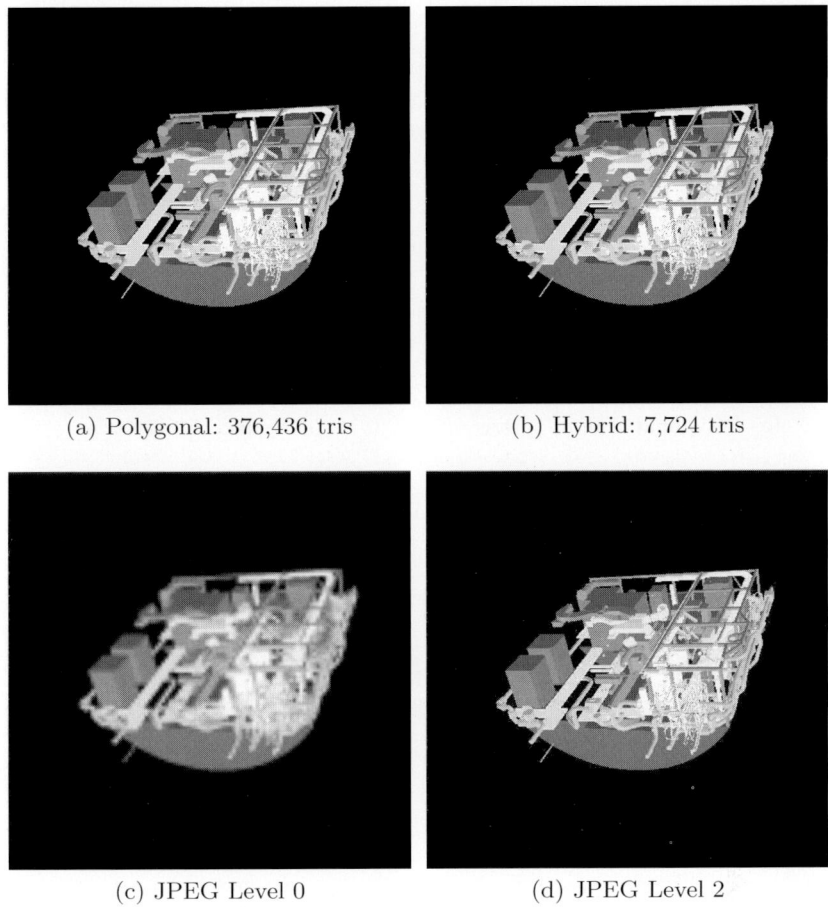

Fig. 5. Different Rendering Methods for the Auxiliary Machine Room of a notional submarine. (See Color Plate 29 on page 366.)

based rendering and image-based rendering. The hybrid rendering method can provide rendering quality comparable to the conventional polygonal rendering at a fraction of the computational cost and has storage requirements that are comparable to image-based rendering methods. The drawback is that it does not permit full navigation capability to the user as in the conventional polygonal method. However, it still retains several other useful features of the polygonal methods such as direct querying to the underlying database and the ability to change illumination and material properties. In future we plan to further explore compression issues for the hybrid method by taking advantage of frame-to-frame coherence in image space and view-dependent geometric hierarchical structures.

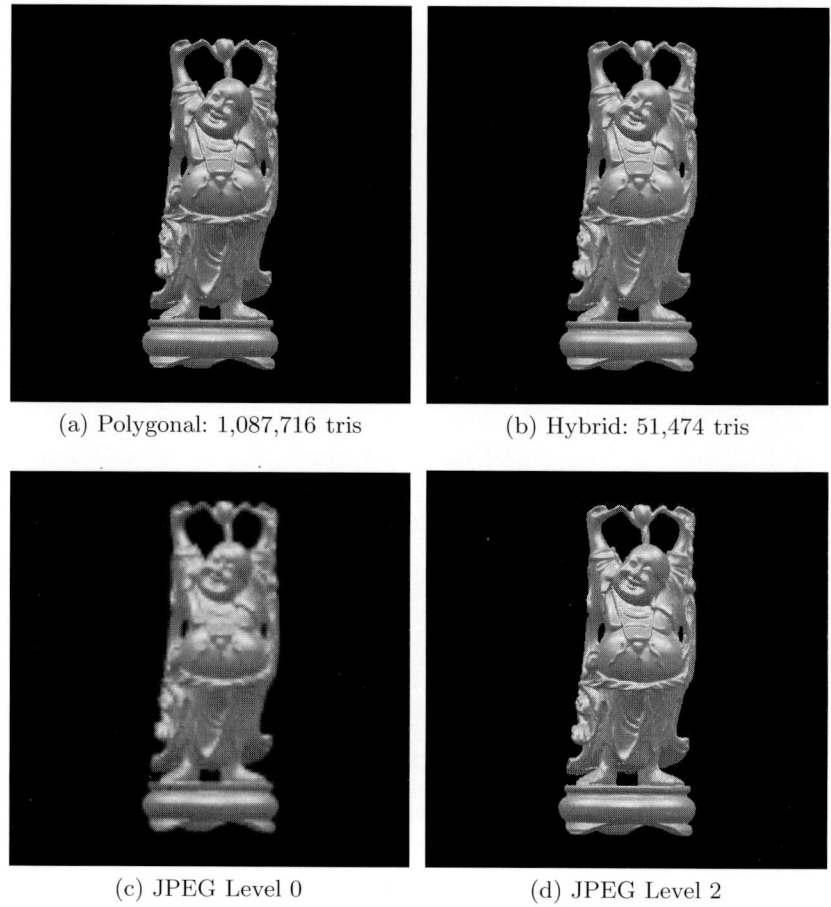

Fig. 6. Different Rendering Methods for the Buddha. (See Color Plate 30 on page 367.)

Acknowledgements

We will like to acknowledge the detailed and useful comments from the reviewers that has improved the presentation of this work. This work has been supported in part by the NSF grants: DMI-9800690, ACR-9812572, and IIS-0081847. We will like to acknowledge Electric Boat Division of General Dynamics for providing us the submarine dataset and the Stanford Graphics lab for providing us the Bunny, the Dragon, and the Buddha models.

References

1. E. H. Adelson and J. R. Bergen. The plenoptic function and the elements of early vision. In M. Landy and J. A. Movshon, editors, *Computational Models of Visual Processing*, chapter 1. The MIT Press, Mass, 1991.

2. Emilio Camahort, Apostolos Lerios, and Donald Fussell. Uniformly sampled light fields. In G. Drettakis and N. Max, editors, *Rendering Techniques '98 (Proceedings of Eurographics Rendering Workshop '98)*, pages 117–130, New York, NY, 1998. Springer Wien.
3. S. E. Chen. Quicktime VR - an image-based approach to virtual environment navigation. In *Computer Graphics Proceedings (SIGGRAPH 95)*, Annual Conference Series, pages 29–38, 1995.
4. S. E. Chen and L. Williams. View interpolation for image synthesis. In *Computer Graphics Proceedings(SIGGRAPH 93)*, Annual Conference Series, pages 279–288, 1993.
5. Daniel Cohen-Or and Eyal Zadicario. Visibility streaming for network-based walkthroughs. In *Graphics Interface*, pages 1–7, June 1998.
6. S. Coorg and S. Teller. Temporally coherent conservative visibility. In *ACM Press*, Proceedings of 20th Annual Symposium on Computational Geometry, pages 78–87, May 1996.
7. S. Coorg and S. Teller. Real time occlusion culling for models with large occluders. In *Proceedings of 1997 Simposium in 3D Interactive Graphics*, pages 83–90, 1997.
8. L. Darsa and B. Costa. Multi-resolution representation and reconstruction of adaptively sampled images. In *Proceedings of SIBGRAPI'96*, pages 321–328, 1996.
9. L. Darsa, B. Costa, and A. Varshney. Navigating static environments using image-space simplication and morphing. In *Proceedings of 1997 Symposium on Interactive 3D Graphics*, pages 25–34, April 1997.
10. P. Debevec, C. Bregler, M. Cohen, L. McMillan, F. Sillion, and R. Szeliski. *SIGGRAPH 2000 Course 35: Image-based Modeling, Rendering, and Lighting*. ACM SIGGRAPH, 2000.
11. S. J. Gortler, R. Grzeszczuk, R. Szeliski, and M. F. Cohen. The lumigraph. In *Computer Graphics Proceedings (SIGGRAPH 96)*, Annual Conference Series, pages 43–54, 1996.
12. Independent JPEG Group. ftp://ftp.uu.net/graphics/jpeg/.
13. Wolfgang Heidrich, Hartmut Schirmacher, Hendrik Kück, and Hans-Peter Seidel. A warping-based refinement of lumigraphs. In N. Thalmann and V. Skala, editors, *Proc. WSCG '99*, 1999.
14. J. T. Kajiya. Anisotropic reflection models. In *Proceedings of SIGGRAPH '85*, volume 19, pages 15–21. Springer Verlag, July 1985.
15. M. Levoy and P. Hanrahan. Light field rendering. In *Computer Graphics Proceedings (SIGGRAPH 96)*, Annual Conference Series, pages 31–42, 1996.
16. D. Luebke, J. Cohen, M. Reddy, A. Varshney, and B. Watson. *SIGGRAPH 2000 Course 41: Advanced Issues in Level of Detail*. ACM SIGGRAPH, 2000.
17. William R. Mark, Leonard McMillan, and Gary Bishop. Post-rendering 3D warping. In Michael Cohen and David Zeltzer, editors, *1997 Symposium on Interactive 3D Graphics*, pages 7–16. ACM SIGGRAPH, April 1997. ISBN 0-89791-884-3.
18. L. McMillan. Computing visibility without depth. Technical Report 95-047, University of North Carolina at Chapel Hill, 1995.
19. L. McMillan. A list-priority rendering algorithm for redisplaying projected surfaces. Technical Report 95-005, University of North Carolina at Chapel Hill, 1995.

20. L. McMillan and G. Bishop. Plenoptic modeling: An image-based rendering system. In *Computer Graphics Proceedings*, Annual Conference Series, pages 39–46. ACM SIGGRAPH, 1995.
21. J. S. Nimeroff, E. Simoncelli, and J. Dorsey. Efficient re-rendering naturally illuminated environments. In *Eurographics*, Fifth Eurographics Workshop on Rendering, pages 359–373, 1994.
22. Manuel M. Oliveira and Gary Bishop. Image-based objects. In Stephen N. Spencer, editor, *Proceedings of the Conference on the 1999 Symposium on interactive 3D Graphics*, pages 191–198, New York, April 26–28 1999. ACM Press.
23. M. Panne and A.J. Stewart. Effective compression techniques for precomputed visibility. In *Springer Computer Science*, Rendering Techniques '99, pages 305–316, 1999.
24. K. Pulli, M. Cohen, T. Duchamp, H. Hoppe, L. Shapiro, and W. Stuetzle. View-based rendering: Visualizing real objects from scanned range and color data. In *Proceedings of Eighth Eurographics Workshop on Rendering*, pages 23–34. Eurographics, June 1997.
25. P. Rademacher and G. Bishop. Multiple-center-of-projection images. In *Computer Graphics Proceedings (SIGGRAPH 98)*, Annual Conference Series, pages 199–206, 1998.
26. Hartmut Schirmacher, Wolfgang Heidrich, and Hans-Peter Seidel. Adaptive acquisition of lumigraphs from synthetic scenes. In Hans-Peter Seidel and Sabine Coquillart, editors, *Eurographics '99*, volume 18, pages C151–C159, Computer Graphics Forum, c/o Mercury Airfreight International Ltd Inc, 365 Blair Road, Avenel, MI 48106, USA, 1999. Eurographics Association, Eurographics Association and Blackwell Publishers Ltd 1999.
27. J. Shade, S. Gortler, L. He, and R. Szeliski. Layer depth images. In *Computer Graphics Proceedings (SIGGRAPH 98)*, Annual Conference Series, pages 231–242, 1998.
28. P. Sloan, M. Cohen, and S. Gortler. Time critical lumigraph rendering. In *Symposium on Interactive 3D Graphics*, pages 17–23, 1997.
29. S. J. Teller and C. H. Sequin. Visibility preprocessing for interactive walkthroughs. In *Computer Graphics Proceedings (SIGGRAPH 91)*, pages 61–69, 1991.
30. Y. Wang, H. Bao, and Q. Peng. Accelerated walkthroughts of virtual environments based on visibility preprocessing and simplification. In *Eurographics*, pages 17(3): 187–194, 1998.
31. T. Wong, P. Heng, S. Or, and W. Ng. Illuminating image-based objects. In *Proc. Pacific Graphics'97, Seoul, Korea*, 1997.
32. T. Wong, P. Heng, S. Or, and W. Ng. Image-based rendering with controllable illumination. In *Proc. 8th Eurographics Workshop on Rendering, St. Etienne, France*, 1997.
33. R. Yagel and W. Ray. Visibility computation for efficient walkthroughs of complex environments. In *Presence*, pages 5(1):45–60, 1995.
34. H. Zhang, D. Manocha, T. Hudson, and K. Hoff. Visibility culling using hierarchical occlusion maps. In *Proceedings of SIGGRAPH '97 (Los Angeles, CA)*, Computer Graphics Proccedings, Annual Conference Series, pages 77–88. ACM SIGGRAPH, ACM Press, August 1997.

Appendix:
Color Plates

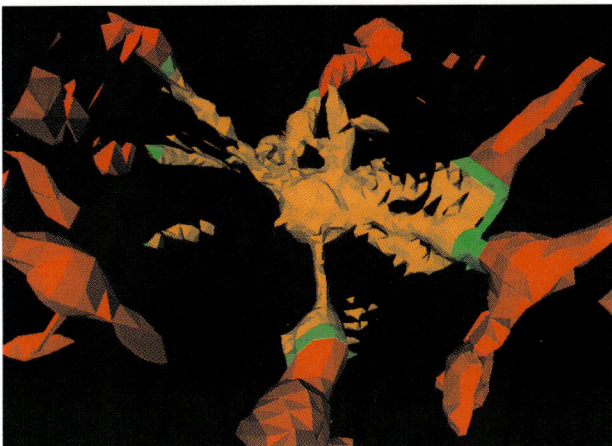

(i) Isosurface obtained when using two of seven levels of AMR hierarchy. Stitch cell generation required approximately 55ms, and isosurface generation required approximately 250ms

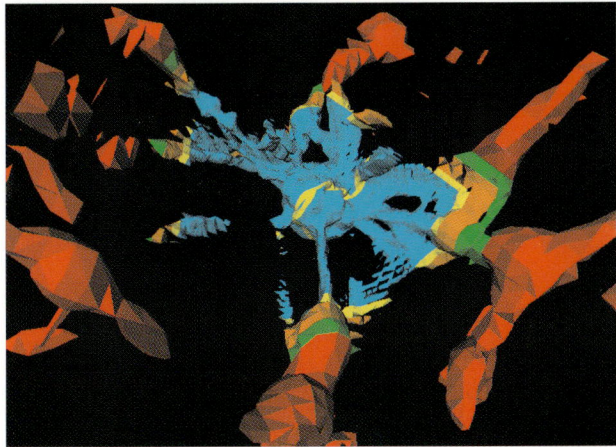

(ii) Isosurface obtained when using three of seven levels of AMR hierarchy. Stitch cell generation required approximately 340ms, and isosurface generation required approximately 600ms

Plate 1. Isosurface extracted from AMR hierarchy simulating star clusters (data set courtesy of Greg Bryan, Massachusetts Institute of Technology, Theoretical Cosmology Group, Cambridge, Massachusetts). (See Fig. 13 on page 37.)

350   Appendix

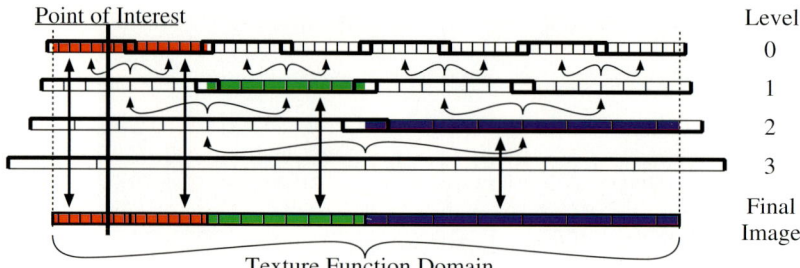

Plate 2. Selecting from a texture hierarchy of four levels. Level 0 is the original texture, broken into eight tiles. The dashed lines show the domain of the texture function over the hierarchy. The bold vertical line represents a point p of interest. Tile selection depends on the width of the tile and the distance from the point. The red, green, and blue shaded regions in levels 0, 1, and 2, respectively, and then in the Final Image, show from which levels of detail the data is used to create the final image. (See Fig. 2 on page 53.)

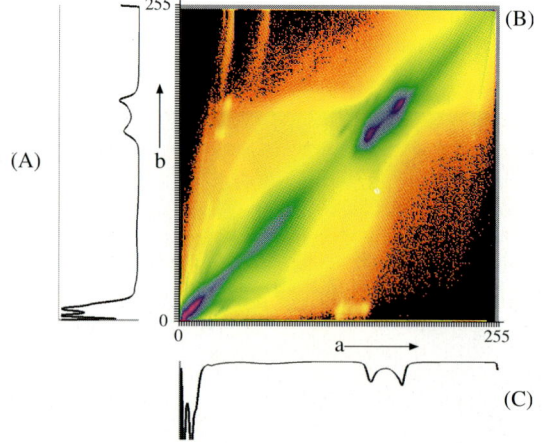

Plate 3. The Visible Female CT data set. Image (B) in the upper-right corner shows the frequency relationship of the original and first approximation of the Visible Female CT data set. The image consists of 256×256 pixels, with a on the horizontal axis and b on the vertical axis; each pixel corresponds to a $Q_{a,b}$ element. This particular table, covering all of the original domain, would not be produced in practice; normally, Q tables associated with the first-level approximation cover fairly small regions of the original domain. Image (B) is shown here to provide the reader with insight into the nature of a typical Q table. The colors are assigned by normalizing the logarithm of the number of occurrences of a $Q_{a,b}$ element, linearly mapped to a rainbow color sequence, where zero maps to red and one maps to violet: $pixel_{a,b} = RainbowColorMap\left[ln\left(Q_{a,b}\right)/ln\left(Max_{a,b}\left\{Q_{a,b}\right\}\right)\right]$. Graph (A), on the left, shows the histogram of the original data (Level 0), with positive frequency pointing left. Graph (C), on the bottom, shows the histogram of the first approximation (Level 1), with positive frequency pointing down. (See Fig. 4 on page 57.)

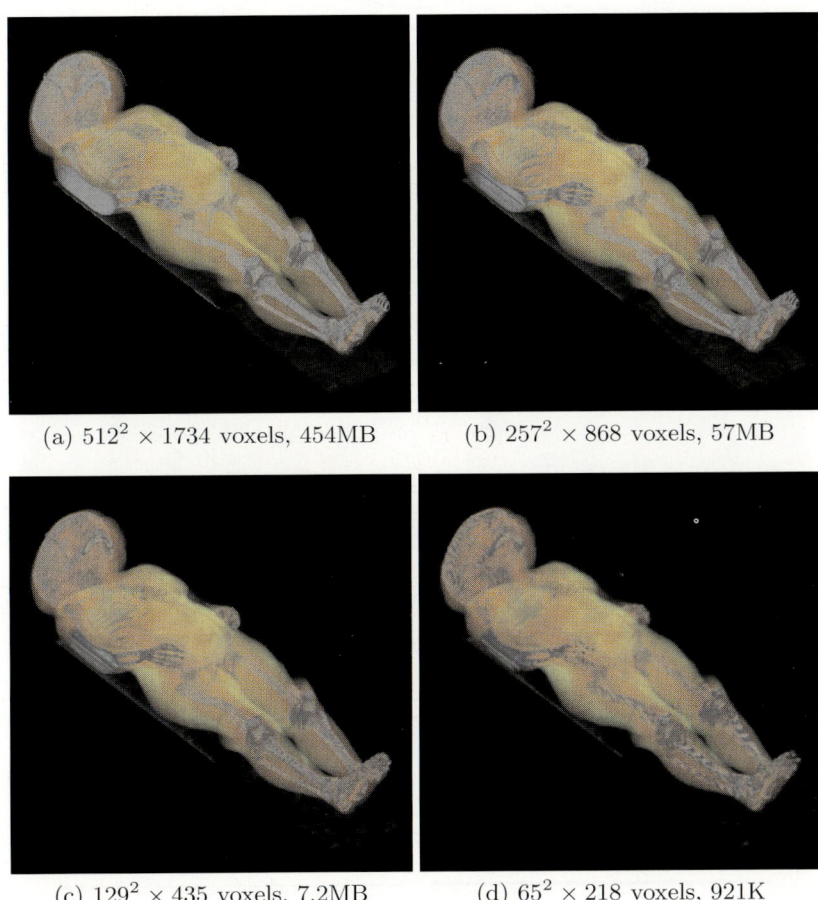

(a) $512^2 \times 1734$ voxels, 454MB (b) $257^2 \times 868$ voxels, 57MB

(c) $129^2 \times 435$ voxels, 7.2MB (d) $65^2 \times 218$ voxels, 921K

Plate 4. Visible Female CT data set rendered at four resolutions (see Table 1). The transfer function shows bones in white, fat in yellow, muscle in red, and internal organs in green. Different spatial scales were used for different sections of the body during data acquisition. (See Fig. 6 on page 60.)

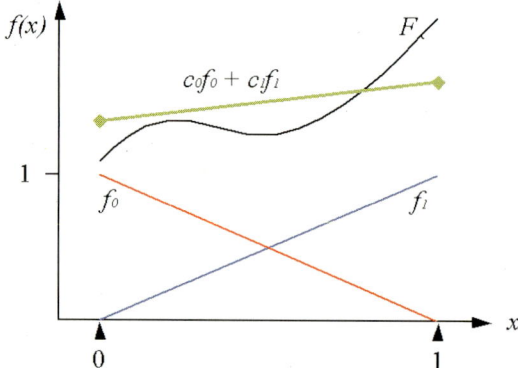

Plate 5. Basis functions f_i, function F to be approximated, and approximation $f(x) = c_0 f_0 + c_1 f_1$. (See Fig. 1 on page 66.)

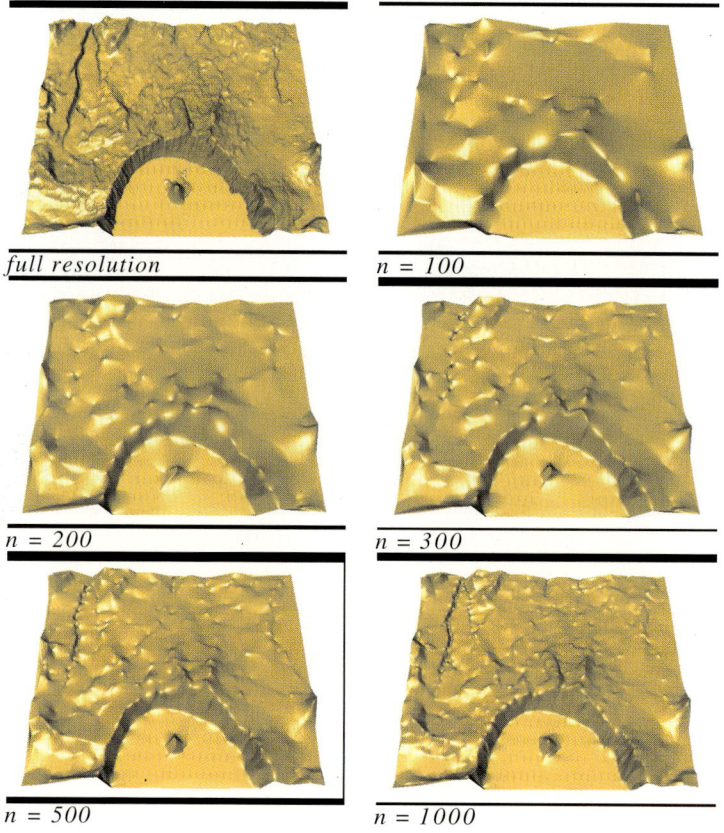

Plate 6. Crater-Lake terrain data set at different levels of resolution (using $p = 2$). The full-resolution data set consists of 159272 points, courtesy of U.S. Geological Survey. (See Fig. 5 on page 96.)

Plate 7. Cross section of a density field approximated using explicit interface representations and separate field representations. The left picture shows the field along with the approximating tetrahedral mesh. (interface error = 0.15). (See Fig. 1 on page 100.)

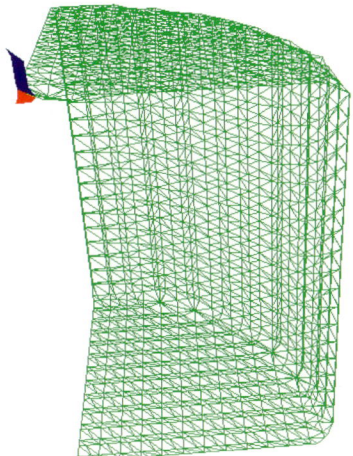

Plate 8. Original triangular meshes representing material interfaces. (See Fig. 7 on page 112.)

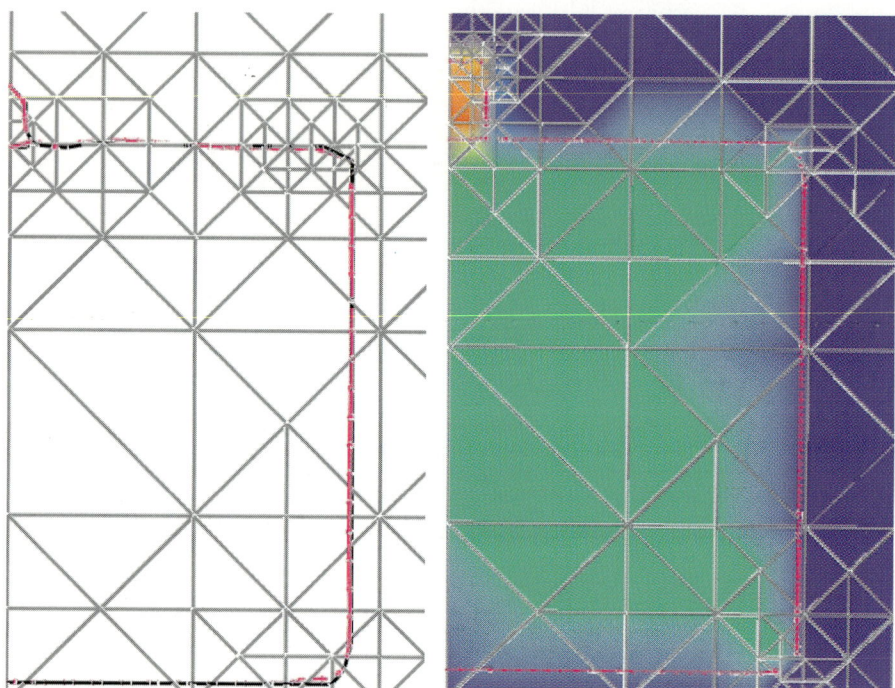

Plate 9. Cross section of the tetrahedral mesh. The left picture shows the original interfaces and their approximations. The picture on the right shows the density field using linear interpolation. (See Fig. 8 on page 114.)

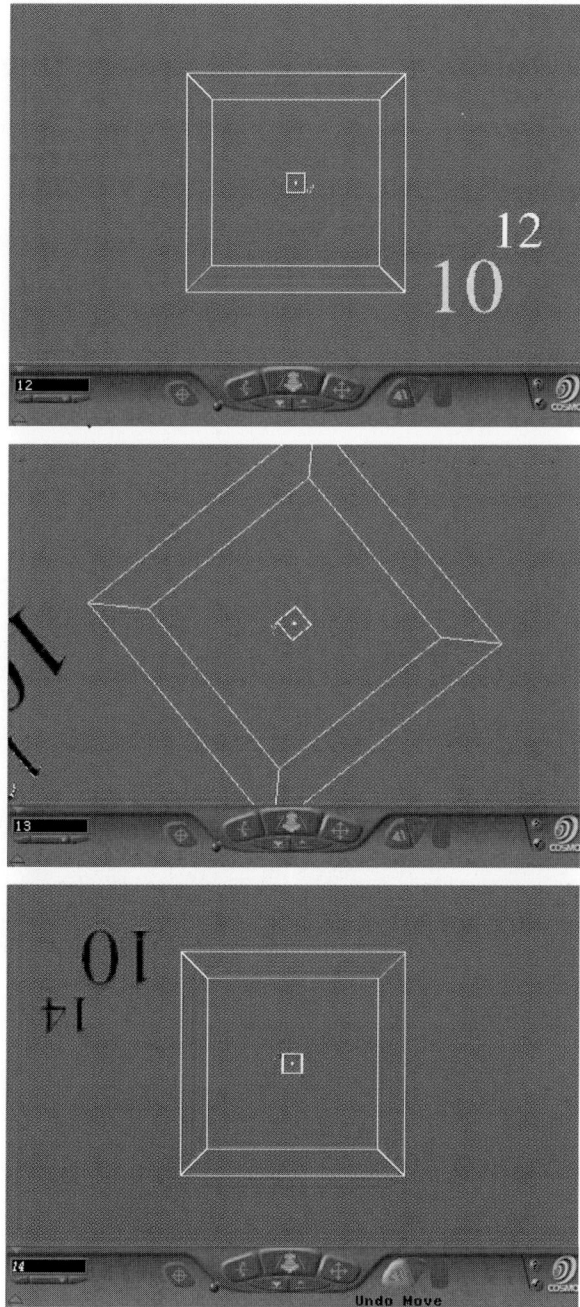

Plate 10. Breakdown of SGI CosmoPlayer running unit VRML cubes near scale parameter values of 10^{13}. (Top) 10^{12}: Normal behavior. (Center) 10^{13}: Lighting fails (blackened lettering) simultaneously with geometry. (Bottom) 10^{14}: Geometric accuracy is lost from here on. (See Fig. 1 on page 123.)

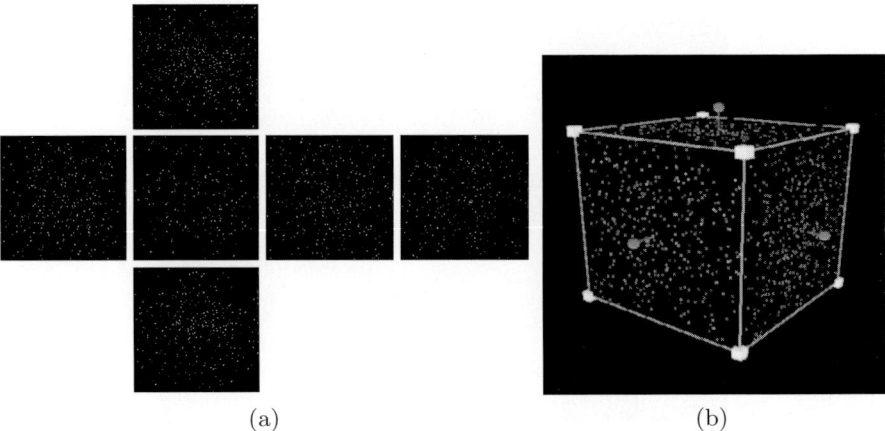

Plate 11. (a) Pieces of cubic environment map used for the nearby stars when $s_{\text{nav}} < 13$. (b) Appearance of wrapped map from outside. (See Fig. 2 on page 125.)

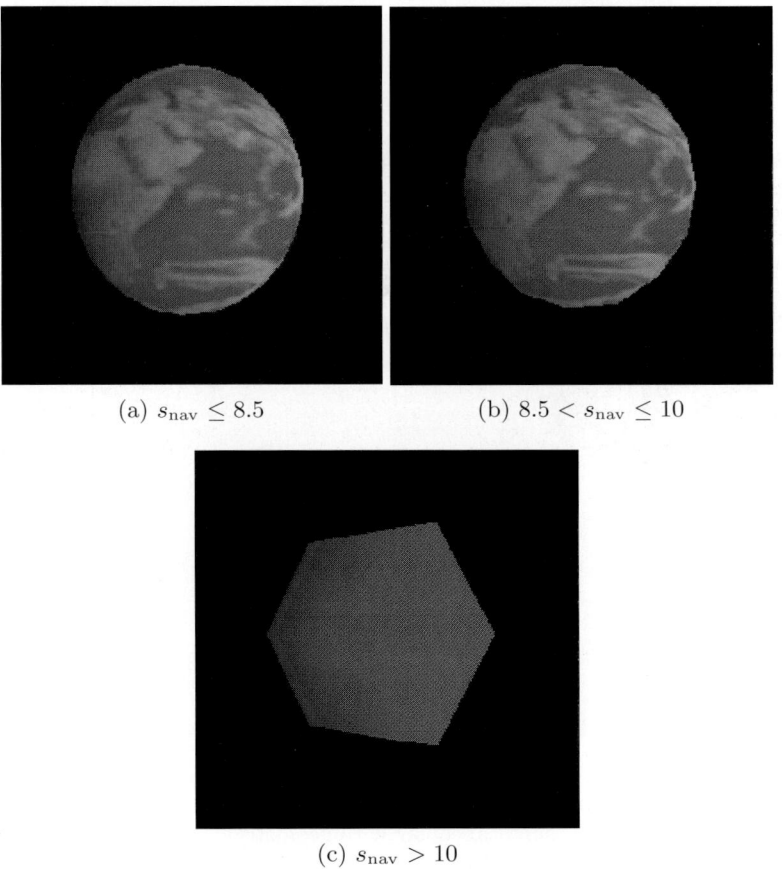

Plate 12. Space-LOD representations of the Earth. (See Fig. 3 on page 126.)

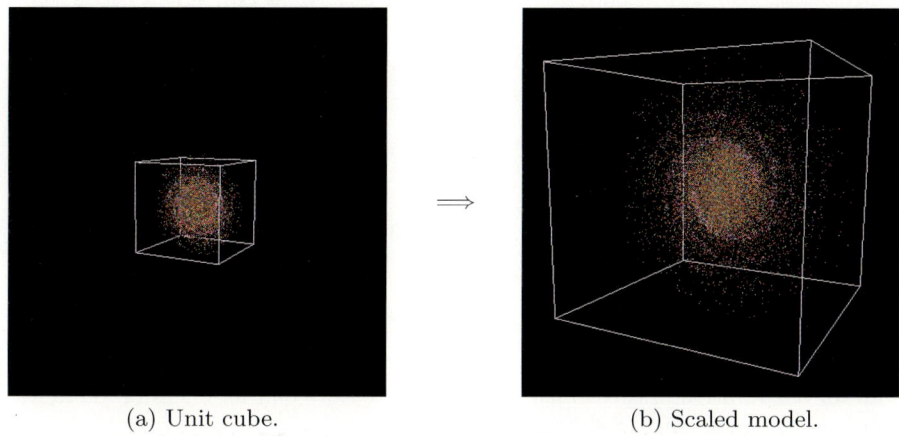

(a) Unit cube.   (b) Scaled model.

Plate 13. Scaling a unit-sized ideal model by 10 to the power $(s_{\text{dataset}} - s_{\text{nav}})$. (See Fig. 4 on page 129.)

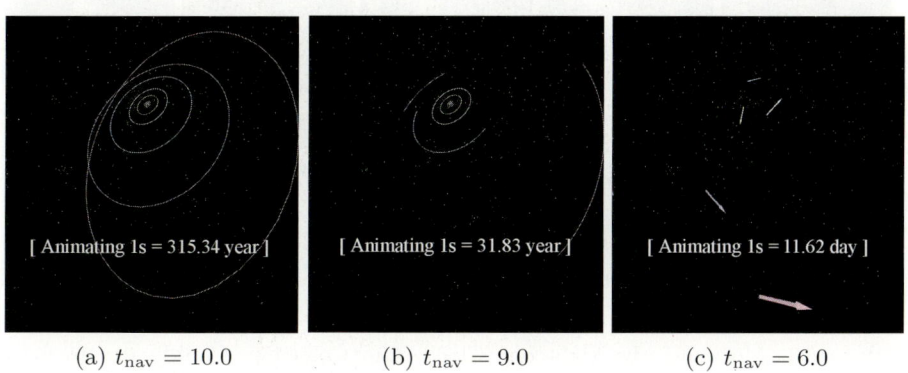

(a) $t_{\text{nav}} = 10.0$ (b) $t_{\text{nav}} = 9.0$ (c) $t_{\text{nav}} = 6.0$

Plate 14. "Too fast" representations at (a) roughly 300 years per screen second and (b) roughly 30 years per screen second. (c) The "too slow" representation for the motion of planets in our Solar System at a scale of approximately 10 days per screen second. (See Fig. 9 on page 139.)

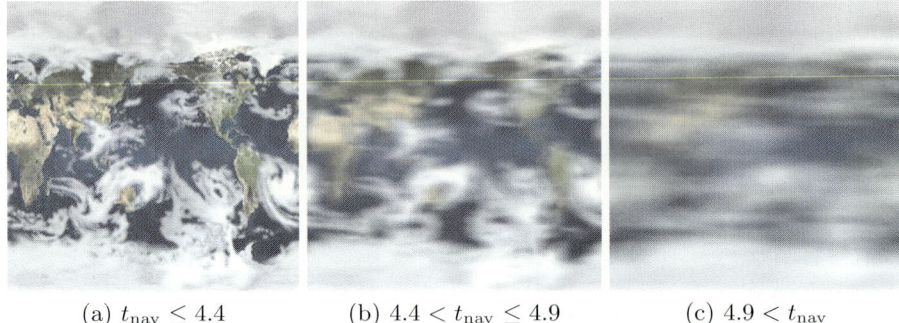

(a) $t_{\text{nav}} \leq 4.4$ (b) $4.4 < t_{\text{nav}} \leq 4.9$ (c) $4.9 < t_{\text{nav}}$

Plate 15. Motion Blur Representation : (a) Normal Earth texture. (b) Texture blur for one rotation in two screen seconds. (c) Texture blur for one rotation in less than one screen second. (See Fig. 10 on page 139.)

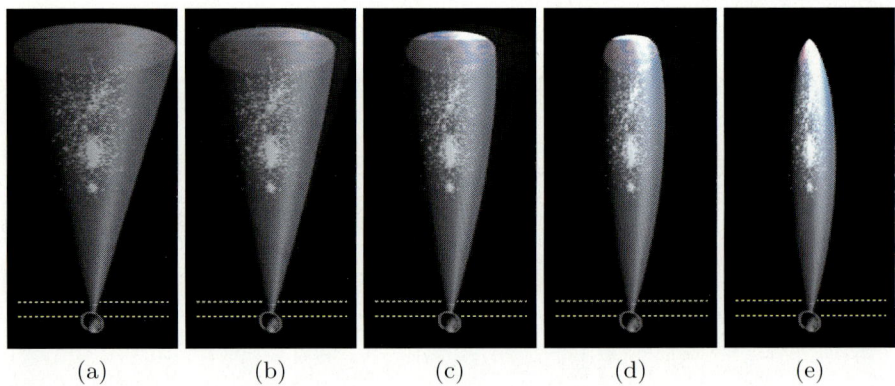

(a)   (b)   (c)   (d)   (e)

Plate 16. Morphing the Lightcone Clock from Comoving Coordinates to $a(t)$-rescaled Physical Coordinates. (See Fig. 11 on page 139.)

Color Plates 359

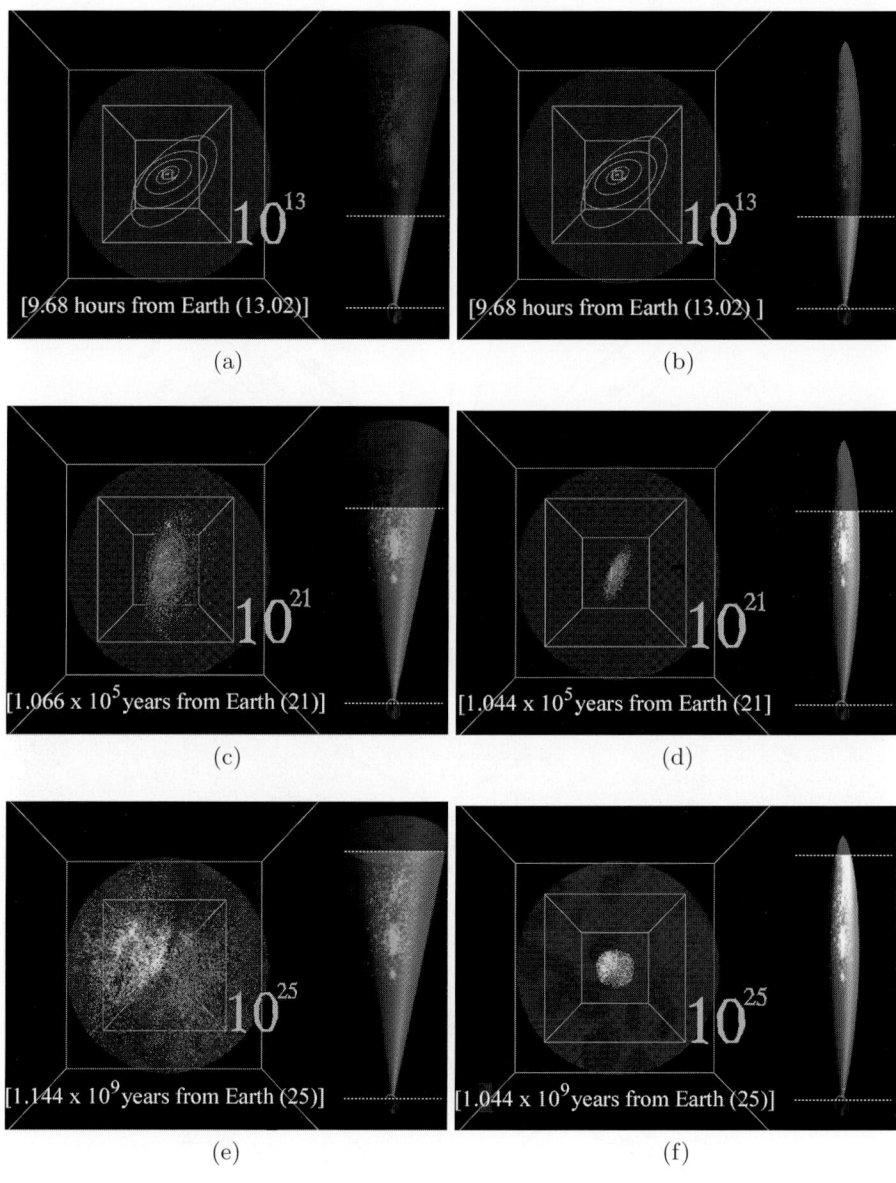

Comoving Coordinates $a(t)$-rescaled Physical Coordinates

Plate 17. Navigation with Comoving Coordinates and $a(t)$-rescaled Physical Coordinates, with the Comoving Coordinate scales superimposed for absolute reference. (See Fig. 12 on page 140.)

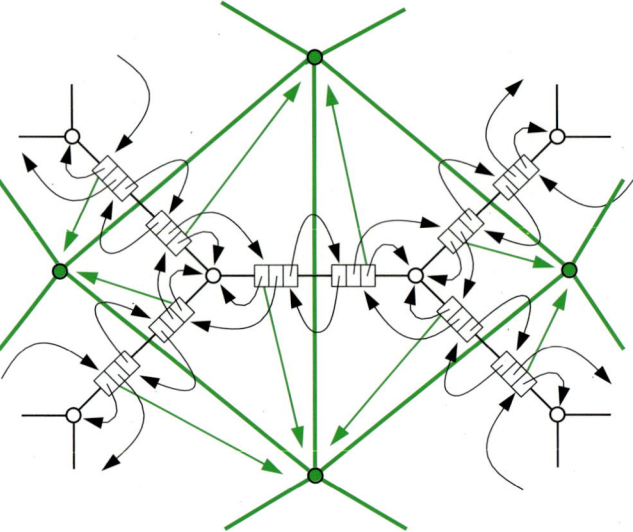

Plate 18. Representing the dual of a mesh with a half-edge data structure. The original mesh and data structure are shown in black with the dual mesh shown in gray. The gray vertex links are the links to the face nodes in the dual structure. Note the counter-clockwise loops about the face, and the clockwise loops about the vertex in the dual mesh. (See Fig. 15 on page 160.)

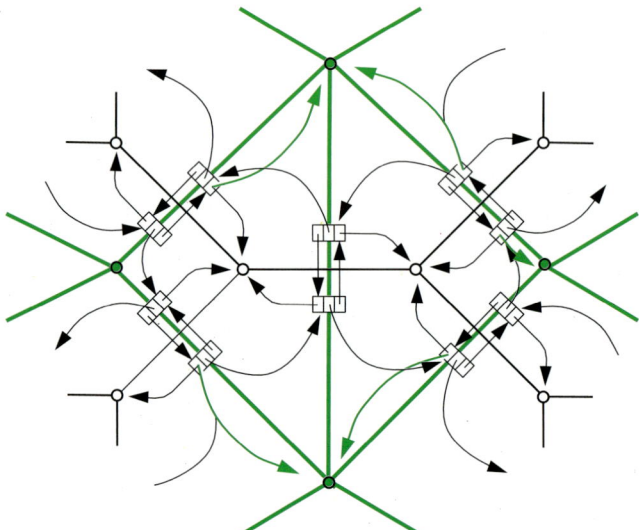

Plate 19. Representing the dual of a mesh with a split-edge data structure. The original mesh and data structure are shown in black with the dual mesh shown in gray. The gray vertex links are the links to the face nodes in the dual structure. Note the counter-clockwise loops about the face and the clockwise loops about the vertex in the dual mesh. (See Fig. 16 on page 161.)

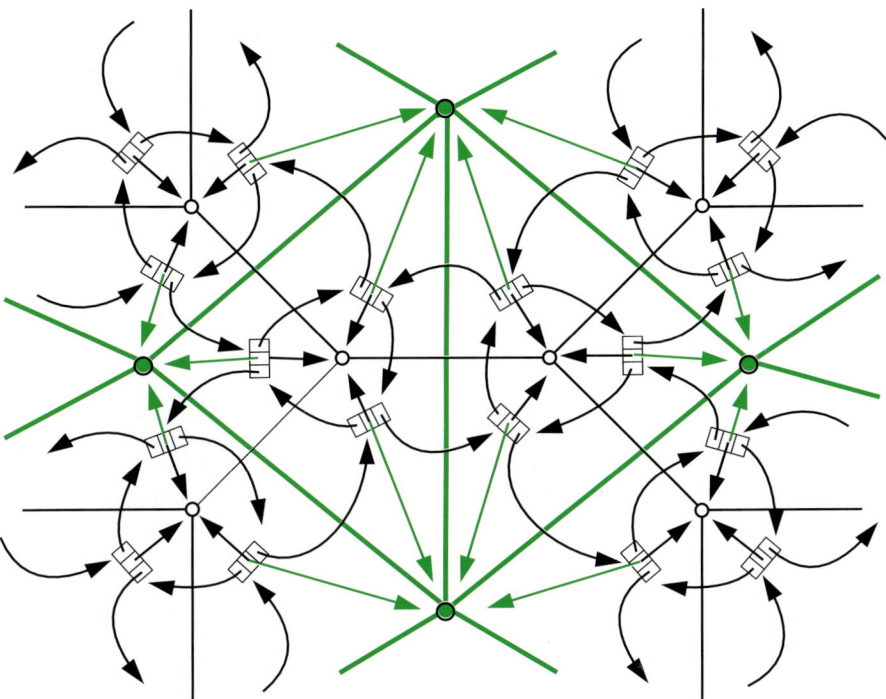

Plate 20. Representing the dual of a mesh with a corner data structure. The original mesh and data structure are shown in black with the dual mesh shown in gray. The gray vertex links are the links to the face nodes in the dual structure. (See Fig. 22 on page 166.)

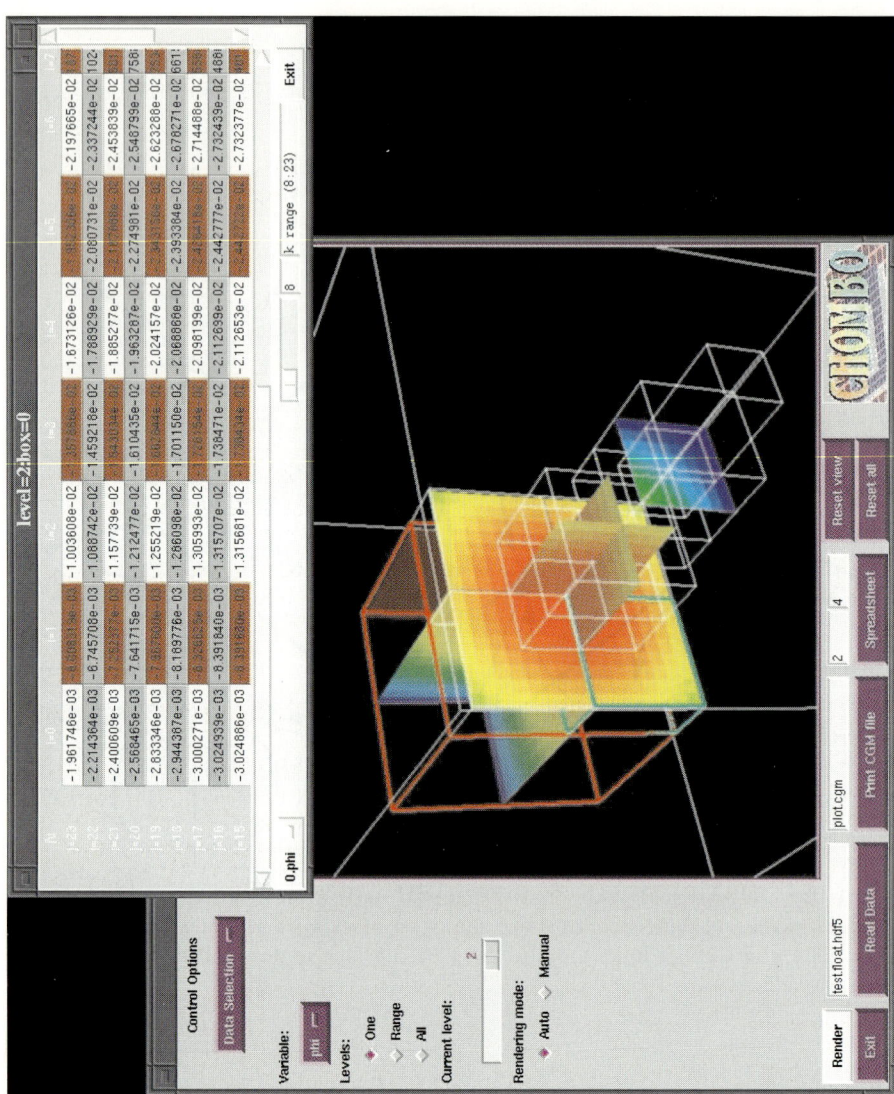

Plate 21. Visualization of an AMR data set using ChomboVis. (See Fig. 1 on page 200.)

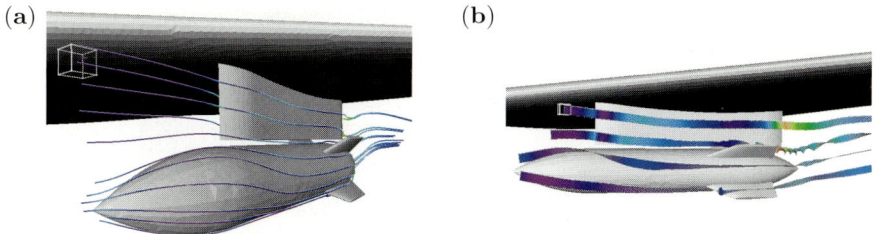

Plate 22. Exploration of vector field on tetrahedral mesh with embedded geometry. (a) Visualization with streamlines (b) Visualization with streamribbons. (See Fig. 1 on page 207.)

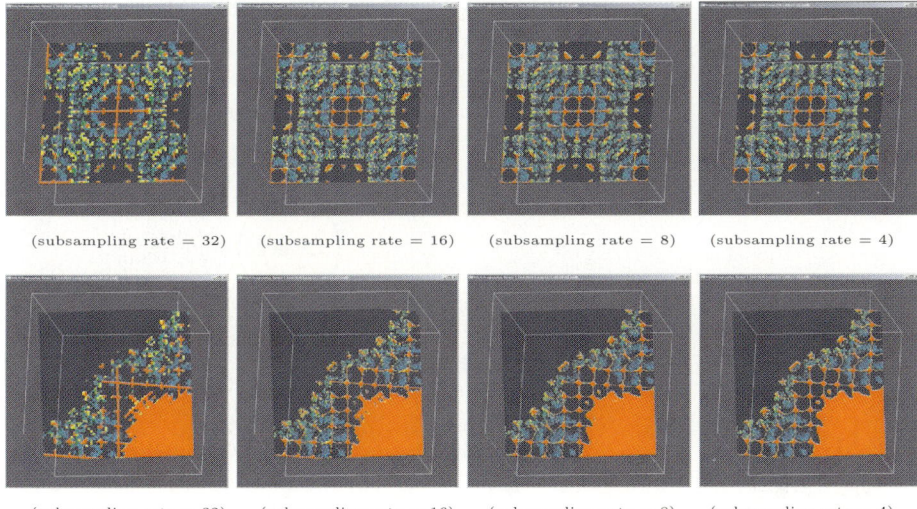

Plate 23. Progressive refinement of two slices of the PPM dataset. (top row) Slice orthogonal to the x axis. (bottom row) Slice at an arbitrary orientation. (See Fig. 11 on page 239.)

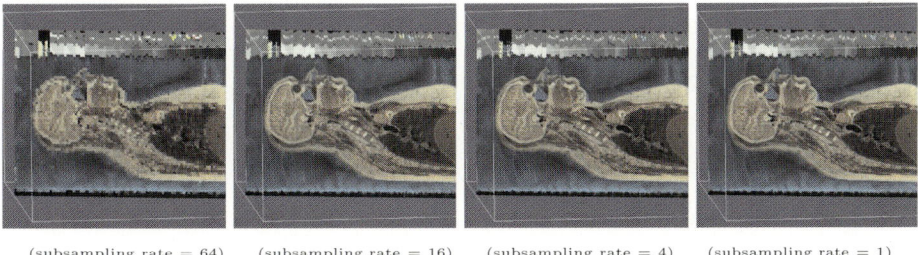

Plate 24. Progressive refinement of one slice of the visible human dataset. Note how the inclination of the slice allows to show at the same time the nose and the eye. This view cannot be obtained using only orthogonal slice. (See Fig. 12 on page 239.)

364   Appendix

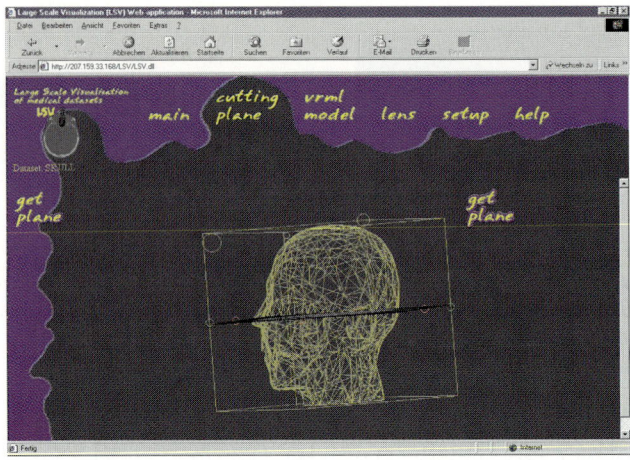

Plate 25. Prototype application. (See Fig. 8 on page 292.)

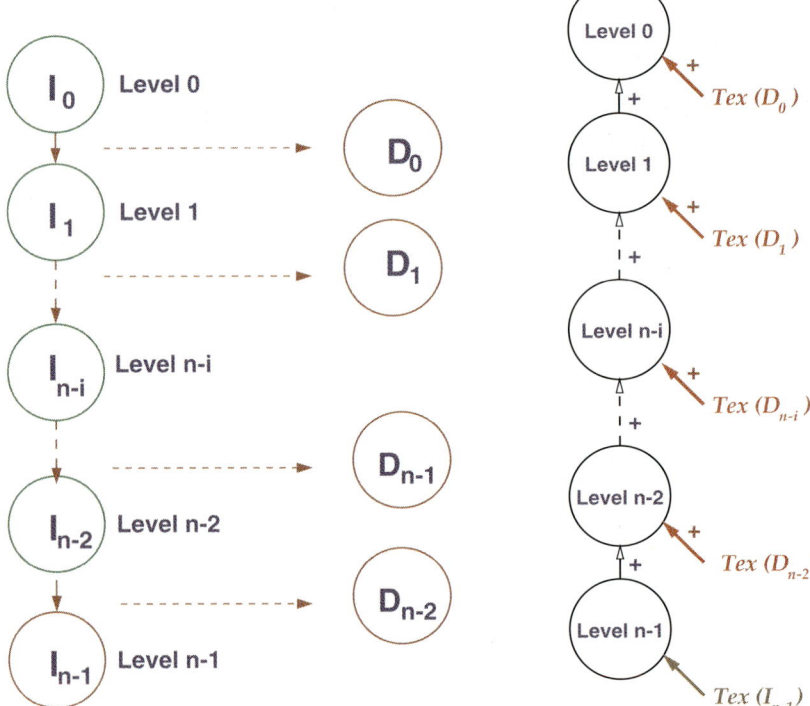

Plate 26. Multi-Level Image Construction. (See Fig. 1 on page 334.)

Plate 27. Hierarchical Accumulation of JPEG Images. (See Fig. 2 on page 336.)

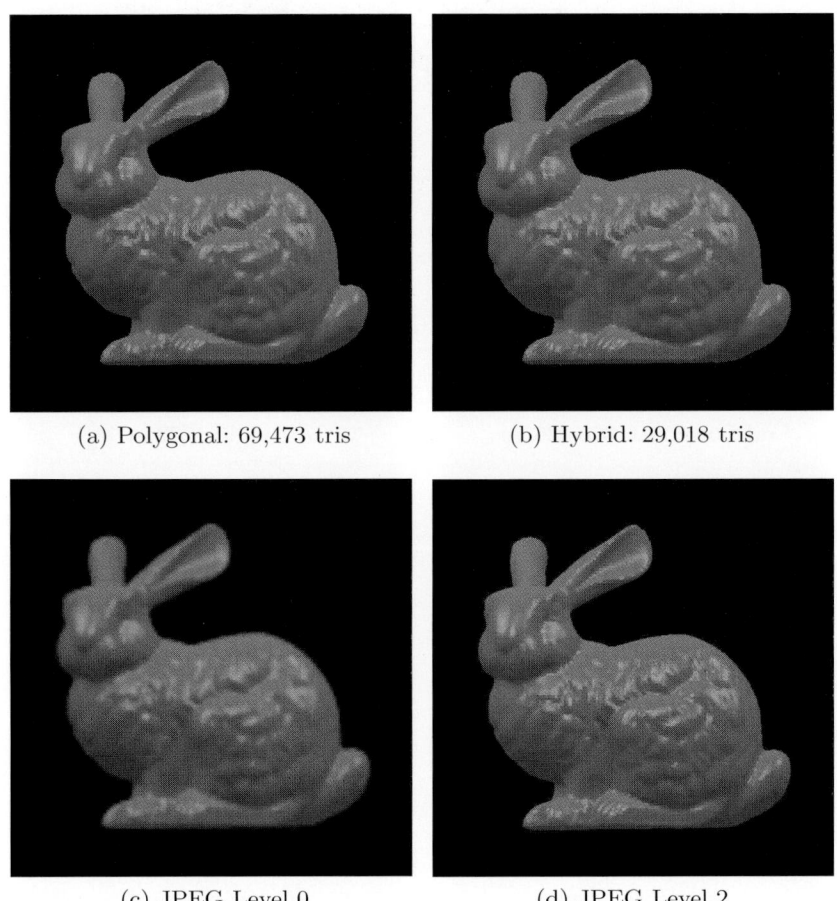

(a) Polygonal: 69,473 tris (b) Hybrid: 29,018 tris

(c) JPEG Level 0 (d) JPEG Level 2

Plate 28. Different Rendering Methods for the Bunny. (See Fig. 4 on page 342.)

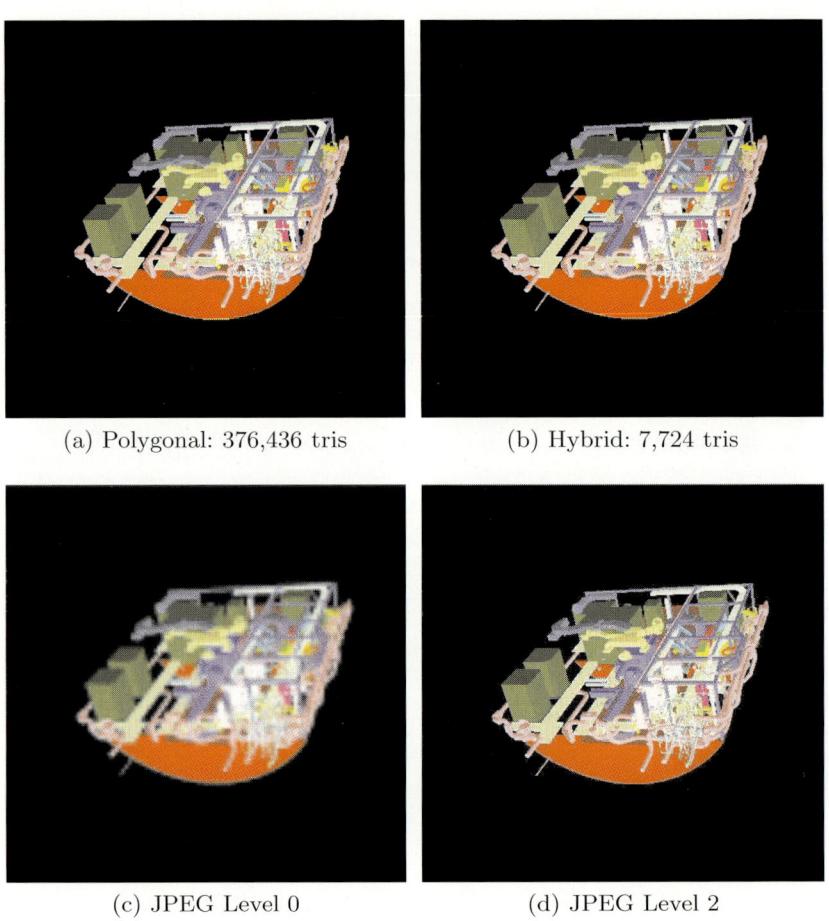

(a) Polygonal: 376,436 tris (b) Hybrid: 7,724 tris

(c) JPEG Level 0 (d) JPEG Level 2

Plate 29. Different Rendering Methods for the Auxiliary Machine Room of a notional submarine. (See Fig. 5 on page 343.)

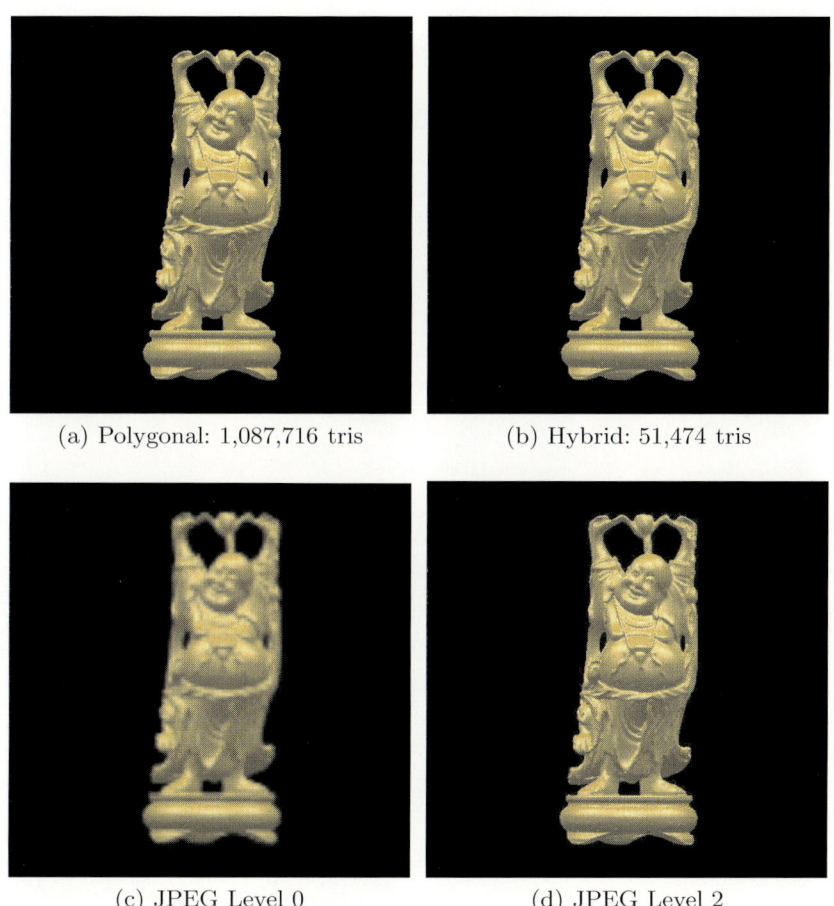

(a) Polygonal: 1,087,716 tris (b) Hybrid: 51,474 tris

(c) JPEG Level 0 (d) JPEG Level 2

Plate 30. Different Rendering Methods for the Buddha. (See Fig. 6 on page 344.)

Druck: Strauss Offsetdruck, Mörlenbach
Verarbeitung: Schäffer, Grünstadt